Pearson Edexcel AS and A level Further Mathematics

Decision Mathematics 1

D1

Series Editor: **Harry Smith**
Authors: Susie Jameson, Peter Sherran, Keith Pledger, Harry Smith

Pearson

Overarching themes

The following three overarching themes have been fully integrated throughout the Pearson Edexcel AS and A level Mathematics series, so they can be applied alongside your learning and practice.

1. Mathematical argument, language and proof

- Rigorous and consistent approach throughout
- Notation boxes explain key mathematical language and symbols
- Dedicated sections on mathematical proof explain key principles and strategies
- Opportunities to critique arguments and justify methods

2. Mathematical problem solving

- Hundreds of problem-solving questions, fully integrated into the main exercises
- Problem-solving boxes provide tips and strategies
- Structured and unstructured questions to build confidence
- Challenge boxes provide extra stretch

The Mathematical Problem-solving cycle

specify the problem

collect information

process and represent information

interpret results

3. Mathematical modelling

- Dedicated modelling sections in relevant topics provide plenty of practice where you need it
- Examples and exercises include qualitative questions that allow you to interpret answers in the context of the model
- Dedicated chapter in Statistics & Mechanics Year 1/AS explains the principles of modelling in mechanics

Finding your way around the book

Access an online digital edition using the code at the front of the book.

Each chapter starts with a list of objectives

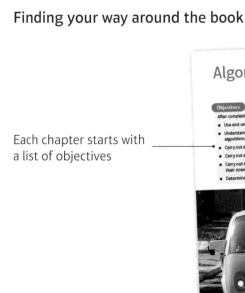

Algorithms **1**

The real world applications of the maths you are about to learn are highlighted at the start of the chapter with links to relevant questions in the chapter

The *Prior knowledge check* helps make sure you are ready to start the chapter

Exercise questions are carefully graded so they increase in difficulty and gradually bring you up to exam standard

Exercises are packed with exam-style questions to ensure you are ready for the exams

A level content is clearly flagged

Each section begins with explanation and key learning points

Exam-style questions are flagged with (E)

Problem-solving questions are flagged with (P)

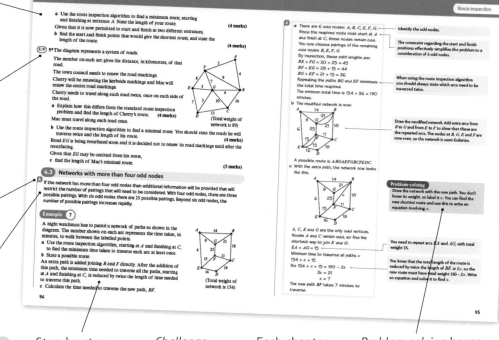

Step-by-step worked examples focus on the key types of questions you'll need to tackle

Challenge boxes give you a chance to tackle some more difficult questions

Each chapter ends with a *Mixed exercise* and a *Summary of key points*

Problem-solving boxes provide hints, tips and strategies, and *Watch out* boxes highlight areas where students often lose marks in their exams

Every few chapters a *Review exercise* helps you consolidate your learning with lots of exam-style questions

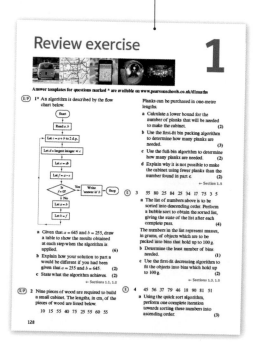

Review exercise

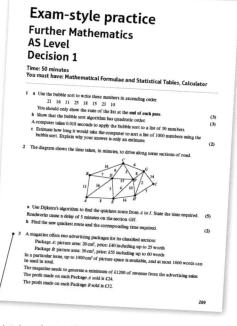

Exam-style practice
Further Mathematics
AS Level
Decision 1

AS and A level practice papers at the back of the book help you prepare for the real thing.

Extra online content

Whenever you see an *Online* box, it means that there is extra online content available to support you.

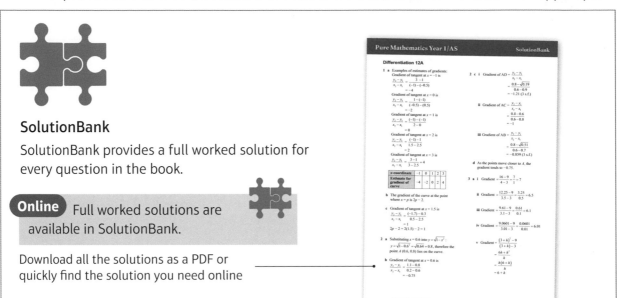

SolutionBank

SolutionBank provides a full worked solution for every question in the book.

Online Full worked solutions are available in SolutionBank.

Download all the solutions as a PDF or quickly find the solution you need online

Use of technology

Explore topics in more detail, visualise problems and consolidate your understanding using pre-made GeoGebra activities.

Online Find the point of intersection graphically using technology.

Ge♦Gebra

GeoGebra-powered interactives

Interact with the maths you are learning using GeoGebra's easy-to-use tools

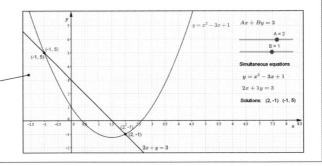

Access all the extra online content for free at:

www.pearsonschools.co.uk/d1maths

You can also access the extra online content by scanning this QR code:

Algorithms

1

Objectives

After completing this chapter you should be able to:

● Use and understand an algorithm given in words → **pages 2–5**

● Understand how flow charts can be used to describe algorithms → **pages 6–10**

● Carry out a bubble sort → **pages 10–13**

● Carry out a quick sort → **pages 13–16**

● Carry out the three bin-packing algorithms and understand their strengths and weaknesses → **pages 16–21**

● Determine the order of an algorithm → **pages 21–24**

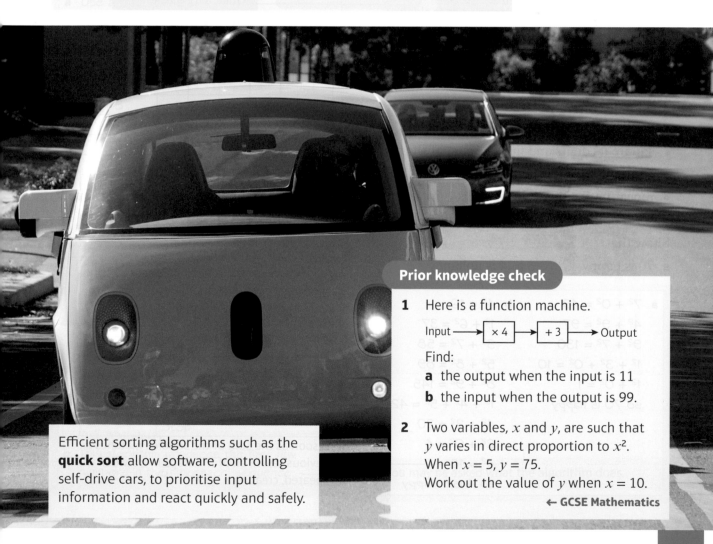

Efficient sorting algorithms such as the **quick sort** allow software, controlling self-drive cars, to prioritise input information and react quickly and safely.

Prior knowledge check

1 Here is a function machine.

Input ⟶ ×4 ⟶ +3 ⟶ Output

Find:
a the output when the input is 11
b the input when the output is 99.

2 Two variables, x and y, are such that y varies in direct proportion to x^2. When $x = 5$, $y = 75$. Work out the value of y when $x = 10$. ← **GCSE Mathematics**

1.2 Flow charts

You need to be able to implement an algorithm given as a flow chart.

- **In a flow chart, the shape of each box tells you about its function.**

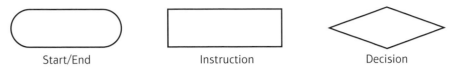

Start/End Instruction Decision

The boxes in a flow chart are linked by arrowed lines. As with an algorithm written in words, you need to follow each step in order.

Example 4

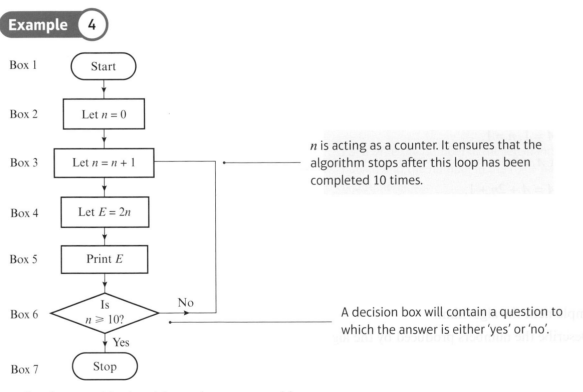

Box 1 Start

Box 2 Let $n = 0$

Box 3 Let $n = n + 1$ — n is acting as a counter. It ensures that the algorithm stops after this loop has been completed 10 times.

Box 4 Let $E = 2n$

Box 5 Print E

Box 6 Is $n \geqslant 10$? No — A decision box will contain a question to which the answer is either 'yes' or 'no'.

Yes

Box 7 Stop

a Implement this algorithm using a trace table.

b Alter box 4 to read 'Let $E = 3n$' and implement the algorithm again.
   How does this alter the algorithm?

a

n	E	Box 6
0		
1	2	no
2	4	no
3	6	no
4	8	no
5	10	no
6	12	no
7	14	no
8	16	no
9	18	no
10	20	yes

Output is 2, 4, 6, 8, 10, 12, 14, 16, 18, 20

b

n	E	Box 6
0		
1	3	no
2	6	no
3	9	no
4	12	no
5	15	no
6	18	no
7	21	no
8	24	no
9	27	no
10	30	yes

Output is 3, 6, 9, 12, 15, 18, 21, 24, 27, 30

This gives the first ten multiples of 3 rather than the first ten multiples of 2.

> In a trace table each step must be made clear.

Example 5

This flow chart can be used to find the roots of an equation of the form $ax^2 + bx + c = 0$.

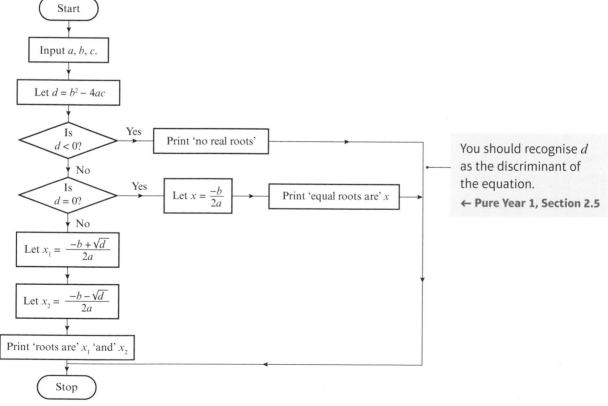

> You should recognise d as the discriminant of the equation.
> ← Pure Year 1, Section 2.5

Demonstrate this algorithm for these equations:

a $6x^2 - 5x - 11 = 0$

b $x^2 - 6x + 9 = 0$

c $4x^2 + 3x + 8 = 0$

a

	a	b	c	d	$d < 0$?	$d = 0$?	x_1	x_2
	6	−5	−11	289	no	no	$\frac{11}{6}$	−1

roots are $\frac{11}{6}$ and −1

b

	a	b	c	d	$d < 0$?	$d = 0$?	x
	1	−6	9	0	no	yes	3

equal roots are 3

c

	a	b	c	d	$d < 0$?
	4	3	8	−119	yes

no real roots

Example 6

Apply the algorithm shown by the flow chart on the right to the data:

$u_1 = 10$, $u_2 = 15$, $u_3 = 9$, $u_4 = 7$, $u_5 = 11$

What does the algorithm do?

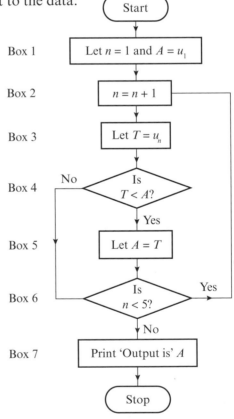

	n	A	T	$T < A$?	$n < 5$?
Box 1	1	10			
Box 2	2				
Box 3			15		
Box 4				No	
Box 6					Yes
Box 2	3				
Box 3			9		
Box 4				Yes	
Box 5		9			
Box 6					Yes
Box 2	4				
Box 3			7		
Box 4				Yes	
Box 5		7			
Box 6					Yes
Box 2	5				
Box 3			11		
Box 4				No	
Box 6					No
Box 7	Output is 7				

The algorithm selects the smallest number from a list.

This is quite complicated because it has questions and a list of data. Tackle one step at a time.

The box numbers have been included to help you to follow the algorithm. You do not need to include them in your exam.

Exercise 1B

1 Apply the flow chart in Example 5 to the following equations.

   a $4x^2 - 12x + 9 = 0$ **b** $-6x^2 + 13x + 5 = 0$ **c** $3x^2 - 8x + 11 = 0$

2 **a** Apply the flow chart in Example 6 to the following sets of data.

    i $u_1 = 28, u_2 = 26, u_3 = 23, u_4 = 25, u_5 = 21$

    ii $u_1 = 11, u_2 = 8, u_3 = 9, u_4 = 8, u_5 = 5$

   b If box 4 is altered to $\langle\, \text{Is } T > A? \,\rangle$, how will this affect the output?

   c Which box would need to be altered if the algorithm had to be applied to a list of 8 numbers?

3 The flow chart describes an algorithm that can be used to find the roots of the equation $2x^3 + x^2 - 15 = 0$.

   a Use $a = 2$ to find a root of the equation.

   b Use $a = 20$ to find a root of the equation. Comment on your answer.

Links   This flow chart implements the iterative

formula $x_{n+1} = \sqrt[3]{\dfrac{15 - x_n^2}{2}}$

← **Pure Year 2, Section 10.2**

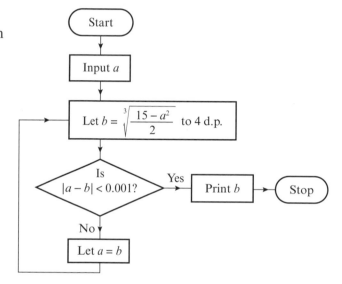

(E/P) 4 The flow chart on the right describes how to apply Euclid's algorithm to two non-zero integers, a and b.

   a Apply Euclid's algorithm to:

    i   507 and 52   **(2 marks)**

    ii   884 and 85   **(2 marks)**

    iii 4845 and 3795   **(2 marks)**

   b Explain what Euclid's algorithm does.   **(2 marks)**

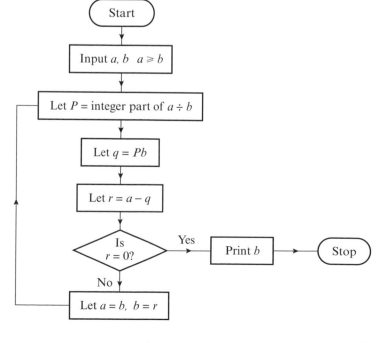

5 The flow chart describes an algorithm.

a Copy and complete this table, using the flow chart with $A = 18$ and $B = 7$.

A	B	$A < B?$	Output

(4 marks)

b Explain what is achieved by this flow chart. **(2 marks)**

c Given that $A = kB$ for some positive integer k, write down the output of the flow chart. **(1 mark)**

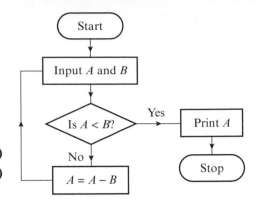

1.3 Bubble sort

A common data processing task is to sort an unordered list into alphabetical or numerical order.

■ **Unordered lists can be sorted using a bubble sort or a quick sort.**

Lists can be sorted into **ascending** or **descending** order.

■ **In a bubble sort, you compare adjacent items in a list:**
  • **If they are in order, leave them.**
  • **If they are not in order, swap them.**
  • **The list is in order when a pass is completed without any swaps.**

As the bubble sort develops, it is helpful to consider the original list as being divided into a **working list**, where comparisons are made, and a **sorted list** containing the items that are in their final positions. To start with, all items are in the working list.

This is the bubble sort algorithm:

1 Start at the beginning of the working list and move from left to right comparing adjacent items.

  a If they are in order, leave them.

  b If they are not in order, swap them.

2 When you get to the end of the working list, the last item will be in its final position. This item is then no longer in the working list.

3 If you have made some swaps in the last pass, repeat step **1**.

4 When a pass is completed without any swaps, every item is in its final position and the list is in order.

You need to learn the bubble sort algorithm.

Notation Each time you get to the end of the working list you complete one **pass** of the algorithm. The length of the working list reduces by 1 with each pass.

Note The elements in the list 'bubble' to the end of the list in the same way as bubbles in a fizzy drink rise to the top. This is how the algorithm got its name.

Example 7

Use a bubble sort to arrange this list into ascending order.

24   18   37   11   15   30

(24 18) 37 11 15 30	1st comparison: swap
18 (24 37) 11 15 30	2nd comparison: leave
18 24 (37 11) 15 30	3rd comparison: swap
18 24 11 (37 15) 30	4th comparison: swap
18 24 11 15 (37 30)	5th comparison: swap
18 24 11 15 30 37	End of first pass

After the second pass the list becomes

18 11   15 24 30 37

After the third pass the list is

11   15 18 24 30 37

After the fourth pass the list is

11   15 18 24 30 37

No swaps were made in the fourth pass, so the list is in order.

Hint In your exam you may be asked to show each comparison for one pass, but generally you will only be required to give the state of the list after each pass.

37 is already in its final position. It is now not in the working list. We now return to the start of the working list for the second pass.

After the third pass, the last three items are guaranteed to be in their final positions. In this particular case, the list is fully ordered but the algorithm requires another pass to be made.

Example 8

A list of n letters is to be sorted into alphabetical order, starting at the left-hand end of the list.

a Describe how to carry out the first pass of a bubble sort on the letters in the list.

b Carry out the first pass of a bubble sort to arrange the letters in the word ALGORITHM into alphabetical order, showing every step of the working.

c Show the order of the letters at the end of the second pass.

a Starting at the beginning of the list, compare the first two letters. If they are in alphabetical order, leave them in position, if not then swap them. Continue through the list, to the end, comparing every pair of letters in the same way.

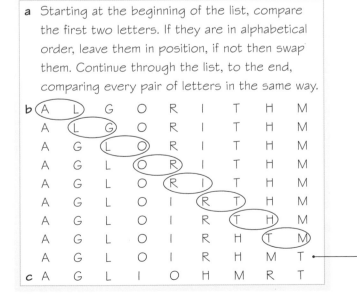

At the end of the first pass, the last letter is guaranteed to be in its correct place.

Example 9

Use a bubble sort to arrange these numbers into descending order.

| 39 | 57 | 72 | 39 | 17 | 24 | 48 |

(39 57) 72 39 17 24 48   39 < 57 so swap

57 (39 72) 39 17 24 48   39 < 72 so swap

57 72 (39 39) 17 24 48   39 ≮ 39 so leave

57 72 39 (39 17) 24 48   39 ≮ 17 so leave

57 72 39 39 (17 24) 48   17 < 24 so swap

57 72 39 39 24 (17 48)   17 < 48 so swap

57 72 39 39 24 48 17

After 1st pass: 57 72 39 39 24 48 17

After 2nd pass: 72 57 39 39 48 24 17

After 3rd pass: 72 57 39 48 39 24 17

After 4th pass: 72 57 48 39 39 24 17

After 5th pass: 72 57 48 39 39 24 17

No swaps in 5th pass, so the list is in order.

Watch out Read the question carefully. You need to sort the list into **descending** order.

Note that the 48 is now between the two 39s. Do not treat the two 39s as one term.

Make sure that you make a statement like this to show that no swaps have been made and you have completed the algorithm.

Exercise 1C

1 Apply a bubble sort to arrange each list into:

  a ascending order   **b** descending order

	i	23	16	15	34	18	25	11	19
	ii	N	E	T	W	O	R	K	S
	iii	A5	D3	D2	A1	B4	C7	C2	B3

For each part, you only need to show the state of the list at the **end** of each pass.

Hint For part **iii,** order alphabetically then numerically. So C2 comes after A5 but before C7.

2 Perform a bubble sort to arrange these place names into alphabetical order.

    Chester   York   Stafford   Bridlington   Burton   Cranleigh   Evesham

(P) 3 A list of n items is to be written in ascending order using the bubble sort.

  a State the minimum number of passes needed.

  b Describe the circumstances in which this number of passes would be sufficient.

  c State the maximum number of passes needed.

  d Describe the circumstances in which this number of passes would be needed.

(E) **4** Here is a list of exam scores:

    63   48   57   55   32   48   72   49   61   39

The scores are to be put in order, highest first, using a bubble sort.

a Describe how to carry out the first pass. **(2 marks)**

b Apply a bubble sort to put the scores in the required order. Only show the state of
the list at the end of each pass. **(4 marks)**

1.4 Quick sort

The quick sort algorithm can be used to arrange a list into alphabetical or numerical order. In many circumstances, it is quicker to implement than the bubble sort algorithm.

■ **In a quick sort, you select a pivot then split the items into two sub-lists:**

  • **One sub-list contains items less than the pivot.**

  • **The other sub-list contains items greater than the pivot.**

  • **You then select further pivots from within each sub-list and repeat the process.**

Note If an item is equal to the pivot it can go in either sub-list.

Here is the quick sort algorithm, used to sort a list into ascending order.

1 Choose the item at the midpoint of the list to be the first pivot.

    If the list has an even number of items, the pivot should be the item to the right of the middle.

2 Write down all the items that are less than the pivot, keeping their order, in a sub-list.

    Do not sort the items as you write them down.

3 Write down the pivot.

4 Write down the remaining items (those greater than the pivot) in a sub-list.

5 Apply steps **1** to **4** to each sub-list.

    This is a recursive algorithm. It is like 'zooming in' on the answer.

6 When all items have been chosen as pivots, stop.

The number of pivots has the potential to double at each pass. There is 1 pivot at the first pass, there could be 2 at the second, 4 at the third, 8 at the fourth, and so on.

Example **10**

Use quick sort to arrange the numbers below into ascending order.

    21   24   42   29   23   13   8   39   38

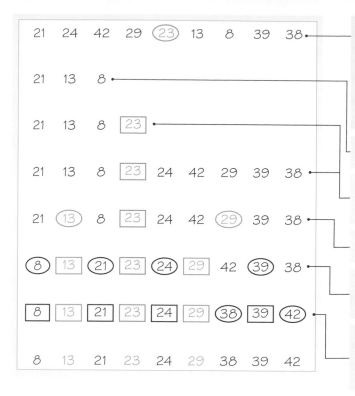

For n items, the pivot will be the $\frac{n+1}{2}$ th item, rounding up if necessary.

There are 9 numbers in the list so the middle will be $\frac{9+1}{2} = 5$, so the pivot is the 5th number in the list. Circle it.

Write all the numbers less than 23.

Write the pivot in a box, then write the remaining numbers.

Now select a pivot in each sub-list.

There are now four sub-lists so we choose 4 pivots (circled).

We can only choose two pivots this time. Each number has been chosen as a pivot, so the list is in order.

Example 11

Use quick sort to arrange the list below into descending order.

37   20   17   26   44   41   27   28   50   17

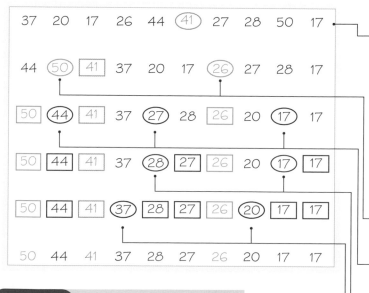

There are 10 items in the list so we choose the number to the right of the middle. This is the 6th number from the left.

Numbers greater than the pivot are to the left of the pivot, those smaller than the pivot are to the right, keeping the numbers in order. Numbers equal to the pivot may go either side, but must be dealt with consistently.

Two pivots are chosen, one for each sub-list.

Now three pivots are selected.

Watch out Colour is used here to make the method clear, but colours should not be used in your exam.

We now choose the next two pivots, even if the sub-list is in order.

The final pivots are chosen to give the list in order.

Exercise **1D**

1 Use a bubble sort to arrange the list

         8   3   4   6   5   7   2

into:

a ascending order   **b** descending order

2 Use quick sort to arrange the list

         22   17   25   30   11   18   20   14   7   29

into:

a ascending order   **b** descending order

3 Sort the letters below into alphabetical order using:

a a bubble sort   **b** a quick sort

     N   H   R   K   S   C   J   E   M   P   L

4 The list shows the test results of a group of students.

Alex	33	Hugo	9
Alison	56	Janelle	89
Amy	93	Josh	37
Annie	51	Lucy	57
Dom	77	Myles	19
Greg	91	Sam	29
Harry	49	Sophie	77

Produce a list of students, in descending order of their marks, using:

a a bubble sort   **b** a quick sort

(E/P) **5** A list of n items is to be written in ascending order using the bubble sort.

a Find an expression, in terms of n, for the maximum number of comparisons to be made.

(2 marks)

b Describe a situation where the bubble sort might be quicker than the quick sort. **(2 marks)**

c Decide whether the bubble sort or the quick sort will be quicker in the following cases:

i 1   2   3   7   4   5   6

ii 2   3   4   5   6   7   1

Explain how you made your decisions. **(4 marks)**

(E) **6** The table shows a list of nine names of students in a dance class.

Hassler	Sauver	Finch	Giannini	Mellor	Clopton	Miranti	Worth	Argi
H	S	F	G	Me	C	Mi	W	A

a Explain how to carry out the first pass of a quick sort algorithm to order the list alphabetically. **(2 marks)**

b Carry out the first two passes of a quick sort on this list, writing down the pivots used in each pass. **(3 marks)**

Challenge

You will need a pack of ordinary playing cards, with any jokers removed.

a Use the quick sort algorithm to sort the cards into ascending order, from Ace to King within each suit and with the suits in the order Hearts, Clubs, Diamonds, Spades. Follow these steps:

1 Shuffle the pack thoroughly and hold it face up.

2 Remove the 27th card and place it face up. This is your pivot card.

3 Go through the pack from the top. Place the cards into two piles depending on whether they are lower or higher than the pivot card.

4 Repeat these steps with each new pile, choosing the card halfway through the pile as the pivot card.

Record the total number of passes needed to sort the deck completely.

b Once the cards are in order, what single change could be made so that a bubble sort would require 51 passes to put the cards back in order?

> **Hint** The final order should be:
> A♥, 2♥, …, K♥, A♣, 2♣, …, K♣, A♦, …, K♦, A♠, …, K♠

1.5 Bin-packing algorithms

Bin packing refers to a whole class of problems.

The easiest is to imagine stacking boxes of fixed width a and length b, but varying heights, into bins of width a and length b, using the minimum number of bins.

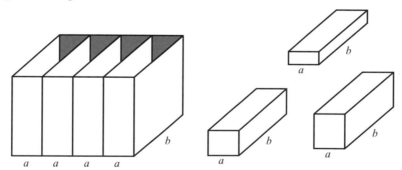

Similar problems are: loading cars of different lengths onto a ferry with several lanes of equal length, a plumber needing to cut sections from lengths of copper pipe, or recording music tracks onto a set of CDs.

You need to be able to implement three different bin-packing algorithms, and be aware of their strengths and weaknesses.

■ **The three bin-packing algorithms are: first-fit, first-fit decreasing and full-bin.**

It is useful first to find a **lower bound** for the number of bins needed. There is no guarantee that you will be able to pack the items into this number of bins, but it will tell you if you have found an optimal solution.

> **Notation** An **optimal** solution is one that cannot be improved upon. For bin packing, an optimal solution will use the smallest possible number of bins.

Example 12

Nine boxes of fixed cross-section have heights, in metres, as follows.

0.3   0.7   0.8   0.8   1.0   1.1   1.1   1.2   1.5

They are to be packed into bins with the same fixed cross-section and height 2 m.
Determine the lower bound for the number of bins needed.

0.3 + 0.7 + 0.8 + 0.8 + 1.0 + 1.1 + 1.1 + 1.2 + 1.5 = 8.5 m $$\frac{8.5}{2} = 4.25 \text{ bins}$$ So a minimum of 5 bins will be needed.

Sum the heights and divide by the bin size. You must always round **up** to determine the lower bound.

Watch out   In practice, it may not be possible to pack these boxes into 5 bins. All that the lower bound is telling us, is that **at least** five bins will be needed.

With small amounts of data it is often possible to 'spot' an optimal answer.

The algorithms you will learn in this chapter will not necessarily find an optimal solution, but can be implemented quickly.

■ **The first-fit algorithm works by considering items in the order they are given.**

First-fit algorithm

1 Take the items **in the order given**.

2 Place each item in the first available bin that can take it. Start from bin 1 each time.

Advantage: It is quick to implement.
Disadvantage: It is not likely to lead to a good solution.

Online   See the operation of the first-fit algorithm using GeoGebra.

Example 13

Use the first-fit algorithm to pack the following items into bins of size 20. (The numbers in brackets are the size of the item.) State the number of bins used and the amount of wasted space.

A(8)   B(7)   C(14)   D(9)   E(6)   F(9)   G(5)   H(15)   I(6)   J(7)   K(8)

Bin 1:	A(8)	B(7)	G(5)
Bin 2:	C(14)	E(6)	
Bin 3:	D(9)	F(9)	
Bin 4:	H(15)		
Bin 5:	I(6)	J(7)	
Bin 6:	K(8)		

This used 6 bins and there are
2 + 5 + 7 + 12 = 26 units of waste of space.

A(8) goes into bin 1, leaving space of 12.
B(7) goes into bin 1, leaving space of 5.
C(14) goes into bin 2, leaving space of 6.
D(9) goes into bin 3, leaving space of 11.
E(6) goes into bin 2, leaving space of 0.
F(9) goes into bin 3, leaving space of 2.
G(5) goes into bin 1, leaving space of 0.
H(15) goes into bin 4, leaving space of 5.
I(6) goes into bin 5, leaving space of 14.
J(7) goes into bin 5, leaving space of 7.
K(8) goes into bin 6, leaving space of 12.

- **The first-fit decreasing algorithm requires the items to be in descending order before applying the algorithm.**

First-fit decreasing algorithm

1 Sort the items so that they are in descending order.

2 Apply the first-fit algorithm to the reordered list.

Advantages: You usually get a fairly good solution.
         It is easy to implement.
Disadvantage: You may not get an optimal solution.

Online See the operation of the first-fit decreasing algorithm using GeoGebra.

Example 14

Apply the first-fit decreasing algorithm to the data given in Example 13.

Sort the data into descending order:
H(15) C(14) D(9) F(9) A(8) K(8) B(7)
J(7) E(6) I(6) G(5)
Bin 1: H(15) G(5)
Bin 2: C(14) E(6)
Bin 3: D(9) F(9)
Bin 4: A(8) K(8)
Bin 5: B(7) J(7) I(6)
This used 5 bins and there are
2 + 4 = 6 units of wasted space.

H(15) goes into bin 1, leaving space of 5.
C(14) goes into bin 2, leaving space of 6.
D(9) goes into bin 3, leaving space of 11.
F(9) goes into bin 3, leaving space of 2.
A(8) goes into bin 4, leaving space of 12.
K(8) goes into bin 4, leaving space of 4.
B(7) goes into bin 5, leaving space of 13.
J(7) goes into bin 5, leaving space of 6.
E(6) goes into bin 2, leaving space of 0.
I(6) goes into bin 5, leaving space of 0.
G(5) goes into bin 1, leaving space of 0.

- **Full-bin packing uses inspection to select items that will combine to fill bins. Remaining items are packed using the first-fit algorithm.**

Full-bin packing

1 Use observation to find combinations of items that will fill a bin. Pack these items first.

2 Any remaining items are packed using the first-fit algorithm.

Advantage: You usually get a good solution.
Disadvantage: It is difficult to do, especially when the numbers are plentiful and awkward.

Example 15

A(8)   B(7)   C(10)   D(11)   E(13)   F(17)   G(4)   H(6)   I(12)   J(14)   K(9)

The items above are to be packed in bins of size 25.

a Determine the lower bound for the number of bins.

b Apply the full-bin algorithm.

c Is your solution optimal? Give a reason for your answer.

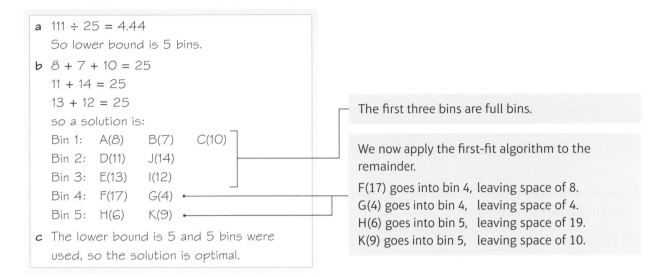

a 111 ÷ 25 = 4.44

So lower bound is 5 bins.

b 8 + 7 + 10 = 25

11 + 14 = 25

13 + 12 = 25

so a solution is:

Bin 1:   A(8)   B(7)   C(10)

Bin 2:   D(11)   J(14)

Bin 3:   E(13)   I(12)

Bin 4:   F(17)   G(4)

Bin 5:   H(6)   K(9)

c The lower bound is 5 and 5 bins were used, so the solution is optimal.

> The first three bins are full bins.

> We now apply the first-fit algorithm to the remainder.
>
> F(17) goes into bin 4, leaving space of 8.
> G(4) goes into bin 4, leaving space of 4.
> H(6) goes into bin 5, leaving space of 19.
> K(9) goes into bin 5, leaving space of 10.

Example 16

A plumber needs to cut the following lengths of copper pipe. (Lengths are in metres.)

A(0.8)   B(0.8)   C(1.4)   D(1.1)   E(1.3)   F(0.9)   G(0.8)   H(0.9)   I(0.8)   J(0.9)

The pipe comes in lengths of 2.5 m.

a Calculate the lower bound of the number of lengths of pipe needed.

b Use the first-fit decreasing algorithm to determine how the required lengths may be cut from the 2.5 m lengths.

c Use full-bin packing to find an optimal solution.

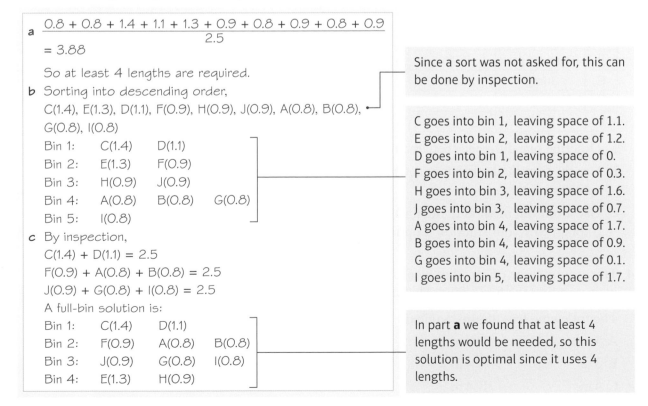

a $\dfrac{0.8 + 0.8 + 1.4 + 1.1 + 1.3 + 0.9 + 0.8 + 0.9 + 0.8 + 0.9}{2.5}$

= 3.88

So at least 4 lengths are required.

b Sorting into descending order,

C(1.4), E(1.3), D(1.1), F(0.9), H(0.9), J(0.9), A(0.8), B(0.8), G(0.8), I(0.8)

Bin 1:   C(1.4)   D(1.1)

Bin 2:   E(1.3)   F(0.9)

Bin 3:   H(0.9)   J(0.9)

Bin 4:   A(0.8)   B(0.8)   G(0.8)

Bin 5:   I(0.8)

c By inspection,

C(1.4) + D(1.1) = 2.5

F(0.9) + A(0.8) + B(0.8) = 2.5

J(0.9) + G(0.8) + I(0.8) = 2.5

A full-bin solution is:

Bin 1:   C(1.4)   D(1.1)

Bin 2:   F(0.9)   A(0.8)   B(0.8)

Bin 3:   J(0.9)   G(0.8)   I(0.8)

Bin 4:   E(1.3)   H(0.9)

> Since a sort was not asked for, this can be done by inspection.

> C goes into bin 1, leaving space of 1.1.
> E goes into bin 2, leaving space of 1.2.
> D goes into bin 1, leaving space of 0.
> F goes into bin 2, leaving space of 0.3.
> H goes into bin 3, leaving space of 1.6.
> J goes into bin 3, leaving space of 0.7.
> A goes into bin 4, leaving space of 1.7.
> B goes into bin 4, leaving space of 0.9.
> G goes into bin 4, leaving space of 0.1.
> I goes into bin 5, leaving space of 1.7.

> In part **a** we found that at least 4 lengths would be needed, so this solution is optimal since it uses 4 lengths.

Exercise 1E

1 18   4   23   8   27   19   3   26   30   35   32

The above items are to be packed in bins of size 50.

a Calculate the lower bound for the number of bins.

b Pack the items into the bins using:
  i the first-fit algorithm   **ii** the first-fit decreasing algorithm   **iii** the full-bin algorithm

2 Laura wishes to record the following television programmes onto DVDs, each of which can hold up to 3 hours of programmes.

Programme	A	B	C	D	E	F	G	H	I	J	K	L	M
Length (minutes)	30	30	30	45	45	60	60	60	60	75	90	120	120

a Apply the first-fit algorithm, in the order A to M, to determine the number of DVDs that need to be used. State which programmes should be recorded on each disc.

b Repeat part **a** using the first-fit decreasing algorithm.

c Is your answer to part **b** optimal? Give a reason for your answer.

Laura finds that her DVDs will only hold up to 2 hours of programmes.

d Use the full-bin algorithm to determine the number of DVDs she needs to use. State which programmes should be recorded on each disc.

E 3 A small ferry loads vehicles into 30 m lanes. The vehicles are loaded bumper to bumper.

	Vehicle	Length (m)
A	car	4 m
B	car 1 trailer	7 m
C	lorry	13 m
D	van	6 m
E	lorry	13 m

	Vehicle	Length (m)
F	car	4 m
G	lorry	12 m
H	lorry	14 m
I	van	6 m
J	lorry	11 m

a Describe one difference between the first-fit and full-bin methods of bin packing.   **(1 mark)**

b Use the first-fit algorithm to determine the number of lanes needed to load all the vehicles.   **(4 marks)**

c Use a full-bin method to obtain an optimal solution using the minimum number of lanes. Explain why your solution is optimal.   **(4 marks)**

E 4 The ground floor of an office block is to be fully recarpeted, with specially made carpet incorporating the firm's logo. The carpet comes in rolls of 15 m.

The following lengths are required.

A 3 m   D 4 m   G 5 m   J 7 m

B 3 m   E 4 m   H 5 m   K 8 m

C 4 m   F 4 m   I 5 m   L 8 m

The lengths are arranged in **ascending** order of size.

a Obtain a lower bound for the number of rolls of carpet needed. **(2 marks)**

b Use the first-fit decreasing bin-packing algorithm to determine the number of rolls needed. State the length of carpet that is wasted using this method. **(3 marks)**

c Give one disadvantage of the first-fit decreasing bin-packing algorithm. **(1 mark)**

d Use a full-bin method to obtain an optimal solution, and state the total length of wasted carpet using this method. **(4 marks)**

(E/P) **5** Eight computer programs need to be copied onto 40 MB discs. The size of each program is given below.

Program	A	B	C	D	E	F	G	H
Size (MB)	8	16	17	21	22	24	25	25

a Use the first-fit decreasing algorithm to determine which programs should be recorded onto each disc. **(3 marks)**

b Calculate a lower bound for the number of discs needed. **(2 marks)**

c Explain why it is not possible to record these programs on the number of discs found in part **b**. **(1 mark)**

Problem-solving

Consider the programs over 20 MB in size.

1.6 Order of an algorithm

The **order** of an algorithm, sometimes called the complexity of the algorithm, tells you how changes in the **size** of a problem affect the approximate time taken for its completion. This is sometimes called the **run time** of the algorithm.

Note The size of a sorting problem, for example, would be given by the number of items to be sorted.

- **The order of an algorithm can be described as a function of its size. If an algorithm has order f(n), then increasing the size of the problem from n to m will increase the run time of the algorithm by a factor of approximately $\dfrac{f(m)}{f(n)}$**

Watch out The exact run time will depend on the exact input data for the algorithm. However, you can use proportion to calculate **estimated** run times of algorithms.

The number of steps needed to complete an algorithm is often used to determine its order.

For the **bubble sort**, most of the steps are to do with making comparisons between pairs of numbers.

If a list has n items, then the first pass will require $(n - 1)$ comparisons.

Assuming that some swaps are made, a second pass will be needed and this will require a further $(n - 2)$ comparisons.

In the worst case, this process continues so that $(n - 3)$ comparisons are needed for the third pass, $(n - 4)$ comparisons for the fourth pass and so on, right down to a single comparison in the final pass.

The total number of comparisons would then be:

$1 + 2 + 3 + \ldots + (n - 4) + (n - 3) + (n - 2) + (n - 1)$

$= \frac{1}{2}(n - 1)n = \frac{1}{2}n^2 - \frac{1}{2}n$

Links This is the sum of the first $(n - 1)$ natural numbers. ← **Core Pure Book 1, Section 3.1**

Since this is a quadratic expression, the bubble sort is taken to have **quadratic order**.

Watch out A different algorithm may require $50n^2 + 11n + 90$ steps to complete a problem of size n. This algorithm would also be described as having quadratic order.

Example 17

An algorithm is defined by this flow diagram, where $n > 2$ and n is an integer.

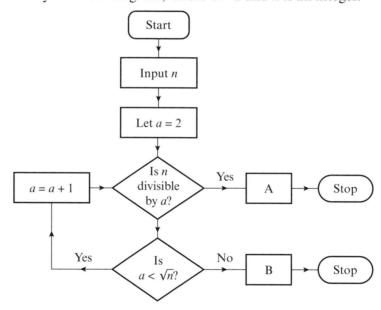

a Describe what the algorithm does.

b Suggest suitable output text for boxes A and B.

c Determine the order of the algorithm.

a The algorithm tests whether or not n is prime.

b Box A: n is not prime.

Box B: n is prime.

c Let the size of the algorithm be n.

At each step the algorithm tests whether n is divisible by a.

If n is prime, the answer at this step will never be 'yes' so the algorithm will continue until $a \geqslant \sqrt{n}$.

The maximum number of steps needed is given by the integer part of $\sqrt{n}$.

So the algorithm has order $\sqrt{n}$.

Problem-solving

The maximum number of steps will not always be needed. If n is even, then the algorithm will only require one step. In general you should consider the worst case scenario when determining the order of an algorithm.

Some common orders you need to recognise:

- Linear order (n)
- Quadratic order (n^2)
- Cubic order (n^3)

You can use these simplified rules for linear, quadratic, and cubic order algorithms:

- **If the size of a problem is increased by some factor k then:**
    - **an algorithm of linear order will take approximately k times as long to complete**
    - **an algorithm of quadratic order will take approximately k^2 times as long to complete**
    - **an algorithm of cubic order will take approximately k^3 times as long to complete.**

Example 18

A given list-searching algorithm has linear order.
To search a list with 500 values, the algorithm takes approximately 0.036 seconds.
Estimate the time taken by this algorithm to search a list containing 1250 values.

$\dfrac{1250}{500} = 2.5$

$0.036 \times 2.5 = 0.09$

The time taken to search a list containing 1250 values is approximately 0.09 seconds.

> The size of the problem has increased from 500 to 1250, which is a factor of 2.5. The algorithm has linear order, so the time taken will increase by the same factor.

Example 19

The bubble sort algorithm has quadratic order.

a Given that it takes a computer 0.2 seconds to sort a list of 1600 numbers using the bubble sort, estimate the time needed to sort a list of 480 000 numbers.

b Comment on the suitability of the bubble sort algorithm for sorting large data sets.

a $\dfrac{480\,000}{1600} = 300$

$0.2 \times 300^2 = 18\,000$

It would take approximately 18 000 seconds or 5 hours to sort 480 000 numbers.

b The length of time needed for bubble sort increases quickly as n increases. This means the bubble sort algorithm is not suitable for large data sets.

> The size of the problem has increased by a factor of 300.

> The algorithm has quadratic order, so the time taken will increase by a factor of $300^2 = 90\,000$.

> The quick sort algorithm has, on average, order $n \log n$. This means that increasing the size of the problem from 1600 to 480 000 would result in an increase in runtime by a factor of $\dfrac{480\,000 \times \log 480\,000}{1600 \times \log 1600} \approx 532$. This is a lot less than the increase of a factor of $300^2 = 90\,000$ for a quadratic order algorithm.

Exercise 1F

1 An algorithm for multiplying two $n \times n$ matrices has cubic order.

A computer program applies this algorithm to multiply two 300×300 matrices, completing the operation in 0.14 seconds.

Estimate the time needed by this computer to apply the algorithm to multiply two:

    a 600×600 matrices   **b** 1000×1000 matrices

(E/P) 2 **a** Explain why the first-fit bin-packing algorithm has quadratic order.   **(2 marks)**

A computer uses the first-fit bin-packing algorithm to determine the number of shipments needed to transport 400 lengths of piping. The total computation time is 0.72 seconds.

    b Estimate the computation time needed to apply this algorithm to 6200 lengths of piping.

                                                            (1 mark)

    c Give a reason why your answer to part **b** is only an estimate.   **(1 mark)**

(E) 3 The first-fit decreasing algorithm has quadratic order. It takes a computer 0.028 seconds to apply the first-fit decreasing algorithm to 50 items.

    a Explain briefly what is meant by a quadratic order algorithm.   **(2 marks)**

    b Estimate how long it would take the computer to apply the algorithm to 500 items.   **(1 mark)**

(P) 4 A student applies a bubble sort followed by a first-fit bin-packing algorithm.
The student states that because both algorithms have order n^2, the complete process must have order n^4.
Explain why the student is incorrect, and state the correct order of the combined process.

(E/P) 5 At a careers day, n students meet with n potential work-experience employers. The employers rate each student out of 10, and the students rate each employer out of 10. An algorithm for matching students to employers is described below.

> **1** Add each student rating for an employer to that employer's rating for the student, to create a score for each employer-student pair.
>
> **2** List the scores for all possible pairs, and order them using bubble sort.
>
> **3** Take the highest score in the list and match the student and employer for work experience. Then delete all other pairings containing either that student or that employer.
>
> **4** Repeat step **3** until all students have been paired with employers.

Determine the order of this algorithm, justifying your answer.   **(3 marks)**

Mixed exercise 1

(E) **1** Use the bubble-sort algorithm to sort, in ascending order, the list

27   15   2   38   16   1

giving the state of the list at each stage. **(4 marks)**

(E/P) **2 a** Use the bubble-sort algorithm to sort, in descending order, the list

25   42   31   22   26   41

giving the state of the list on each occasion when two or more values are interchanged. **(4 marks)**

b Find the **maximum** number of interchanges needed to sort a list of six pieces of data using the bubble-sort algorithm. **(2 marks)**

(E) **3** 8   4   13   2   17   9   15

This list of numbers is to be sorted into ascending order.
Perform a quick sort to obtain the sorted list, giving the state of the list after each rearrangement. **(5 marks)**

(E) **4** 111   103   77   81   98   68   82   115   93

a The list of numbers above is to be sorted into descending order. Perform a quick sort to obtain the sorted list, giving the state of the list after each rearrangement and indicating the pivot elements used. **(5 marks)**

b i Use the first-fit decreasing bin-packing algorithm to fit the data into bins of size 200. **(3 marks)**

ii Explain how you decided in which bin to place the number 77. **(1 mark)**

(E) **5** Trishna wishes to record eight television programmes. The lengths of the programmes, in minutes, are:

75   100   52   92   30   84   42   60

Trishna decides to use 2-hour (120 minute) DVDs only to record all of these programmes.

a Explain how to implement the first-fit decreasing bin-packing algorithm. **(2 marks)**

b Use this algorithm to fit these programmes onto the smallest number of DVDs possible, stating the total amount of unused space on the DVDs. **(3 marks)**

Trishna wants to record an additional two 25-minute programmes.

c Determine whether she can do this using only 5 DVDs, giving reasons for your answer. **(3 marks)**

E **6** A DIY enthusiast requires the following 14 pieces of wood as shown in the table.

Length in metres	0.4	0.6	1	1.2	1.4	1.6
Number of pieces	3	4	3	2	1	1

The DIY store sells wood in 2 m and 2.4 m lengths. He considers buying six 2 m lengths of wood.

a Explain why he will not be able to cut all of the lengths he requires from these six 2 m lengths. **(2 marks)**

b He eventually decides to buy 2.4 m lengths. Use a first-fit decreasing bin-packing algorithm to show how he could use six 2.4 m lengths to obtain the pieces he requires. **(4 marks)**

c Obtain a solution that requires only five 2.4 m lengths. **(4 marks)**

E/P **7** The algorithm described by the flow chart below is to be applied to the five pieces of data below.

$$u_1 = 6.1, u_2 = 6.9, u_3 = 5.7, u_4 = 4.8, u_5 = 5.3$$

Hint This question uses the modulus function. If $x \neq y$, $|x - y|$ is the positive difference between x and y, e.g. $|5 - 6.1| = 1.1$.

a Obtain the final output of the algorithm using the five values given for u_1 to u_5. **(4 marks)**

b In general, for any set of values u_1 to u_5, explain what the algorithm achieves. **(2 marks)**

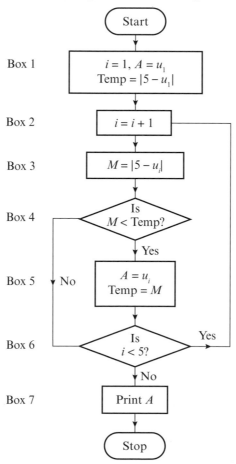

Box 1 — $i = 1$, $A = u_1$, Temp $= |5 - u_1|$

Box 2 — $i = i + 1$

Box 3 — $M = |5 - u_i|$

Box 4 — Is $M <$ Temp?

Box 5 — $A = u_i$, Temp $= M$

Box 6 — Is $i < 5$?

Box 7 — Print A

c If Box 4 in the flow chart is altered to
'Is $M >$ Temp?'
state what the algorithm achieves now. **(1 mark)**

E **8** A plumber is cutting lengths of PVC pipe for a bathroom installation. The lengths needed, in metres, are:

0.3 2.0 1.3 1.6 0.3 1.3 0.2 0.1 2.0 0.5

The pipe is sold in 2 m lengths.

a Carry out a bubble sort to produce a list of the lengths needed in **descending** order. Give the state of the list after each pass. **(4 marks)**

b Apply the first-fit decreasing bin-packing algorithm to your ordered list to determine the total number of 2 m lengths of pipe needed. **(3 marks)**

c Does the answer to part **b** use the minimum number of 2 m lengths? You must justify your answer. **(2 marks)**

E/P **9** Here are the names of eight students in an A level group:

Maggie, Vivien, Cath, Alana, Daisy, Beth, Kandis, Sara

a Use a quick sort to put the names in alphabetical order. Show the result of each pass and identify the pivots. **(5 marks)**

The quick sort algorithm has order $n \log n$.
A computer program can sort a list of 100 names in 0.3 seconds using a quick sort.

b Estimate the time needed for this computer program to apply a quick sort to a list of 1000 names. **(2 marks)**

> ### Challenge
>
> An algorithm for factorising an n-digit integer is found to have order 1.1^n. A computer uses the algorithm to factorise 8 788 751, taking 0.734 seconds.
>
> **a** Estimate the time needed for the computer to factorise:
>   **i** 3 744 388 667   **ii** a number with 100 digits
>
> Internet security is based on large, hard-to-factorise numbers. A cryptographer wants to choose a number which will take at least one year to factorise using this algorithm.
>
> **b** Determine the minimum number of digits the cryptographer should use for their number.
>
> **c** Suggest a reason why the run time of this algorithm might vary widely depending on the choice of number to be factorised.

Summary of key points

1 An **algorithm** is a finite sequence of step-by-step instructions carried out to solve a problem.

2 In a **flow chart**, the shape of each box tells you about its function.

3 Unordered lists can be sorted using a bubble sort or a quick sort.

4 In a **bubble sort**, you compare adjacent items in a list:
   - If they are in order, leave them.
   - If they are not in order, swap them.
   - The list is in order when a pass is completed without any swaps.

5 In a **quick sort**, you select a pivot then split the items into two sub-lists:
   - One sub-list contains items less than the pivot.
   - The other sub-list contains items greater than the pivot.
   - You then select further pivots from within each sub-list and repeat the process.

6 The three bin-packing algorithms are first-fit, first-fit decreasing, and full-bin:
   - The **first-fit algorithm** works by considering items in the order they are given.
   - The **first-fit decreasing** algorithm requires the items to be in descending order before applying the algorithm.
   - **Full-bin packing** uses inspection to select items that will combine to fill bins. Remaining items are packed using the first-fit algorithm.

7 The three bin-packing algorithms have the following advantages and disadvantages:

Type of algorithm	Advantage	Disadvantage
First-fit	Quick to implement	Not likely to lead to a good solution
First-fit decreasing	Usually a good solution Easy to implement	May not get an optimal solution
Full-bin	Usually a good solution	Difficult to do, especially when the numbers are plentiful or awkward

8 The **order** of an algorithm can be described as a function of its size. If an algorithm has order $f(n)$, then increasing the size of the problem from n to m will increase the run time of the algorithm by a factor of approximately $\dfrac{f(m)}{f(n)}$

9 If the size of a problem is increased by some factor k then:
   - an algorithm of linear order will take approximately k times as long to complete
   - an algorithm of quadratic order will take approximately k^2 times as long to complete
   - an algorithm of cubic order will take approximately k^3 times as long to complete.

Graphs and networks

Objectives

After completing this chapter you should be able to:

● Know how graphs and networks can be used to create mathematical models   → pages 30–33

● Be familiar with basic terminology used in graph theory
                                                                → pages 34–38

● Know some special types of graph   → pages 38–40

● Understand how graphs and networks can be represented using matrices   → pages 41–43

● Use the planarity algorithm to determine whether or not a given graph is planar   → pages 43–47

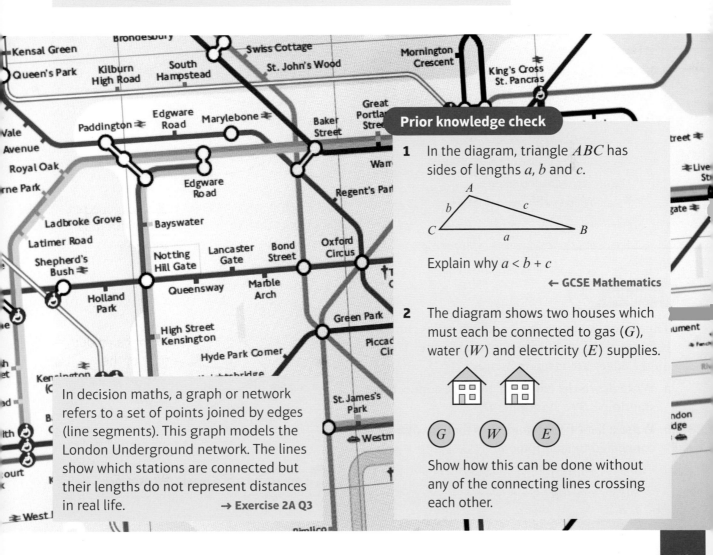

Prior knowledge check

1   In the diagram, triangle ABC has sides of lengths a, b and c.

Explain why $a < b + c$

                                ← GCSE Mathematics

2   The diagram shows two houses which must each be connected to gas (G), water (W) and electricity (E) supplies.

Show how this can be done without any of the connecting lines crossing each other.

In decision maths, a graph or network refers to a set of points joined by edges (line segments). This graph models the London Underground network. The lines show which stations are connected but their lengths do not represent distances in real life.   → Exercise 2A Q3

2.1 Modelling with graphs

You need to know how graphs and networks can be used to create mathematical models. In decision maths, the word **graph** has a very specific meaning.

- **A graph consists of points (called vertices or nodes) which are connected by lines (edges or arcs).**

- **If a graph has a number associated with each edge (usually called its weight), then the graph is known as a weighted graph or network.**

> **Notation**
> - A **vertex** is a point on a graph. **Vertices** are sometimes called **nodes**.
> - An **edge** is a line segment joining vertices. Edges are sometimes called **arcs**.

Example 1

This graph shows the routes flown by an airline.

a Explain why this is a graph.

b Describe what is represented by

  i the vertices

  ii the edges.

c Describe two possible flight routes from Johannesburg to Liberville.

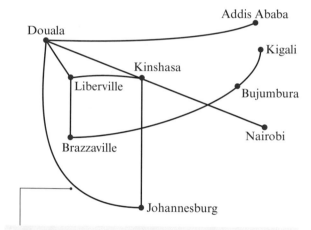

> a It has vertices connected by edges.
> b i Cities served by the airline
>   ii Flight routes connecting cities
> c Johannesburg to Douala to Liberville
>   Johannesburg to Kinshasa to Liberville

The graph shows only which cities are connected by a flight route. There is no scale and the graph is not weighted, so it does not show any distances or the correct geographical position of the cities.

Example 2

This network shows the lengths of pipes running between water standpipes on a camp site. The lengths on the edges are given in metres.

a Write down the length of the pipe connecting standpipe B to standpipe D.

b Write a list of the standpipes that are directly connected to standpipe E.

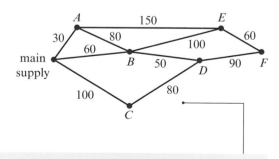

> a 50 m
> b A, B, F

This is an example of a **weighted network**.

Example (3)

The network shows the times taken, in minutes, by a car to travel along some sections of road.

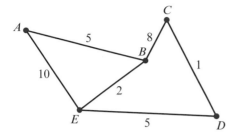

Calculate the minimum time needed to travel from A to D and give the route taken.

Route	Time needed (mins)
$ABCD$	$5 + 8 + 1 = 14$
$ABED$	$5 + 2 + 5 = 12$
AED	$10 + 5 = 15$

The minimum time required is 12 minutes along the route $ABED$.

Watch out A weighted network is not usually drawn to scale. The quickest route may not be the route which appears to be the shortest – you need to consider the weight on each arc.

Exercise (2A)

1 This graph represents the friendships within a group of students.

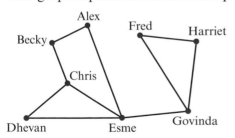

a State what is represented by each

  i vertex

  ii arc

b List all of Chris's friends within the group.

c State two new friendships that could be created to ensure that every pair of friends has at least one mutual friend.

2 This table gives information about the A level subjects studied by some students.

Student	Subjects
A	Maths, Chemistry
B	Maths, Chemistry, Biology, Art
C	Physics, Chemistry, Biology
D	Maths, Physics, Art, English
E	Biology, English, Art
F	English, Maths, Art, Physics

a Copy and complete this graph to represent this information. Use the arcs to show which subjects each student studies.

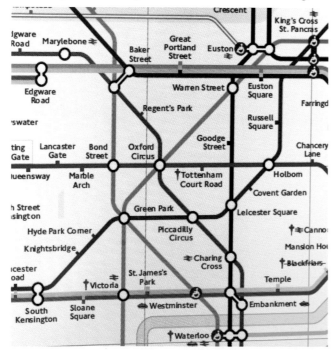

b Which subjects are studied by the most students?

P **3** The diagram shows part of the London Underground map.

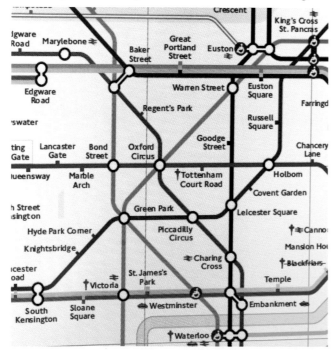

a Work out a possible route from Marylebone to Victoria

  i that requires a minimum number of changes

  ii that passes through a minimum number of stations.

A student estimates that each stop on the London Underground takes 80 seconds, and each change takes 3 minutes.

b Work out the quickest possible journey time from
  i Kings Cross St Pancras to Waterloo
  ii Holborn to St James's Park
  iii Victoria to Baker Street.

In real life the quickest way to get from Kings Cross St Pancras to Waterloo is usually to change at Oxford Circus.

c Compare this to your answer to part **b i** and suggest a potential problem with the student's model.

(P) **4** This weighted network shows the flight times of a particlar airline, in minutes, between some different airports in the UK and Ireland.

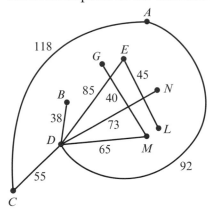

A	Aberdeen
B	Belfast
C	Cork
D	Dublin
E	Edinburgh
G	Glasgow
L	Leeds-Bradford
M	Manchester
N	Newcastle

a Write down the flight time between Glasgow and Manchester.

b Name the route with the longest flight time.

c Suggest which city might be the home base of this airline. Give a reason for your answer.

(P) **5** This network shows the lengths, in km, of a network of mountain bike trails.

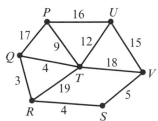

A student states that the shortest possible route to bike from P to V is 27 km long.

a State which route the student has chosen to calculate this distance.

b State, with a reason, whether the student is correct.

Challenge

The diagram shows a wire network in the shape of a cube.
An ant walks from vertex A to vertex G along the edges of the cube.
Find the total number of possible different routes which:
a are of minimum length
b do not pass through any vertex more than once.

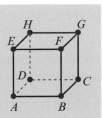

2.2 Graph theory

You need to be familiar with some basic terminology used to describe graphs in decision maths. There are two different types of notation used to describe graphs, as follows:

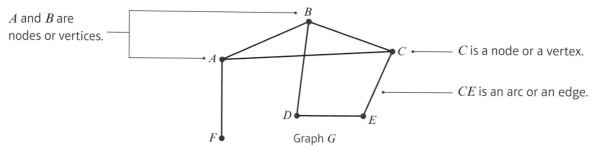

A and *B* are nodes or vertices.

C is a node or a vertex.

CE is an arc or an edge.

Graph *G*

In **graph** *G*, above

- the **vertices** (or **nodes**) are: *A*, *B*, *C*, *D*, *E* and *F* (this list of vertices is sometimes called the **vertex set**)
- the **edges** (or **arcs**) are: *AB*, *AC*, *AF*, *BC*, *BD*, *CE* and *DE* (this list of edges is sometimes called the **edge set**).

Watch out

The intersection of the edges *AC* and *BD* is not a vertex.

- **A subgraph of *G* is a graph, each of whose vertices belongs to *G* and each of whose edges belongs to *G*. It is simply a part of the original graph.**

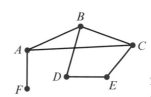

Some possible subgraphs.

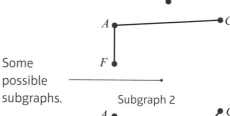

Subgraph 1

Subgraph 2

- **The degree or valency or order of a vertex is the number of edges incident to it.**

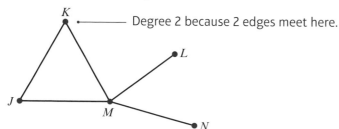

Degree 2 because 2 edges meet here.

Vertex	Degree
J	2
K	2
L	1
M	4
N	1

Notation

A vertex and an edge are **incident** if the vertex is at either end of the edge.

- **If the degree of a vertex is even, you say it has even degree, so *J*, *K* and *M* have even degree. Similarly vertices *L* and *N* have odd degree.**

- **A walk is a route through a graph along edges from one vertex to the next.**

- **A path is a walk in which no vertex is visited more than once.**

Notation

- A **subgraph** is a part of the original graph.
- The **degree** of a vertex is sometimes called the **valency** or **order**.
- **Even degree** means that a vertex has an even number of incident edges. **Odd degree** means the vertex has an odd number of incident edges.

- **A trail** is a walk in which no edge is visited more than once.
- **A cycle** is a walk in which the end vertex is the same as the start vertex and no other vertex is visited more than once.
- **A Hamiltonian cycle** is a cycle that includes every vertex.

Links A trail which traverses every arc and starts and ends at the same vertex is called an **Eulerian circuit**. → Section 4.1

In the graph below, an example of:

- a **walk** is $RSUWVU$ ———— It is ok to include a vertex or an edge more than once on a walk.
- a **path** is $RSUVW$
- a **trail** is $RUSVUW$
- a **cycle** is $RSUR$ ———— It is not necessary to include every vertex in a cycle.
- a **Hamiltonian cycle** is $RSUVWTR$.

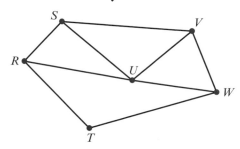

- **Two vertices are connected** if there is a path between them. **A graph is connected** if all its vertices are connected.

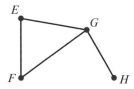 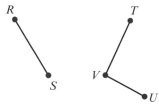

This is a **connected** graph.
A path can be found between any two vertices.

This graph is **not connected**.
There is no path from R to V, for example.

- **A loop** is an edge that starts and finishes at the same vertex.

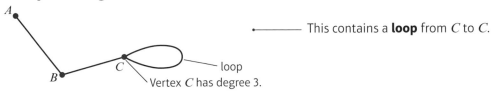

———— This contains a **loop** from C to C.

loop
Vertex C has degree 3.

- **A simple graph** is one in which there are no loops and there is at most one edge connecting any pair of vertices.

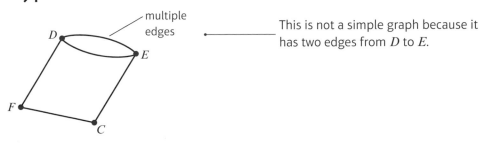

multiple edges

This is not a simple graph because it has two edges from D to E.

■ If the edges of a graph have a direction associated with them they are known as **directed edges** and the graph is known as a **directed graph**, often abbreviated to **digraph**.

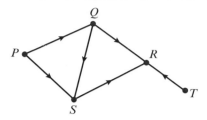

■ In any undirected graph, the sum of the degrees of the vertices is equal to **2 × the number of edges**. As a consequence, **the number of odd vertices must be even, including possibly zero**. This result is known as **Euler's handshaking lemma**.

Example 4

a Find the sum of the valencies of the vertices in this graph.

b Verify that your answer to part **a** is twice the number of edges in the graph.

a

Vertex	Valency
A	3
B	3
C	3
D	3
E	4
F	2
Total	18

Notice that the number of vertices of odd degree is even (there are 4 of them).

If each vertex represents a person and each edge indicates that two people have shaken hands, then the number of people who have shaken hands an odd number of times must be even.

b The graph has 9 edges.
   Sum of valencies = 18 = 2 × 9
                          = 2 × number of edges

Example 5

A graph has 5 nodes and 8 edges. The valencies of the nodes are x, $x - 1$, $x + 1$, $2x - 1$ and $x - 1$.
Find the value of x.

Sum of valencies
   $= x + x - 1 + x + 1 + 2x - 1 + x - 1$
   $= 6x - 2$
Sum of valencies = 2 × number of edges
                      = 2 × 8 = 16
$6x - 2 = 16$
   $6x = 18$
   $x = 3$

Problem-solving

Use Euler's handshaking lemma to formulate an equation in x.

Exercise **2B**

1 Draw a connected graph with:

    a one vertex of degree 4 and four vertices of degree 1

    b three vertices of degree 2, one of degree 3 and one of degree 1

    c two vertices of degree 2, two of degree 3 and one of degree 4.

2 Which of the graphs below are not simple?

3 In question **2**, which graphs are not connected?

4 For the graph on the right, state:

    a four paths from *F* to *D*

    b a cycle passing through *F* and *D*

    c the degree of each vertex.

Use the graph to:

    d draw a subgraph

    e confirm the handshaking lemma.

Notation A **lemma** is a mathematical fact used as a stepping stone to more important results.

5 Repeat question **4** parts **c**, **d** and **e** using this graph:

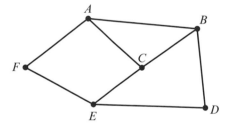

P **6** Show that it is possible to draw a graph with:

    a an even number of vertices of even degree

    b an odd number of vertices of even degree.

    c It is not possible to draw a graph with an odd number of vertices of odd degree. Explain why not.

7 Copy the diagram and complete a digraph for the relationship 'is a factor of'.

2 •   16 •

4 •   8 •

(E) 8 a Explain what is meant by a Hamiltonian cycle. **(2 marks)**

   b For this graph, list all of the Hamiltonian cycles that start from the vertex P. **(4 marks)**

   c Explain what is meant by a subgraph. **(2 marks)**

   d Draw a subgraph of this graph with 4 vertices. **(1 mark)**

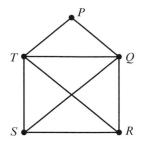

2.3 Special types of graph

You need to know about the following special types of graph.

- **A tree is a connected graph with no cycles.**

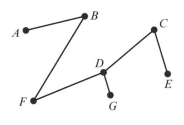

A tree

This is not a tree.
It contains cycle $RTUSR$.

- **A spanning tree of a graph, G, is a subgraph which includes all the vertices of G and is also a tree.**

For example, starting with this graph:

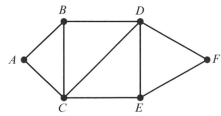

possible spanning trees are:

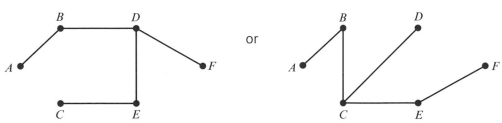

or

There are many other possible spanning trees.

- **A complete graph** is a graph in which every vertex is directly connected by a single edge to each of the other vertices.

K_3

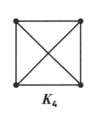

K_4

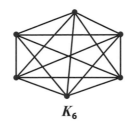

K_6

> **Notation** The complete graph with n vertices is written as K_n.

- **Isomorphic graphs** are graphs which show the same information but may be drawn differently.

For example,

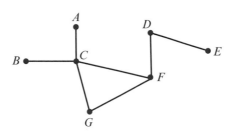

is isomorphic to:

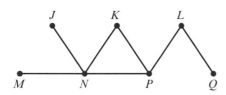 and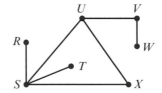

For two graphs to be isomorphic, they must have the same numbers of vertices of the same degrees, and these vertices must also be connected together in the same ways: if there is an edge between two vertices in one graph, then there is also an edge between the two corresponding vertices in the other graph.

If graphs are isomorphic it is possible to pair equivalent vertices.
In this case,

- A can be paired with J in the first graph and R in the second.
- B can be paired with M in the first graph and T in the second.
- C can be paired with N in the first graph and S in the second.
- D can be paired with L in the first graph and V in the second.
- E can be paired with Q in the first graph and W in the second.
- F can be paired with P in the first graph and U in the second.
- G can be paired with K in the first graph and X in the second.

Exercise

1 State which of the following graphs are trees.

a  **b**   **c** 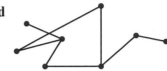 **d**

2 There are 11 spanning trees for the graph shown on the right. Draw them all.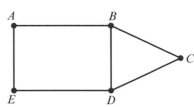

3 Identify a complete subgraph of this graph that has 4 vertices.

4 Identify which of graphs *A*, *B* and *C* are isomorphic to the graph shown on the right.

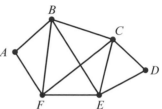

A 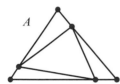 *B* 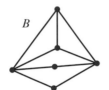 *C*

5 a Define the terms:

  i tree (1 mark)

  ii spanning tree. (1 mark)

 b Explain why it is not possible to construct a spanning tree for this graph. (1 mark)

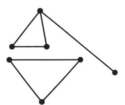

6 a Draw K_4. (2 marks)

 b State the degree of each vertex in the graph K_n. (1 mark)

 c Find the total number of edges in the graph K_{20}. (2 marks)

Challenge

Draw all possible connected graphs with three edges.

Hint Not counting isomorphic graphs, there are 11 such graphs.

2.4 Representing graphs and networks using matrices

You can use an **adjacency matrix** to represent a graph or network. The adjacency matrix provides information about the connections between the vertices in a graph.

- **Each entry in an adjacency matrix describes the number of arcs joining the corresponding vertices.**

Example 6

Use an adjacency matrix to represent this graph.

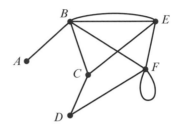

	A	B	C	D	E	F
A	0	1	0	0	0	0
B	1	0	1	0	2	1
C	0	1	0	1	1	0
D	0	0	1	0	0	1
E	0	2	1	0	0	1
F	0	1	0	1	1	2

Watch out You should be able to write down the adjacency matrix given the graph, and draw a graph given the adjacency matrix.

This indicates that there are 2 direct connections between B and E.

This indicates a loop from F to F. It could be travelled in either direction, and hence counts as 2.

The matrix associated with a weighted graph is called a **distance matrix**.

- **In a distance matrix, the entries represent the weight of each arc, not the number of arcs.**

Example 7

Use a distance matrix to represent this network.

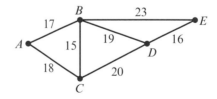

	A	B	C	D	E
A	—	17	18	—	—
B	17	—	15	19	23
C	18	15	—	20	—
D	—	19	20	—	16
E	—	23	—	16	—

Notice that the matrix is symmetrical about the leading diagonal (top left to bottom right). This will be the case for any non-directed network.

Watch out You should be able to write down the distance matrix given the network and draw the network given the distance matrix.

If the network has directed edges the distance matrix will not be symmetrical.

Example 8

Use a distance matrix to represent this directed network.

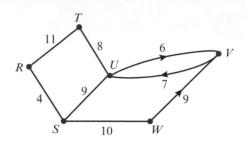

	R	S	T	U	V	W
R	—	4	11	—	—	—
S	4	—	—	9	—	10
T	11	—	—	8	—	—
U	—	9	8	—	6	—
V	—	—	—	7	—	—
W	—	10	—	—	9	—

Watch out This matrix is not symmetrical about the leading diagonal.

This indicates a direct link of weight 6 from U to V.

This indicates a direct link of weight 7 from V to U.

This shows a direct link of weight 9 from W to V.

Exercise 2D

1

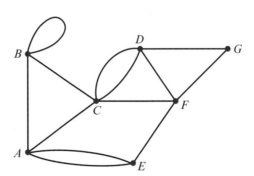

Use an adjacency matrix to represent the graph above.

2 Draw a graph corresponding to each adjacency matrix.

a

	A	B	C	D	E
A	0	1	0	1	0
B	1	0	1	1	1
C	0	1	0	2	0
D	1	1	2	0	1
E	0	1	0	1	0

b

	A	B	C	D
A	0	1	0	1
B	1	0	0	1
C	0	0	2	1
D	1	1	1	0

c

	A	B	C	D
A	0	1	0	1
B	1	0	1	1
C	0	1	0	1
D	1	1	1	0

d

	A	B	C	D
A	0	2	0	1
B	2	0	1	0
C	0	1	0	1
D	1	0	1	0

3 Draw the network corresponding to each distance matrix.

a

	A	B	C	D	E
A	—	21	—	20	23
B	21	—	17	23	—
C	—	17	—	18	41
D	20	23	18	—	22
E	23	—	41	22	—

b

	A	B	C	D	E	F
A	—	—	—	—	15	8
B	—	—	9	13	17	11
C	—	9	—	8	—	—
D	—	13	8	—	10	—
E	15	17	—	10	—	—
F	8	11	—	—	—	—

4 Use a distance matrix to represent the directed network below.

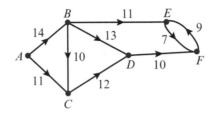

E **5 a** Draw a graph to represent this adjacency matrix.

	A	B	C	D	E
A	0	2	0	0	1
B	2	0	1	0	1
C	0	1	0	1	1
D	0	0	1	0	1
E	1	1	1	1	0

Watch out This is an adjacency matrix, not a distance matrix.

(3 marks)

b For this graph, a spanning tree may be drawn such that one vertex connects to every edge. Which vertex is this? **(1 mark)**

c Draw the spanning tree referred to in part **b**. **(2 marks)**

2.5 The planarity algorithm

A
You need to be able to determine whether a graph is planar.

- **A planar graph is one that can be drawn in a plane such that no two edges meet except at a vertex.**

Both of these graphs have edges crossing, so it appears that neither is planar. However, K_4 can be drawn with no edges crossing.

K_4

K_5

The same cannot be done with K_5. It follows that K_4 is planar, but K_5 is not.

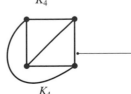

K_4

This version of K_4 shows the same connections, but this time no edges cross.

43

A If a graph contains a Hamiltonian cycle then it is possible to apply an algorithm to determine whether the graph is planar.

- **The planarity algorithm can be applied to graphs that contain a Hamiltonian cycle. The steps are as follows:**

1 Identify a Hamiltonian cycle.

2 Draw a polygon with vertices labelled to match the ones in the Hamiltonian cycle.

Try to draw a regular polygon as this makes things clearer.

3 Draw edges inside the polygon to match the edges of the original graph not already represented by the polygon itself.

4 Make a list of all edges **inside** the polygon, in any order.

This is a list of all the edges in the original graph that are not part of the Hamiltonian cycle. You will attempt to show how the graph can be drawn in a planar way by choosing edges to draw inside ('I') and outside ('O') the Hamiltonian cycle.

5 Choose any unlabelled edge in your list and label it 'I'. If all edges are now labelled then the graph is **planar**.

6 Look at any **unlabelled** edges that cross the edge(s) just labelled:

- If there are none, then go back to step **5**
- If any of these edges cross each other, then the graph is **non-planar**.
- If none of these edges cross each other, give them the **opposite label** to the edge(s) just labelled.
- If all edges are now labelled, the graph is **planar**. Otherwise, go back to the start of step **6**.

If the last set of edges were labelled 'I', then the 'opposite label' would be 'O', and vice versa.

A pair of edges that cross inside the Hamiltonian cycle would also cross if they were both drawn **outside** the Hamiltonian cycle. This algorithm attempts to partition the edges not in the Hamiltonian cycle into two sets: those that can be drawn inside the cycle, and those that can be drawn outside it. If the algorithm determines that the graph is planar, you can draw it without crossings by drawing any edges labelled 'I' inside the Hamiltonian cycle, and any edges labelled 'O' outside.

Example 9

Use the planarity algorithm to determine whether this graph is planar.

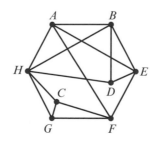

A

Step 1: A Hamiltonian cycle is *ABDEFGCHA*.

Steps 2, 3:

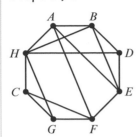

Draw your Hamiltonian cycle as a regular octagon, then add the remaining edges to create a graph that is isomorphic to the given graph. Be careful when labelling your vertices. The order around the polygon should be the same as the order in the Hamiltonian cycle.

Step 4: *AE, AF, BH, BE, DH, FC, GH*

Count to check: there are 7 edges inside the polygon so there should be 7 edges in your list.

Step 5: *AE*(I), *AF, BH, BE, DH, FC, GH*

Choose any edge to start and label it 'I'.

Step 6: Edge just labelled 'I': *AE*
Unlabelled edges that cross *AE: BH* and *DH*
AE(I), *AF, BH*(O), *BE, DH*(O), *FC, GH*

Edges *BH* and *DH* cross *AE* and are unlabelled. They don't cross each other, so label them 'O'. The graph still contains unlabelled edges, so go back to the start of step **6**.

Step 6: Edges just labelled 'O': *BH* and *DH*
Unlabelled edges that cross *BH* or *DH*: *AF* and *BE*
AE(I), *AF*(I), *BH*(O), *BE*(I), *DH*(O), *FC, GH*

Consider **all** the edges that cross **any** of those just labelled. *AE* also crosses, but it is labelled already, so you don't need to include it. *AF* and *BE* do not cross each other, so label them 'I', and go back to the start of step **6**.

Step 6: Edges just labelled 'I': *AF* and *BE*
Unlabelled edges that cross *AF* or *BE*: none

BH, BE and *DH* all cross the edges just labelled, but they are already labelled. Go back to step **5**.

Step 5:
AE(I), *AF*(I), *BH*(O), *BE*(I), *DH*(O), *FC*(I), *GH*

Choose either of the unlabelled edges and label it 'I'.

Step 6: Edge just labelled 'I': *FC*
Unlabelled edges that cross *FC: GH*
AE(I), *AF*(I), *BH*(O), *BE*(I), *DH*(O), *FC*(I), *GH*(O)
All edges are labelled so the graph is planar:

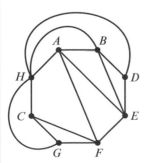

Watch out You can demonstrate the completed algorithm by giving each edge inside the Hamiltonian cycle with its label, but the safest way to check your answer and show that the graph is planar is to draw an isomorphic graph in a planar way. Draw any edges labelled 'I' inside the Hamiltonian cycle, and any edges labelled 'O' outside the Hamiltonian cycle.

Example 10

A An electronics company attaches components A, B, C, D, E and F to a circuit board. These components are to be connected by wires that are not allowed to cross. The table gives information about the connections.

Component	Connected to
A	B, D, E, F
B	C, D, E, F
C	D, E
D	E, F
E	F

In the table, each connection is only represented once. For example, the connection from A to B is shown, but not the connection from B to A.

Using the Hamiltonian cycle $ABCDEFA$, apply the planarity algorithm to determine whether the connections can be made without any wires crossing.

Step 1: A Hamiltonian cycle is $ABCDEFA$.

Steps 2, 3:

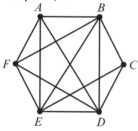

You should use the given Hamiltonian cycle. Draw a graph showing all the connections in the table.

Step 4: AD, AE, BD, BE, BF, CE, DF

Step 5: $AD(I)$, AE, BD, BE, BF, CE, DF

Step 6: Edge just labelled 'I': AD
Unlabelled edges that cross AD: BE, BF, CE
$AD(I)$, AE, BD, $BE(O)$, $BF(O)$, $CE(O)$, DF

Step 6: Edges just labelled 'O': BE, BF, CE
Unlabelled edges that cross BE, BF or CE: AE, BD, DF

AE and DF cross each other, so the graph is non-planar. The connections cannot be made without wires crossing.

Problem-solving

You could produce a written argument to mirror the working shown in the algorithm:
BE, BF and CE must all be in the same set (inside or outside), as they are all crossed by AD.
AE and DF must therefore also be in the same set (inside or outside), as they each cross at least one of BE, BF and CE.
But AE and DF cross, so the graph cannot be drawn without crossings.

Exercise 2E

E **1 a** Explain why there is no Hamiltonian cycle that begins $ABC...$ **(2 marks)**

   b Complete a Hamiltonian cycle that starts with $ABE...$ **(1 mark)**

   c Use the planarity algorithm to show that this graph is planar. **(3 marks)**

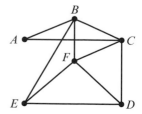

A **2 a** Define what is meant by a planar graph. **(2 marks)**

E

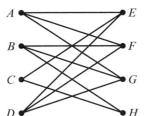

    b Starting at A, find a Hamiltonian cycle for the graph shown. **(1 mark)**

    c Use the planarity algorithm to determine whether or not the graph is planar.
You must make your working clear and justify your answer. **(4 marks)**

E/P **3** Electronic components P, Q, R, S, T and U are attached to a circuit board.
The table shows the connections to be made with wires between the components.
Each connection is shown only once.

Component	Connects directly to
P	Q, R, S, T, U
Q	R, U
R	T, U
S	T, U
T	U

    a Draw a graph to represent the connections between the components. **(2 marks)**

    b Determine whether or not the connections can be made without crossing any wires. **(4 marks)**

    A new connection is required between R and S.

    c Apply the planarity algorithm a second time to determine whether the addition of this
connection will affect your answer to part **b**. **(4 marks)**

E/P **4** The diagram shows three houses that must each be connected to gas (G), water (W) and
electricity (E) supplies.

    Use the planarity algorithm on a suitable graph to show that it is not possible to do this in
such a way that no two supply lines cross each other. **(6 marks)**

Challenge

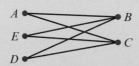

    a Explain why the planarity algorithm cannot be used to
determine whether or not this graph is planar.

    b Show that the graph is planar.

Mixed exercise 2

1 *A*, *B*, *C* and *D* are four towns. The distance matrix shows the direct distances by road between the towns in kilometres.

	A	*B*	*C*	*D*
A	—	11	15	9
B	11	—	—	8
C	15	—	—	16
D	9	8	16	—

Draw a weighted graph to show this information.

2 Which of these graphs are isomorphic?

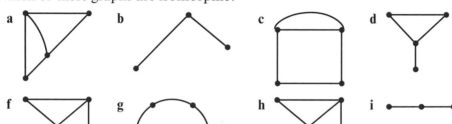

3 a Draw a graph with eight vertices, all of degree 1.

  b Draw a graph with eight vertices, all of degree 2, so that the graph is:

    i connected and simple   **ii** not connected and simple   **iii** not connected and not simple.

4 a Describe the difference between a distance matrix and an adjacency matrix.

  b Use a distance matrix to represent the network below.

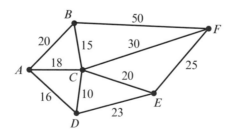

  c Draw a spanning tree for this network and state the weight of your spanning tree.

(P) 5 Write an expression for the number of edges in a spanning tree containing *v* vertices.

6 Write down all the possible routes from *P* to *R* in this digraph.

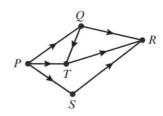

(E) 7 The diagram shows a graph.

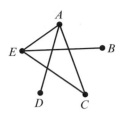

    a Write down the valency of vertex A. **(1 mark)**

    b Identify:

       i a path from B to D **(1 mark)**

       ii a cycle. **(1 mark)**

    c Sketch:

       i a spanning tree for this graph **(1 mark)**

      ii a complete subgraph of this graph. **(2 marks)**

(E/P) 8 **a** Explain why it is not possible to draw a graph with exactly four vertices with degrees 3, 1, 2 and 1. **(1 mark)**

    A connected graph has exactly four vertices and 10 edges. The degrees of the vertices are $k^2 - 3k$, $k + 1$, $8 - k$ and $k - 4$ respectively.

    b Find the value of k. **(3 marks)**

(A)
(E) 9 A Hamiltonian cycle for this graph begins $AGB…$

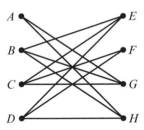

    a Complete the Hamiltonian cycle. **(2 marks)**

    b Use the planarity algorithm to determine whether the graph is planar. **(4 marks)**

(E/P) 10

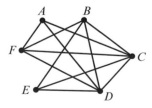

    a Write down a Hamiltonian cycle starting from A. **(2 marks)**

    b Show that this graph is planar by using the planarity algorithm. **(4 marks)**

    Arc AB is added to the graph.

    c Use the planarity algorithm to show that the new graph is non-planar. **(4 marks)**

Challenge

A The diagram shows a connected planar graph. The **finite regions** of the graph are shaded in different colours.

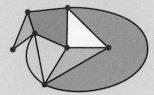

Let V denote the number of vertices, E denote the number of edges, and R denote the number of finite regions in a connected, planar graph.

a For the graph above, verify that $V + R - E = 1$.

b Draw a graph with one vertex and four edges, and verify that $V + R - E = 1$.

c Explain why the relationship $V + R - E = 1$ holds for any graph with one vertex.

A **contraction** of an edge in a graph glues together adjacent vertices, creating a graph with one fewer vertex. This diagram shows the above graph with one edge contracted.

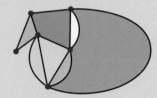

d Given that G is a planar connected graph, and that G' is the graph formed by contracting one edge of G, show that if the relationship $V + R - E = 1$ holds for G', then it must hold for G.

e Using your results from parts **c** and **d**, show that the relationship $V + R - E = 1$ holds for all planar connected graphs.

Links In part **e**, use proof by induction.

← **Core Pure Book 1, Chapter 8**

Summary of key points

1 A **graph** consists of points (called **vertices** or **nodes**) which are connected by lines (**edges** or **arcs**).

2 If a graph has a number associated with each edge (usually called its weight), then the graph is known as a **weighted graph** or **network**.

3 A **subgraph** is a graph, each of whose vertices belongs to the original graph and each of whose edges belongs to the original graph. It is part of the original graph.

4 The **degree** (or **valency** or **order**) of a vertex is the number of edges incident to it.

5 If the degree of a vertex is even, you say that it has **even degree**. If the degree of a vertex is odd it has **odd degree**.

6 A **walk** is a route through a graph along edges from one vertex to the next.

7 A **path** is a walk in which no vertex is visited more than once.

8 A **trail** is a walk in which no edge is visited more than once.

9 A **cycle** is a walk in which the end vertex is the same as the start vertex and no other vertex is visited more than once.

10 A **Hamiltonian cycle** is a cycle that includes every vertex.

11 Two vertices are **connected** if there is a path between them. A graph is connected if all its vertices are connected.

12 A **loop** is an edge that starts and finishes at the same vertex.

13 A **simple graph** is one in which there are no loops and there is at most one edge connecting any pair of vertices.

14 If the edges of a graph have a direction associated with them they are known as **directed edges** and the graph is known as a **directed graph** (or **digraph**).

15 In any undirected graph, the sum of the degrees of the **vertices** is equal to 2 × the number of edges. As a consequence, the number of odd nodes must be even. This result is called **Euler's handshaking lemma**.

16 A **tree** is a connected graph with no cycles.

17 A **spanning tree** of a graph is a subgraph which includes all the vertices of the original graph and is also a tree.

18 A **complete graph** is a graph in which every vertex is directly connected by a single edge to each of the other vertices.

19 **Isomorphic graphs** are graphs which show the same information but may be drawn differently.

20 Each entry in an **adjacency matrix** describes the number of arcs joining the corresponding vertices.

21 In a **distance matrix**, the entries represent the weight of each arc, not the number of arcs.

22 A **planar graph** is one that can be drawn in a plane such that no two edges meet except at a vertex.

23 The **planarity algorithm** may be applied to any graph that contains a Hamiltonian cycle. It provides a method of redrawing the graph in such a way that it becomes clear whether or not it is planar.

3 Algorithms on graphs

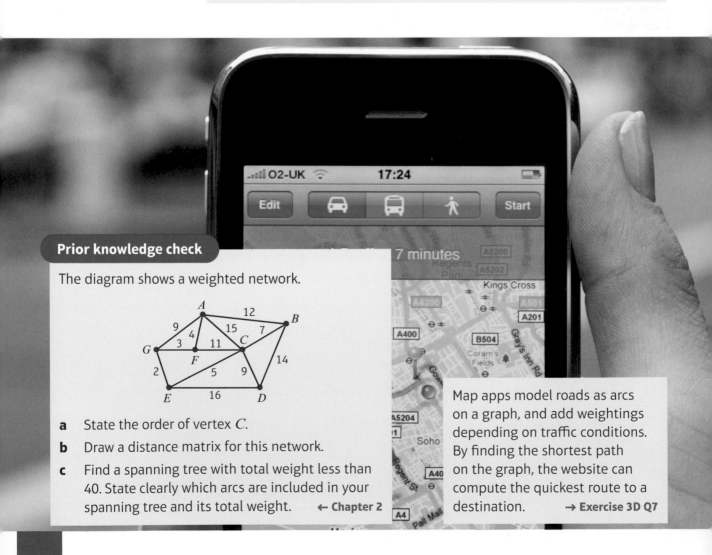

Prior knowledge check

The diagram shows a weighted network.

a State the order of vertex C.

b Draw a distance matrix for this network.

c Find a spanning tree with total weight less than 40. State clearly which arcs are included in your spanning tree and its total weight. ← **Chapter 2**

Map apps model roads as arcs on a graph, and add weightings depending on traffic conditions. By finding the shortest path on the graph, the website can compute the quickest route to a destination. → **Exercise 3D Q7**

3.1 Kruskal's algorithm

You can use Kruskal's algorithm to find a minimum spanning tree. This can tell you the shortest, cheapest or fastest way of linking all the nodes in a network.

- **A minimum spanning tree (MST) is a spanning tree such that the total length of its arcs (edges) is as small as possible.**

- **Kruskal's algorithm can be used to find a minimum spanning tree:**

Notation An MST is sometimes called a **minimum connector**.

1 Sort all the arcs (edges) into ascending order of weight.
2 Select the arc of least weight to start the tree.
3 Consider the next arc of least weight.
  - If it would form a cycle with the arcs already selected, reject it.
  - If it does not form a cycle, add it to the tree.
  If there is a choice of equal arcs, consider each in turn.
4 Repeat step **3** until all vertices are connected.

Online Explore Kruskal's algorithm using GeoGebra.

Example 1

Use Kruskal's algorithm to find a minimum spanning tree for this network. List the arcs in the order that you consider them. State the weight of your tree.

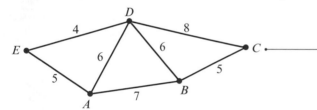

In a network of n vertices a spanning tree will always have $(n - 1)$ arcs. In this case there will be 4 arcs in the spanning tree.

By inspection the order of the arcs is
$DE(4)$, $AE(5)$, $BC(5)$, $AD(6)$, $BD(6)$, $AB(7)$, $CD(8)$.
Start with DE.

AE and BC could have been written in either order, as both have weight 5. In the same way, AD and BD could have been written in either order.

Add AE.

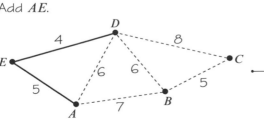

You do not need to draw each of these diagrams. The list of arcs, in order, with your decision about rejecting or adding them, is sufficient to make your method clear.

Add *BC*.

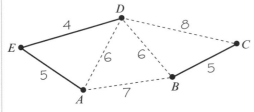

Reject *AD* (it would make the cycle *AEDA*).
Add *BD*.

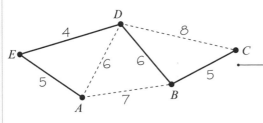

All vertices are connected so this is a
minimum spanning tree.
Its weight is 5 + 4 + 6 + 5 = 20

Watch out The spanning tree does not
necessarily remain connected as it grows, but the
finished spanning tree must be connected.

All the vertices are now connected so you can stop.

If you continued to work through the list of arcs,
each would be rejected because they would make
cycles.

Example 2

Use Kruskal's algorithm and show that there are
four minimum spanning trees for this network.
State their weight.

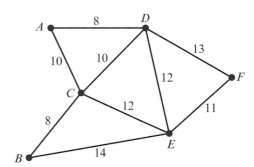

Order of arcs

length 8:	*AD* and *BC*
length 10:	*AC* and *CD*
length 11:	*EF*
length 12:	*CE* and *DE*
length 13:	*DF*
length 14:	*BE*

Spanning tree

Start with *AD*.

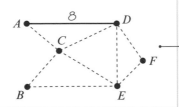

Online Explore Kruskal's algorithm using
GeoGebra.

Remember that you can choose *AD* and *BC* in
either order, *AC* and *CD* in either order, and *CE*
and *DE* in either order.

You may find it helpful to draw out the tree as
you go. It makes it easier to check for cycles.

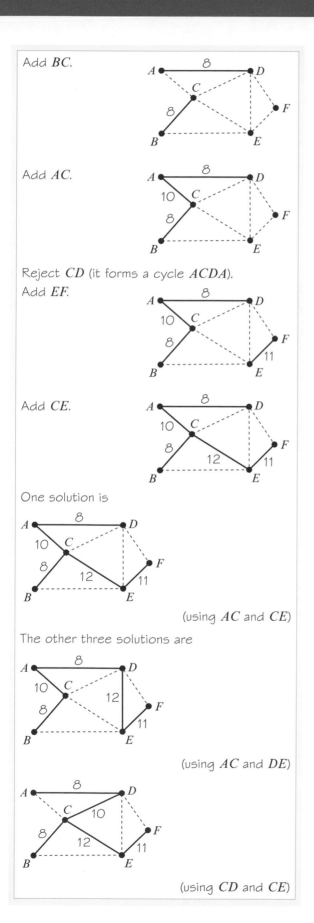

Add *BC*.

Add *AC*.

Reject *CD* (it forms a cycle *ACDA*).
Add *EF*.

Add *CE*.

One solution is

(using *AC* and *CE*)

The other three solutions are

(using *AC* and *DE*)

(using *CD* and *CE*)

Watch out You need to consider all the arcs in turn, even when they have equal weight. If you start with *AD* then you have to consider *BC* next.

Notice that, when using Kruskal's algorithm, the chosen arcs seem to jump around the network. Remember **K**ruskal's algorithm seems **c**haotic.

Looking at the list of arcs, you can see that, although you included both arcs of length 8, you only included one arc of length 10 and one arc of length 12. You can never select both length 10 or both length 12 arcs. However, you must always choose one of each. This helps to determine the other three solutions.

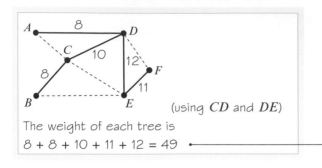

(using CD and DE)

The weight of each tree is

$8 + 8 + 10 + 11 + 12 = 49$

All four solutions are minimum spanning trees, so they will all have the same total weight.

Exercise 3A

1 Use Kruskal's algorithm to find minimum spanning trees for each of these networks. State the weight of each tree. You must list the arcs in the order in which you consider them.

a 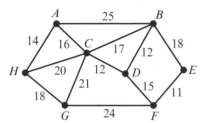   b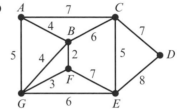

c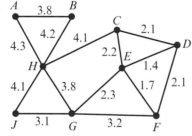

(E/P) 2 a State what is meant by:

    i a tree (1 mark)

    ii a minimum spanning tree. (1 mark)

b Use Kruskal's algorithm to find a minimum spanning tree for this network. (3 marks)

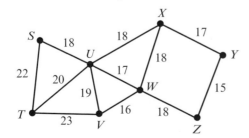

c Draw the minimum spanning tree found in part **b**. (1 mark)

d State, giving a reason, whether this minimum spanning tree is unique. (1 mark)

(P) **3** Draw a network in which:

    a the three shortest edges form part of the minimum connector (MST)

    b not all of the three shortest edges form part of the minimum connector.

(E) **4** The diagram shows nine estates and the distances between them in kilometres. A cable TV company plans to link up the estates.

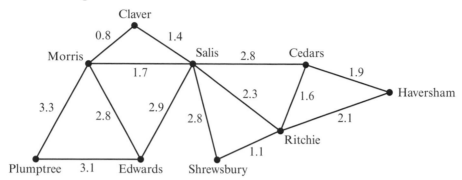

    a Find a minimum spanning tree for the network using Kruskal's algorithm. List the arcs in the order that they were added to the tree. **(4 marks)**

    b Use your answer to part **a** to find the minimum length of cable required to link all the estates together. **(1 mark)**

3.2 Prim's algorithm

■ **Prim's algorithm can be used to find a minimum spanning tree:**

> **1** Choose any vertex to start the tree.
>
> **2** • Select an arc of least weight that joins a vertex already in the tree to a vertex not yet in the tree.
>
>   • If there is a choice of arcs of equal weight, choose any of them.
>
> **3** Repeat step **2** until all the vertices are connected.

Note Prim's algorithm considers **vertices**, whereas Kruskal's algorithm considers **edges**.

Example **3**

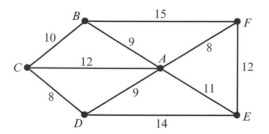

Online Explore Prim's algorithm using GeoGebra.

Watch out In your exam, you may be asked to use Prim's or Kruskal's algorithm. You must know which is which.

Use Prim's algorithm to find a minimum spanning tree for the network above. List the arcs in the order in which you add them to your tree.

Choose to start the tree at A.
Add the arc of least weight, AF, to the tree.

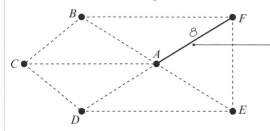

The arcs to consider are those linking A to another vertex: $AF(8)$, $AB(9)$, $AD(9)$, $AE(11)$ and $AC(12)$. Add the one of least weight, AF, to the tree.

Add AD to the tree.

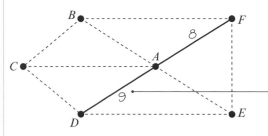

Now consider arcs that link either A or F to a vertex not in the tree: $AB(9)$, $AD(9)$, $AE(11)$, $AC(12)$, $FE(12)$ and $FB(15)$.

Add the arc of least weight, from A or F, that introduces a new vertex to the tree. In this case there are two arcs of least weight, AD and AB. You can choose either. In this case, AD is chosen.

Add DC to the tree.

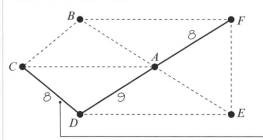

The arcs linking A, F and D to the remaining vertices are $AB(9)$, $AE(11)$, $AC(12)$, $FE(12)$, $FB(15)$, $DC(8)$, $DE(14)$. The one of least weight is DC.

Add AB to the tree.

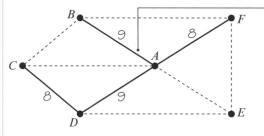

The arcs linking A, F, D and C to the remaining vertices (B and E) are $AB(9)$, $AE(11)$, $FE(12)$, $FB(15)$, $DE(14)$, $CB(10)$. The one of least weight is AB.

Add AE to the tree.

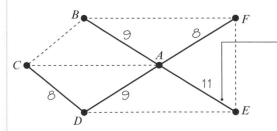

The arcs linking A, F, D, C and B to the remaining vertex, E, are $AE(11)$, $FE(12)$, $DE(14)$. The one of least weight is AE.

Arcs added in this order: AF, AD, DC, AB, AE.

Hint Notice that with Prim's algorithm the tree always grows in a connected fashion. The arcs do not jump around as they sometimes do in Kruskal's algorithm.

Exercise 3B

1 Repeat Question **1** in Exercise 3A using Prim's algorithm. Start at vertex *A* each time.

2 Describe two differences between Prim's algorithm and Kruskal's algorithm.

(E) 3 The network shows the distances, in kilometres, between eight weather monitoring stations. The eight stations need to be linked together with underground cables.

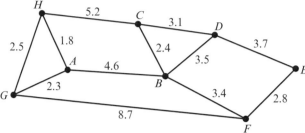

a Use Prim's algorithm, starting at *A*, to find a minimum spanning tree. You must make your order of arc selection clear. **(3 marks)**

b Given that cable costs £850 per kilometre to lay, find the minimum cost of linking these weather stations. **(1 mark)**

(E/P) 4

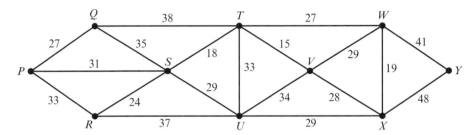

The network shows ten villages and the costs, in thousands of pounds, of connecting them with a new energy supply.

a Use Prim's algorithm, starting at *P*, to find the energy supply network that would connect all ten villages for minimum cost. **(3 marks)**

b Draw your minimum connector and state its cost. **(2 marks)**

c Unforeseen problems with the link between villages *W* and *X* mean that the cost of connecting them rises to £34 000. Explain how this affects your minimum spanning tree. **(2 marks)**

(E/P) 5 a Explain why it is not necessary to check for cycles when using Prim's algorithm. **(1 mark)**

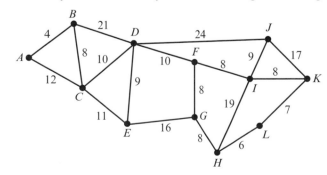

b Use Prim's algorithm, starting at *A*, to find a minimum spanning tree for this network. You must make your order of arc selection clear. **(3 marks)**

c State, with a reason, whether this minimum spanning tree is unique. **(2 marks)**

3.3 Applying Prim's algorithm to a distance matrix

Networks (especially large ones) are often described using distance matrices. This is a convenient way to input a large network into a computer. You can apply Prim's algorithm directly to a distance matrix, which makes it more convenient for computer applications.

- **The distance matrix form of Prim's algorithm is:**

  1 Choose any vertex to start the tree.

  2 Delete the **row** in the matrix for the chosen vertex.

  3 Number the **column** in the matrix for the chosen vertex.

  4 Put a ring round the lowest undeleted entry in the numbered columns. (If there is an equal choice, choose randomly.)

  5 The ringed entry becomes the next arc to be added to the tree.

  6 Repeat steps **2**, **3**, **4** and **5** until all rows have been deleted.

Example 4

Apply Prim's algorithm to the distance matrix to find a minimum spanning tree.
Start at A.

	A	B	C	D	E
A	–	27	12	23	74
B	27	–	47	15	71
C	12	47	–	28	87
D	23	15	28	–	75
E	74	71	87	75	–

Delete row A, and number column A.

The lowest undeleted entry in column A is 12, so put a ring round it. The first arc is AC.

The first arc is AC.

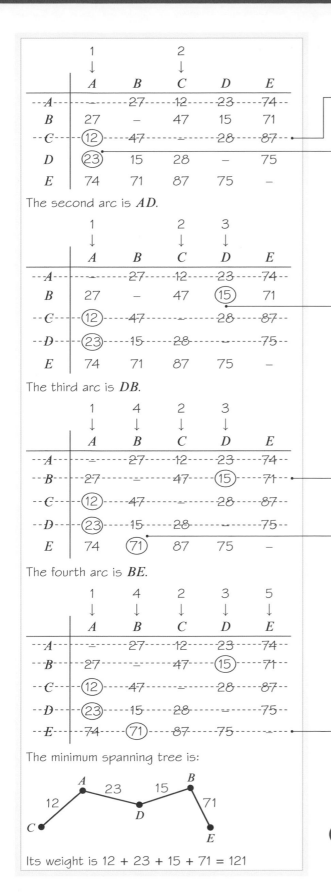

	1 ↓ A		2 ↓ C		
	A	B	C	D	E
A	–	27	12	23	74
B	27	–	47	15	71
C	(12)	47	–	28	87
D	(23)	15	28	–	75
E	74	71	87	75	–

The second arc is AD.

	1 ↓ A		2 ↓ C	3 ↓ D	
	A	B	C	D	E
A	–	27	12	23	74
B	27	–	47	(15)	71
C	(12)	47	–	28	87
D	(23)	15	28	–	75
E	74	71	87	75	–

The third arc is DB.

	1 ↓ A	4 ↓ B	2 ↓ C	3 ↓ D	
	A	B	C	D	E
A	–	27	12	23	74
B	27	–	47	(15)	71
C	(12)	47	–	28	87
D	(23)	15	28	–	75
E	74	(71)	87	75	–

The fourth arc is BE.

	1 ↓ A	4 ↓ B	2 ↓ C	3 ↓ D	5 ↓ E
	A	B	C	D	E
A	–	27	12	23	74
B	27	–	47	(15)	71
C	(12)	47	–	28	87
D	(23)	15	28	–	75
E	74	(71)	87	75	–

The minimum spanning tree is:

```
        A   23   15   B
   12  /  \       /    \ 71
  C       D            
                        E
```

Its weight is $12 + 23 + 15 + 71 = 121$

The new vertex is C. Delete row C, and number column C.

The lowest undeleted entry in columns A and C is 23, so put a ring round it. The second arc is AD.

The new vertex is D. Delete row D, and number column D.

The lowest undeleted entry in columns A, C and D is 15. Put a ring round it. The third arc is DB.

The new vertex is B. Delete row B, and number column B.

The lowest undeleted entry in columns A, C, D and B is 71. Put a ring round it. The fourth arc is BE.

The new vertex is E. Delete row E and number column E.

All rows have now been deleted, so the algorithm is complete.

Watch out You do not need to show all of these tables. The final annotated table, plus a list of arcs in order, is sufficient to make your method clear.

Example 5

Here is the distance matrix used in Example 4.

a Work out the number of comparisons required to complete Prim's algorithm.

b Work out the number of comparisons required for an $n \times n$ distance matrix.

c State the order of this implementation of Prim's algorithm in terms of the number of vertices, n.

	A	B	C	D	E
A	–	27	12	23	74
B	27	–	47	15	71
C	12	47	–	28	87
D	23	15	28	–	75
E	74	71	87	75	–

a Once the 1st vertex has been selected, you need to choose the least of $5 - 1 = 4$ values in a single column.

```
        1
        ↓
       A    B    C    D    E
 A ----------27----12----23----74--
 B    27    –    47   15   71
 C    12   47    –    28   87
 D    23   15   28    –    75
 E    74   71   87   75    –
```

This requires $4 - 1 = 3$ comparisons.
Once the 2nd vertex has been selected, there are 2 columns to consider.

```
        1         2
        ↓         ↓
       A    B    C    D    E
 A ----------27----12----23----74--
 B    27    –    47   15   71
 C --(12)----47----------28----87--
 D   (23)   15   28    –    75
 E    74   71   87   75    –
```

The number of values in each column is $5 - 2 = 3$ so there are a further $3 \times 2 - 1 = 5$ comparisons to make.
Once the 3rd vertex has been selected, there are 3 columns to consider.
The number of values in each column is $5 - 3 = 2$ so there are a further $2 \times 3 - 1 = 5$ comparisons to make.
Once the 4th vertex has been selected, there are 4 columns to consider.

The three comparisons are:
1. Compare the values in rows B and C, and select the smallest.
2. Compare this to the value in row D, and select the smallest.
3. Compare this to the value in row E and select the smallest.

In general, you can choose the smallest of n different numbers by making $n - 1$ comparisons.

The number of values in each column is
$5 - 4 = 1$ so there are a further
$1 \times 4 - 1 = 3$ comparisons to make.
The total number of comparisons is
$3 + 5 + 5 + 3 = 16$

This corresponds roughly to the number of steps a computer program would have to carry out in order to execute Prim's algorithm on this distance matrix.

b Starting with an $n \times n$ matrix, the row corresponding to the starting vertex is crossed out, leaving $n - 1$ values in the corresponding column. The smallest of these is to be found. This requires $(n - 1) - 1$ comparisons.

At each stage, the number of columns included increases by 1 and the number of values remaining in each of those columns reduces by 1.

The total number of comparisons required to complete the algorithm is given by

$((n - 1) \times 1 - 1) + ((n - 2) \times 2 - 1) + ((n - 3) \times 3 - 1) + \ldots + ((n - (n - 1)) \times (n - 1) - 1)$

$= \sum_{r=1}^{n-1} ((n - r) \times r - 1)$

$= n\sum_{r=1}^{n-1} r - \sum_{r=1}^{n-1} r^2 - \sum_{r=1}^{n-1} 1$

$= n\left(\frac{1}{2}(n - 1)n\right) - \frac{1}{6}(n - 1)n(2n - 1) - (n - 1)$

$= \frac{3n^2(n - 1) - n(n - 1)(2n - 1) - 6(n - 1)}{6}$

$= \frac{n^3 - 7n + 6}{6}$

c The order of Prim's algorithm is n^3.

Problem-solving

An $n \times n$ distance matrix corresponds to a network with n vertices. By considering the number of comparisons needed in an $n \times n$ distance matrix, you can determine the order of Prim's algorithm when applied to a network with n vertices.

After the rth vertex has been selected you will need to find the least of $(n - r) \times r$ values, which requires $(n - r) \times r - 1$ comparisons.

Use $\sum_{r=1}^{n} 1 = n$, $\sum_{r=1}^{n} r = \frac{1}{2}n(n + 1)$ and
$\sum_{r=1}^{n} r^2 = \frac{1}{6}n(n + 1)(2n + 1)$

← **Core Pure Book 1, Chapter 3**

As a check, when $n = 5$, this gives
$\frac{5^3 - 7 \times 5 + 6}{6} = \frac{96}{6} = 16$
which is the result found in part **a**.

When determining the order of an algorithm, you only need to consider the highest power of n. As n gets large, the n^3 term will have the largest effect on the runtime of the algorithm.

Exercise 3C

Answer templates for questions marked * are available at www.pearsonschools.co.uk/d1maths

1* Apply Prim's algorithm to the distance matrices below. List the arcs in order of selection and state the weight of your minimum spanning tree.

a

	A	B	C	D	E	F
A	–	15	20	34	25	9
B	15	–	36	38	28	14
C	20	36	–	43	38	22
D	34	38	43	–	26	40
E	25	28	38	26	–	31
F	9	14	22	40	31	–

b

	R	S	T	U	V
R	–	28	30	31	41
S	28	–	16	19	43
T	30	16	–	22	41
U	31	19	22	–	37
V	41	43	41	37	–

2* The table shows the distance, in miles, between five cities. It is intended to link these five cities by a transit system.

	Birmingham	Nottingham	Lincoln	Stoke	Manchester
Birmingham	–	164	100	49	88
Nottingham	164	–	37	56	74
Lincoln	100	37	–	90	86
Stoke	49	56	90	–	44
Manchester	88	74	86	44	–

Use Prim's algorithm, starting at Birmingham, to find the minimum total length of a transit system linking all five cities. You must list the arcs in order of selection and state the weight of your tree.

(**E/P**) **3*** The matrix shows the costs, in euros per 1000 words, of translating appliance instruction manuals between eight languages.

	A	B	C	D	E	F	G	H
A	–	84	53	35	–	47	–	42
B	84	–	71	113	142	61	75	–
C	53	71	–	–	–	–	59	–
D	35	113	–	–	58	67	151	–
E	–	142	–	58	–	168	159	48
F	47	61	–	67	168	–	–	73
G	–	75	59	151	159	–	–	52
H	42	–	–	–	48	73	52	–

a Use Prim's algorithm, starting from language D, to find the cost of translating an instruction manual of 3000 words from D into the seven other languages. **(4 marks)**

b Draw your minimum spanning tree. **(1 mark)**

A manual is written in language E and needs to be translated into language G.
The table shows that it costs 159 euros per 1000 words to translate from language E to G.

c Give a reason why:
   i it might be decided not to translate directly from E to G
   ii it might be decided to translate directly. **(2 marks)**

E/P **4*** The table shows the distances, in miles, between nine oil rigs and the depot X.
Pipes are to be laid to connect the rigs and the depot.

	X	A	B	C	D	E	F	G	H	I
X	–	65	80	89	74	26	71	41	41	74
A	65	–	27	41	22	37	20	29	25	43
B	80	27	–	30	24	55	16	46	40	42
C	89	41	30	–	50	84	24	70	49	26
D	74	22	24	50	–	51	35	34	47	63
E	26	37	55	84	51	–	52	18	23	68
F	71	20	16	24	35	52	–	45	31	27
G	41	29	46	70	34	18	45	–	25	64
H	41	25	40	49	47	23	31	25	–	44
I	74	43	42	26	63	68	27	64	44	–

a Use Prim's algorithm, starting at X, to find a minimum spanning tree for
the network. You must make the order of arc selection clear. **(3 marks)**

A computer program finds this minimum
spanning tree in 0.7 seconds. Given that
the computer program implements Prim's
algorithm with cubic order,

> **Watch out** The specific way in which an algorithm
> is implemented can affect its efficiency. There are
> alternative implementations of Prim's algorithm
> that have quadratic complexity. In your exam you
> will be told what order you need to use in your
> calculations. **→ Mixed exercise, Challenge**

b estimate the time it would take this
computer program to find a minimum
spanning tree for a different network
of 24 oil rigs and the depot. **(2 marks)**

Oil rig A exhausts its supply and is closed down.

c The distance between the depot and oil rig I is found to be incorrect.
Give a possible value for this distance that:
i will not affect the minimum spanning tree you found in part **a**
ii will affect the minimum spanning tree you found in part **a**.

Give a reason for each of your choices. **(4 marks)**

3.4 Using Dijkstra's algorithm to find the shortest path

You can use Dijkstra's algorithm to find the shortest path between two vertices in a network. This could be useful for finding the cheapest, shortest or quickest transportation route between two locations.

- **Dijkstra's algorithm can be used to find the shortest path from S to T through a network:**

> **Note** Dijkstra is pronounced 'Dike-Stra'.

1 Label the start vertex, S, with the final label, 0.

2 Record a working value at every vertex, Y, that is directly connected to the vertex, that has just received its final label, X.
- Working value at Y = final label at X + weight of arc XY
- If there is already a working value at Y, it is only replaced if the new value is smaller.
- Once a vertex has a final label, it is not revisited and its working values are no longer considered.

3 Look at the working values at all vertices without final labels. Select the smallest working value. This now becomes the final label at that vertex. (If two vertices have the same smallest working value, either may be given its final label first.)

4 Repeat steps **2** and **3** until the destination vertex, T, receives its final label.

5 To find the shortest path, trace back from T to S. Given that B already lies on the route, include arc AB whenever:
final label of B − final label of A = weight of arc AB

The algorithm makes use of labels. Start at the initial vertex and move through the network, putting working values (often called temporary labels) on each vertex. Each pass finds the shortest route to one of the vertices and records its final label (also called its permanent label). Once a vertex has its final label it is 'sealed' and its working values are no longer considered. Continue in this way until the destination vertex is reached.

> **Online** Explore Dijkstra's algorithm using GeoGebra.

Example 6

Use Dijkstra's algorithm to find the shortest route from S to T in the network below.

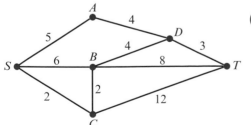

> **Notation** To make the working clear you replace the vertices with boxes like this:
>
> | Vertex | Order of labelling | Final label |
> |--------|--------------------|-------------|
> | | Working values | |

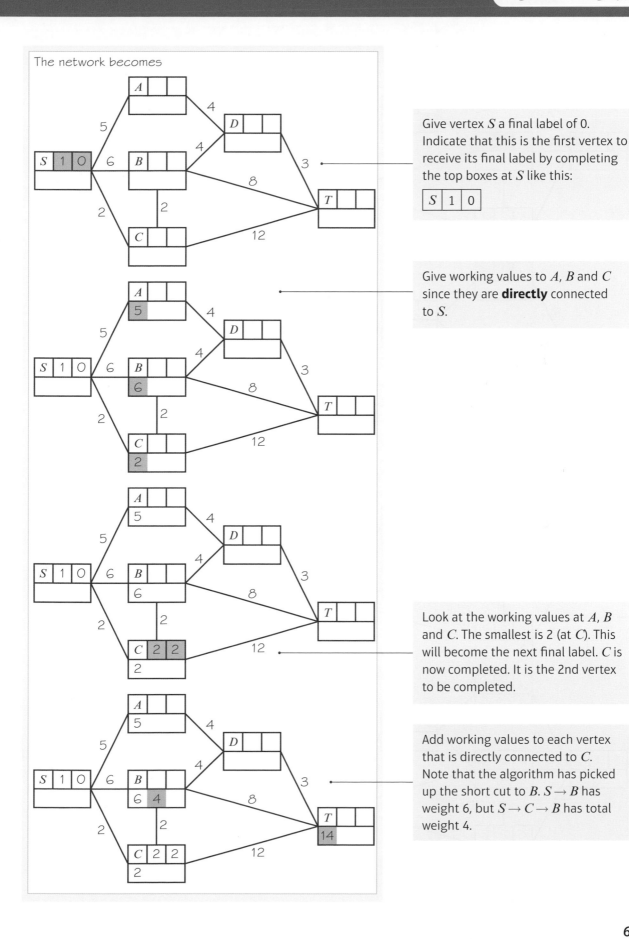

The network becomes

Give vertex S a final label of 0. Indicate that this is the first vertex to receive its final label by completing the top boxes at S like this:

| S | 1 | 0 |

Give working values to A, B and C since they are **directly** connected to S.

Look at the working values at A, B and C. The smallest is 2 (at C). This will become the next final label. C is now completed. It is the 2nd vertex to be completed.

Add working values to each vertex that is directly connected to C. Note that the algorithm has picked up the short cut to B. $S \rightarrow B$ has weight 6, but $S \rightarrow C \rightarrow B$ has total weight 4.

67

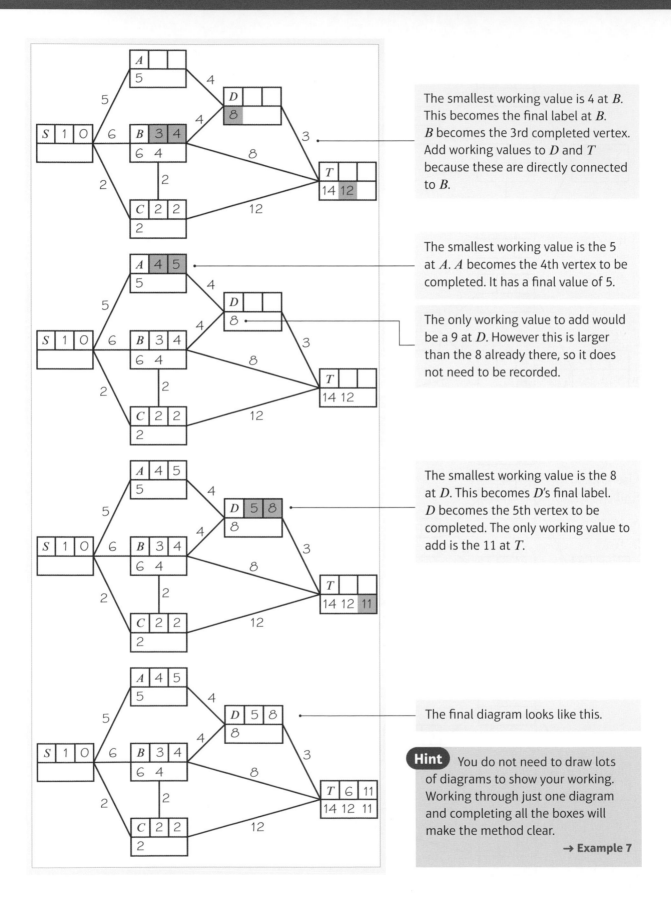

The smallest working value is 4 at *B*. This becomes the final label at *B*. *B* becomes the 3rd completed vertex. Add working values to *D* and *T* because these are directly connected to *B*.

The smallest working value is the 5 at *A*. *A* becomes the 4th vertex to be completed. It has a final value of 5.

The only working value to add would be a 9 at *D*. However this is larger than the 8 already there, so it does not need to be recorded.

The smallest working value is the 8 at *D*. This becomes *D*'s final label. *D* becomes the 5th vertex to be completed. The only working value to add is the 11 at *T*.

The final diagram looks like this.

Hint You do not need to draw lots of diagrams to show your working. Working through just one diagram and completing all the boxes will make the method clear.

→ **Example 7**

The length of the shortest route from S to T is 11.

To find the shortest route, start at T and trace back looking at final values and arc lengths.

Check the arcs into T.

$11 - 2 \neq 12$ so CT is <u>not</u> on the route

$11 - 4 \neq 8$ so BT is <u>not</u> on the route

$11 - 8 = 3$ so DT <u>is</u> on the route

To get to T in 11 you must have come from D. Continue working back from D.

The working is

$11 - 8 = 3$	DT
$8 - 4 = 4$	BD
$4 - 2 = 2$	CB
$2 - 0 = 2$	SC

So the shortest route is

$SCBDT$, length 11

Watch out Unless you are asked to explain how you obtained your final route, you do not need to show all of this working. You can identify the shortest route from your fully labelled diagram by inspection.

- **You can use Dijkstra's algorithm to find the shortest route between the start vertex and any other vertex with a final label.**

Example 7

Use Dijkstra's algorithm in this network to find the length of the shortest route:

a from A to H

b from A to G.

List the routes you use.

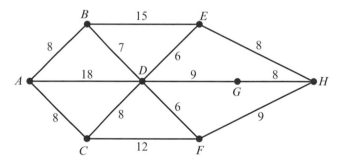

a The final diagram looks like this. It shows all the working needed to make the method clear.

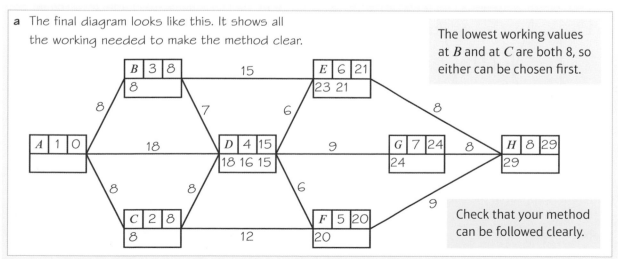

The lowest working values at B and at C are both 8, so either can be chosen first.

Check that your method can be followed clearly.

The length of the shortest route from A to H is 29.

There are two shortest routes:

$ABDEH$ and $ACFH$.

b From the diagram, the length of the shortest route from
A to G is 24.

The route is $ABDG$.

either	29 − 21 = 8 *EH*
	21 − 15 = 6 *DE*
	15 − 8 = 7 *BD*
	8 − 0 = 8 *AB*
or	29 − 20 = 9 *FH*
	20 − 8 = 12 *CF*
	8 − 0 = 8 *AC*

Problem-solving

G already has a final value, so you do not need to implement the algorithm again. Just work backwards from G to A.

- **It is possible to use Dijkstra's algorithm on networks with directed arcs. This is like trying to find a driving route where some of the roads are one-way streets.**

Example 8

The network below represents part of a road system in a city. Some roads are one-way and these are indicated by directed arcs. The number on each arc represents the time, in minutes, to travel along that arc.

a Show that there are two quickest routes from A to I. Explain how you found your routes from your labelled diagram.

Road HI is closed due to roadworks.

b Find the quickest route from A to I, avoiding HI.

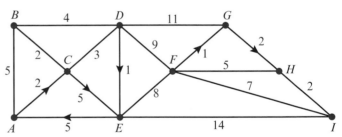

a The final diagram looks like this:

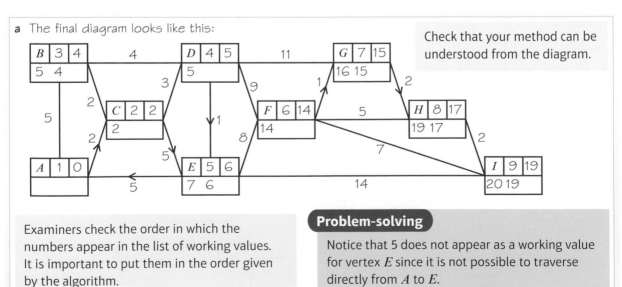

Check that your method can be understood from the diagram.

Examiners check the order in which the numbers appear in the list of working values. It is important to put them in the order given by the algorithm.

Problem-solving

Notice that 5 does not appear as a working value for vertex E since it is not possible to traverse directly from A to E.

The two quickest routes are

$A\ C\ D\ F\ G\ H\ I$ and $ACDEFGHI$

since

			since		
19 − 17 = 2	*HI*		19 − 17 = 2	*HI*	
17 − 15 = 2	*GH*		17 − 15 = 2	*GH*	
15 − 14 = 1	*FG*		15 − 14 = 1	*FG*	
14 − 5 = 9	*DF*		14 − 6 = 8	*EF*	
5 − 2 = 3	*CD*		6 − 5 = 1	*DE*	
2 − 0 = 2	AC		5 − 2 = 3	*CD*	
			2 − 0 = 2	AC	

> **Watch out** In this question you are told to explain how you found the route from your diagram, so you need to show these calculations.

Both are of length 19 minutes.

b Removing *HI* from the network would leave a final value of 20 at *I*. Start at *I* and find the route of length 20.

20 − 6 = 14	*EI*
6 − 5 = 1	*DE*
5 − 2 = 3	*CD*
2 − 0 = 2	AC

So the quickest route from *A* to *I*, avoiding *HI*, is *ACDEI*, of length 20 minutes.

Exercise 3D

Answer templates for questions marked * are available at www.pearsonschools.co.uk/d1maths

1* Use Dijkstra's algorithm to find the shortest route from *S* to *T* in each of the following networks. In each case, explain how you determined the shortest path from your fully labelled diagram.

a

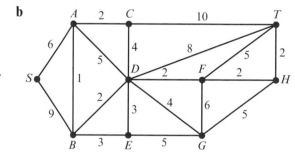

b

(P) **2*** The network shows part of a road system in a city. The number on each arc gives the time, in minutes, it takes to travel along that arc.

Use Dijkstra's algorithm to find:

a the quickest route from *A* to *Q* and its length

b the quickest route from *A* to *L* and its length

c the quickest route from *M* to *A* and its length

d the quickest route from *P* to *A* and its length.

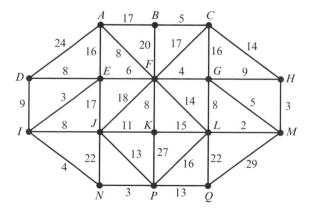

71

Chapter 3

3 * Use Dijkstra's algorithm to find the shortest route, and its length, from A to F in this directed network.

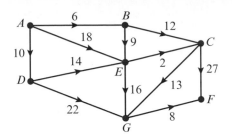

(E/P) **4** * The network represents the lengths, in metres, of all the roads in a building site. A crane is needed for one day at T. There are two cranes available on-site, one at S_1, and the other at S_2.
One of these two cranes will be moved to T. In order to minimise the cost it is decided to move the crane that is closest to T. Use Dijkstra's algorithm to determine which crane should be moved.

(7 marks)

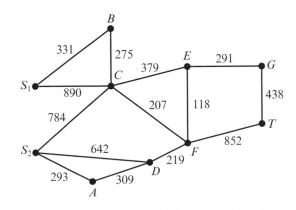

Problem-solving

It is possible to solve this problem with only one application of Dijkstra's algorithm. Think carefully about the starting point.

(E/P) **5** * **a** Use Dijkstra's algorithm to find the shortest route from A to H. Indicate how you obtained your shortest route from your labelled diagram. **(6 marks)**

b Find the shortest route from A to H via G. **(2 marks)**

c Find the shortest route from A to H, not using CE. **(2 marks)**

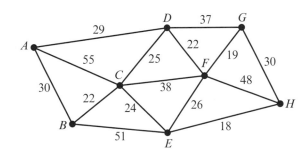

(E) **6** * Use Dijkstra's algorithm to find the shortest route from S to T in this directed network. State the length of your route. **(6 marks)**

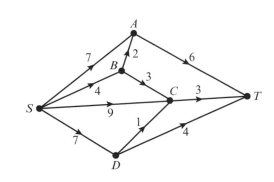

72

(E/P) **7*** A navigation app uses real-time data to determine driving times, between different locations. The diagram shows a network of driving times, in minutes, along roads joining ten locations.

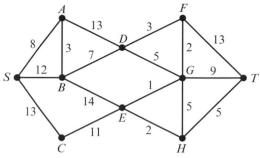

   a Use Dijkstra's algorithm to find the quickest route from S to T, and state the total driving time for this route.
   (6 marks)

A driver is following this quickest route. Due to traffic congestion, the travel time between H and T increases from 5 minutes to 13 minutes.

   b State how this will affect the quickest route if this information is discovered:
   i before the start of the journey
   ii when the driver reaches point H.
   (4 marks)

Dijkstra's algorithm has quadratic order.

   c If the app takes 0.026 seconds to apply Dijkstra's algorithm to the above network, estimate the time it would take the app to compute the quickest route on a network with roads connecting 40 different locations.
   (2 marks)

   d Explain why your answer to part **c** is only an estimate.
   (1 mark)

3.5 Floyd's algorithm

A
You can use Floyd's algorithm to find the shortest path between every pair of vertices in a network. The algorithm produces two tables. The first shows the minimum distance between each pair of vertices and the second can be used to find the corresponding route.

Note Unlike Dijkstra's algorithm, Floyd's algorithm will tell you the shortest distance between **any pair** of vertices in the network.

- **Floyd's algorithm can be used to find the shortest path between every pair of vertices in a network:**

   1 Complete an initial **distance table** for the network. If there is no direct route from one vertex to another, label the distance with the infinity symbol.

   2 Complete an initial **route table** by making every entry in the first column the same as the label at the top of the first column, making every entry in the second column the same as the label at the top of the second column and so on.

   3 In the first iteration, copy the first row and the first column values of the distance table into a new table. Lightly shade these values.

	A	B	C	D
A	–			
B		–		
C			–	
D				–

A

4 Consider each unshaded position in turn. Compare the value in this position in the previous table with the sum of the corresponding shaded values.

Distance table

	A	B	C	D
A	–		Y	
B	X	–	Z	
C			–	
D				–

Route table

	A	B	C	D
A	A	B	C	D
B	A	B	[A]	D
C	A	B	C	D
D	A	B	C	D

Watch out You compare Z with the sum of the shaded values in the **same row and column**. You are always comparing an unshaded value with the sum of two shaded values.

If $X + Y \geqslant Z$ then copy Z into the new table, i.e. there is no change – the object is to keep the smallest values in the table.

If $X + Y < Z$, copy $X + Y$ into the new table and write A in the corresponding position in the route table. Once all values in the unshaded area have been considered, the first iteration is complete.

A is written into the route table to show that the direct distance BC has been replaced by $BA + AC$, i.e. the route has taken a detour through A. Any other changes in this iteration will also result in a detour through A and so A is written in the route table, in each case, in the position corresponding to the changed value.

5 For the second iteration, copy the second row and the second column from the last iteration into a new distance table. Lightly shade these values.

	A	B	C	D
A	–			
B		–		
C			–	
D				–

6 Repeat step **4** with the new unshaded positions. This time any changes will result in a detour through B and so you should write B in the new route table, in each case, in the position corresponding to the changed value.

7 If there are n vertices then completing the algorithm will require n iterations continuing in the same way.

Example 9

This distance graph shows the direct distances, by road, between four towns A, B, C and D, in miles. The road from D to A is a one-way road as shown by the arrow.

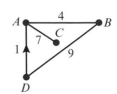

a Use Floyd's algorithm to produce a table of shortest distances. You should give the distance table and route table for each iteration.

b Find the route of minimum length from C to D.

A

a Initial tables:

	A	B	C	D
A	–	4	7	∞
B	4	–	∞	9
C	7	∞	–	∞
D	1	9	∞	–

	A	B	C	D
A	A	B	C	D
B	A	B	C	D
C	A	B	C	D
D	A	B	C	D

> The table on the left is the initial distance table but, where there is no direct route, the ∞ symbol is used.

> The table on the right is the initial route table.

> The first row and column of the distance table are copied into a blank table and these values are now shaded. In each unshaded position, the sum of the shaded values is compared with the value in the initial table. The smaller value is written into this table. The figures in brackets are the ones that have been changed. For example, $4 + 7 = 11 < ∞$ so two of the ∞ symbols are replaced with 11. Also, $1 + 4 = 5 < 9$ so 5 replaces the 9 in the bottom row. Finally, $1 + 7 = 8 < ∞$ so 8 replaces the ∞ symbol in the bottom row.

First iteration:

	A	B	C	D
A	–	4	7	∞
B	4	–	[11]	9
C	7	[11]	–	∞
D	1	[5]	[8]	–

	A	B	C	D
A	A	B	C	D
B	A	B	[A]	D
C	A	[A]	C	D
D	A	[A]	[A]	D

> The four letters in the matching positions in the route table are changed to the letter A.

> The second row and column of the previous table are now copied into a blank table and these values are shaded. The two remaining ∞ symbols are replaced with $4 + 9 = 13$ and $11 + 9 = 20$. The letters in the corresponding positions in the route table are changed to the letter B. No other value in an unshaded position needs to be changed from the previous table.

Second iteration:

	A	B	C	D
A	–	4	7	[13]
B	4	–	11	9
C	7	11	–	[20]
D	1	5	8	–

	A	B	C	D
A	A	B	C	[B]
B	A	B	A	D
C	A	A	C	[B]
D	A	A	A	D

> In a new distance table, the third row and the third column are now shaded. The values in those positions are copied from the previous table. In each unshaded position, the value in the previous table is smaller than the sum of the shaded values, so no changes are made in the third iteration. It follows that no changes are made to the route table.

Third iteration:

	A	B	C	D
A	–	4	7	13
B	4	–	11	9
C	7	11	–	20
D	1	5	8	–

	A	B	C	D
A	A	B	C	B
B	A	B	A	D
C	A	A	C	B
D	A	A	A	D

A

Fourth iteration:

	A	B	C	D
A	–	4	7	13
B	4	–	11	9
C	7	11	–	20
D	1	5	8	–

	A	B	C	D
A	A	B	C	B
B	A	B	A	D
C	A	A	C	B
D	A	A	A	D

The final distance and route tables are:

	A	B	C	D
A	–	4	7	13
B	4	–	11	9
C	7	11	–	20
D	1	5	8	–

	A	B	C	D
A	A	B	C	B
B	A	B	A	D
C	A	A	C	B
D	A	A	A	D

b The route of minimum length from C to D is $CABD$.

In the final iteration, once again, no changes are made because each value in the previous table is smaller than the sum of the shaded values.

To find this route, look at row C and column D in the route table. The letter in this position is B, which means that a detour is taken through B. So far, the route is CBD.
Now look at row C and column B. The letter in this position is A, which means that the route passes through A in going from C to B. So far, the route is $CABD$.
Since there are only four vertices in the graph, there can be no further detours. You can verify this with the route table by checking CA, AB, and BD:
• CA gives A (no detour)
• AB gives B (no detour)
• BD gives D (no detour)

Problem-solving

The corresponding entry in the final distance table tells you the length of this route. The entry in row C and column D is 20, so the total length of this route is 20 miles.

Online Explore Floyd's algorithm using GeoGebra.

Example 10

8 departure gates in an airport, which are linked by a system of travellators and escalators, are modelled using the directed network shown. The weight of each edge represents the travel time, in minutes, from one gate to another.

7 iterations of Floyd's algorithm are applied to the network, resulting in the following distance and route tables.

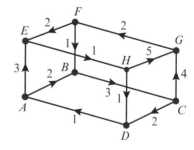

Distance table

	A	B	C	D	E	F	G	H
A	–	2	5	7	3	11	9	4
B	6	–	3	5	9	9	7	10
C	3	5	–	2	6	6	4	7
D	1	3	6	–	4	12	10	5
E	∞	∞	∞	∞	–	∞	∞	1
F	7	1	4	6	2	–	8	3
G	9	3	6	8	4	2	–	5
H	2	4	7	1	5	7	5	–

Route table

	A	B	C	D	E	F	G	H
A	A	B	B	C	E	G	C	E
B	D	B	C	C	D	G	C	E
C	D	D	C	D	D	G	G	E
D	A	A	B	D	A	G	C	E
E	A	B	C	D	E	F	G	H
F	D	B	B	C	E	F	C	E
G	F	F	F	F	F	F	G	F
H	D	D	D	D	D	G	G	H

A **a** Apply the final iteration of Floyd's algorithm to give the final distance and route tables for the network.

b Sarah needs to get from gate D to gate F. State the minimum time needed to make this journey, and determine the route she should take.

a Final iteration:

Distance table

	A	B	C	D	E	F	G	H
A	–	2	5	[5]	3	11	9	4
B	6	–	3	5	9	9	7	10
C	3	5	–	2	6	6	4	7
D	1	3	6	–	4	12	10	5
E	[3]	[5]	[8]	[2]	–	[8]	[6]	1
F	[5]	1	4	[4]	2	–	8	3
G	[7]	3	6	[6]	4	2	–	5
H	2	4	7	1	5	7	5	–

Route table

	A	B	C	D	E	F	G	H
A	A	B	B	[H]	E	G	C	E
B	D	B	C	C	D	G	C	E
C	D	D	C	D	D	G	G	E
D	A	A	B	D	A	G	C	E
E	[H]	[H]	[H]	[H]	E	H	[H]	[H]
F	[H]	B	B	[H]	E	F	C	E
G	[H]	F	F	[H]	F	F	G	F
H	D	D	D	D	D	G	G	H

Shade the final row and column of the distance table. For each unshaded position, compare the value in this position in the previous table (Z) with the sum of the corresponding two shaded numbers ($X + Y$), and enter the minimum of $X + Y$ and Z as the new entry. In this case, 11 values are changed as part of the final iteration.

For each entry that has changed in the distance table, change the corresponding entry in the distance table to 'H'.

b The distance table shows the minimum travel times between gates, so the minimum travel time from gate D to gate F is 12 minutes.
Using the route table,

$D \rightarrow G \rightarrow F$

$D \rightarrow C \rightarrow G \rightarrow F$

$D \rightarrow B \rightarrow C \rightarrow G \rightarrow F$

$D \rightarrow A \rightarrow B \rightarrow C \rightarrow G \rightarrow F$

Sarah's quickest route from gate D to gate F is $DABCGF$.

Looking at row D and column F, the entry in this position is G, so the quickest route goes via gate G.

Watch out You need to check the quickest route from D to G, and from G to F. The route from D to G must detour via C. No detour is required on the route from G to F.

Every stage in this route is now direct, so this is the quickest route from D to F.

Exercise **3E**

A
E

1* This network shows the distances, in miles, between four locations *P*, *Q*, *R* and *S*.

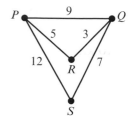

a Use Floyd's algorithm to produce a table of shortest distances. You should give the distance table and route table for each iteration. **(7 marks)**

b Explain how you can use your route table to show that the shortest route from *R* to *S* is *RQS*. **(2 marks)**

2* In this network, the number next to each edge denotes the length of the edge. Use Floyd's algorithm to determine the shortest distance between all pairs of nodes.

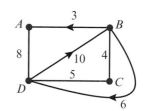

3* This network shows the times needed in minutes to transport materials between storage points on a building site. Use Floyd's algorithm to determine the shortest time needed to transport materials between all possible pairs of storage points.

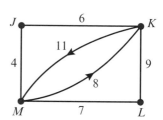

E/P **4** Floyd's algorithm is applied to a weighted graph. The tables shown below are the final distance table and final route table.

	A	*B*	*C*	*D*			*A*	*B*	*C*	*D*
A	–	7	3	11		*A*	*A*	*B*	*C*	*D*
B	*x*	–	9	17		*B*	*A*	*B*	*A*	*A*
C	12	*y*	–	*z*		*C*	*B*	*B*	*C*	*B*
D	8	15	11	–		*D*	*A*	*A*	*A*	*D*

a Explain how you know that the graph contains directed edges. **(1 mark)**

b Use the tables to work out the values of *x*, *y* and *z*. **(4 marks)**

E **5* a** Explain the difference between the output of Floyd's algorithm and the output of Dijkstra's algorithm. **(1 mark)**

b This network shows the direct distances in miles by road between seven towns.

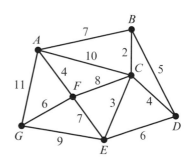

   i Show the initial distance table for the network.

   ii Show the distance and route tables after one iteration of Floyd's algorithm. **(6 marks)**

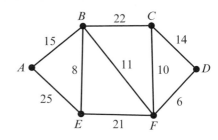

A **6*** Jared and Amy are planning a school trip to a theme park.
E/P They have shown the main attractions and the walking
times between them on a network.

Jared wants to enter the park at location A, and would like
to know the shortest distance from there to each of the
main attractions. Amy would like to give her students
information about the shortest distances and corresponding
routes between any pair of the main attractions.

a Suggest a suitable algorithm for each situation, giving reasons for your choices. **(2 marks)**

b Apply two iterations of Floyd's algorithm to this network, showing both the distance table
and the route table after each iteration. **(5 marks)**

P **7 a** Floyd's algorithm is applied to an $n \times n$ distance matrix. State the number of comparisons that
are made with each iteration.

b Give the order of Floyd's algorithm.

c A computer program takes 0.012 seconds to apply Floyd's algorithm to a 30×30 distance
matrix. Estimate the time required for the same computer program to apply the algorithm to a
100×100 distance matrix.

Mixed exercise 3

Answer templates for questions marked * are available at www.pearsonschools.co.uk/d1maths

E **1** The network represents a theme park
with seven zones. The number on each
arc shows a distance in metres.

Tramways are to be built to link the
seven zones and the entrance.

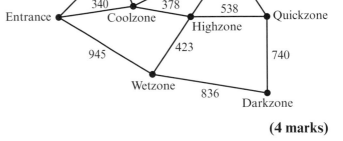

a Find a minimum connector using:

i Kruskal's algorithm **(4 marks)**

ii Prim's algorithm, starting at the
entrance. **(4 marks)**

You must make your order of arc selection clear.

b Draw your tree and state its weight. **(2 marks)**

E/P **2** The network represents eight observation
points in a wildlife reserve and the possible
paths connecting them. The number on
each arc is the distance, in kilometres,
along that path. It is decided to link the
observation points by paths but, in order to
minimise the impact on the wildlife reserve,
the least total length of path is to be used.

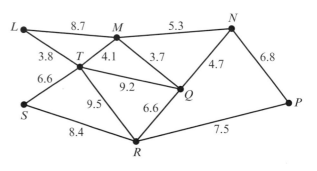

    a Find a minimum spanning tree for the network using:

      i Prim's algorithm, starting at L **(4 marks)**

      ii Kruskal's algorithm.   **(4 marks)**

    In each case list the arcs in the order in which you consider them.

    b Given that paths TQ and RP already exist, and so must form part of the tree, state which algorithm, Prim's or Kruskal's, you would select to complete the spanning tree. Give a reason for your answer.   **(2 marks)**

(E/P)   **3*** The table shows the distances, in mm, between six nodes A to F in a network.

	A	B	C	D	E	F
A	–	124	52	87	58	97
B	124	–	114	111	115	84
C	52	114	–	67	103	98
D	87	111	67	–	41	117
E	58	115	103	41	–	121
F	97	84	98	117	121	–

    a Use Prim's algorithm, starting at A, to solve the minimum connector problem for this table of distances. You must explain your method carefully and indicate clearly the order in which you selected the arcs.   **(3 marks)**

    b Draw a sketch showing the minimum connector and find its length.   **(2 marks)**

    A computer applies Prim's algorithm to a network containing 80 vertices in 0.02 seconds.

    c Given that Prim's algorithm has order n^3, estimate the time required for the computer to apply the algorithm to a network containing 240 vertices.   **(2 marks)**

    d Explain why your answer to part **c** is only an estimate.   **(1 mark)**

(E)   **4*** It is intended to network five computers at a large race track. There is one computer at the office and one at each of the four different entrances. Cables need to be laid to link the computers. Cable laying is expensive, so a minimum total length of cable is required. The table shows the shortest distances, in metres, between the various sites.

	Office	**Entrance 1**	**Entrance 2**	**Entrance 3**	**Entrance 4**
Office	–	1514	488	980	945
Entrance 1	1514	–	1724	2446	2125
Entrance 2	488	1724	–	884	587
Entrance 3	980	2446	884	–	523
Entrance 4	945	2125	587	523	–

    a Starting at Entrance 2, demonstrate the use of Prim's algorithm and hence find a minimum spanning tree. You must make your method clear, indicating the order in which you selected the arcs in your final tree.   **(4 marks)**

    b Calculate the minimum total length of cable required.   **(2 marks)**

E/P 5

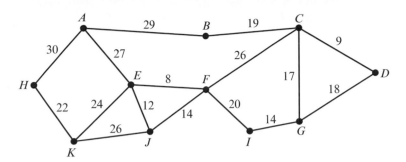

The diagram above shows a weighted network. The weights of each arc are given in the following table:

AB	AE	AH	BC	CD	CF	CG	DG	EF	EJ	EK	FI	FJ	GI	HK	JK
29	27	30	19	9	26	17	18	8	12	24	20	14	14	22	26

a Use a quick sort to write these arcs in order of weight, smallest first. **(4 marks)**

b Use Kruskal's algorithm to find a minimum spanning tree for this network. **(3 marks)**

c Draw your minimum spanning tree and state its total weight. **(2 marks)**

For any connected network,
e = number of edges in the minimum spanning tree
v = number of vertices in the network.

d Write down the relationship between e and v. **(2 marks)**

E/P 6 A company is to install power lines to buildings on a large industrial estate. The lines are to be laid by the side of the roads on the estate. The estate is shown here as a network. The buildings are designated A, B, C, ..., N and the distances between them are given in hundreds of metres. The manager wants to minimise the total length of power line to be used.

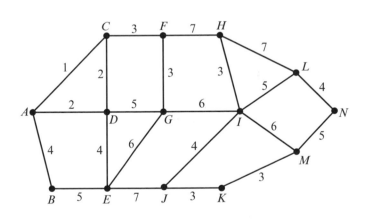

a Use Kruskal's algorithm to obtain a minimum spanning tree for the network and hence determine the minimum length of power line needed. **(4 marks)**

Owing to a change of circumstances, the company modifies its plans for the estate. The result is that the road from F to G now has a length of 700 metres.

b Determine the new minimum total length of power line. **(3 marks)**

7* A weighted network is shown. The number on each arc indicates the weight of that arc.

a Use Dijkstra's algorithm to find a path of least weight from A to K.

State clearly:

i the order in which the vertices were labelled

ii how you determined the path of least weight from your labelling.

b List all alternative paths of least weight.

c Describe a practical problem that could be modelled by the network and solved using Dijkstra's algorithm.

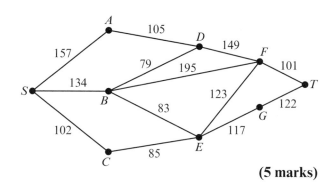

E **8*** The network above shows the distances, in miles, between nine cities. Use Dijkstra's algorithm to determine the shortest route, and its length, between cities S and T. You must indicate clearly the order in which the vertices are labelled and how you used your labelled diagram to decide which cities to include in the shortest route.

(5 marks)

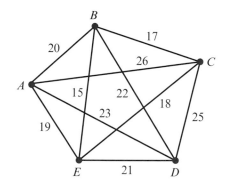

P **9 a** Find a minimum spanning tree for this network using:

i Prim's algorithm

ii Kruskal's algorithm.

b Compare the application of the two algorithms in this case. State which is easier.

c Describe a condition under which Prim's algorithm may be quicker to apply than Kruskal's algorithm.

A **10** The diagram represents part of a road network with distances shown in kilometres. The road from S to Q is a one-way road, as indicated by the arrow.

E

a Use Floyd's algorithm to find the table of least distances, showing both the distance table and the route table after each iteration. **(7 marks)**

b Use the route table to determine the shortest route from R to P, explaining your method clearly. **(2 marks)**

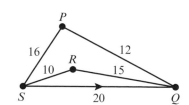

11 The diagram shows a network of roads between some towns. The weight shown on each edge represents the distance in miles. The local council wants to designate a network of 'snow emergency roads' between the towns.

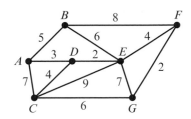

One councillor proposes choosing a set of roads that link all of the towns, such that the total length of the included roads should be as small as possible. These roads would then be cleared as a priority.

a Name a suitable algorithm to find this network. **(1 mark)**

b Apply the algorithm and draw the selected roads, stating the total length of the roads that would need to be cleared under this proposal. **(4 marks)**

The snowplough depot is located in town A. Another councillor proposes that the shortest distance and corresponding route to each town from town A should be found.

c Name a suitable algorithm that will provide this information. **(1 mark)**

d Apply the algorithm, and state the minimum distance from town A to town G, and the corresponding route. **(4 marks)**

A third councillor questions whether the depot is in the best location, and suggests that a table is produced showing the minimum distance between every pair of towns.

e Name a suitable algorithm that will provide this information. **(1 mark)**

Challenge

A computer scientist wants to develop a more efficient implementation of Prim's algorithm. For each vertex Y not contained in the growing tree, she stores a value $\min(Y)$, which is the weight of the shortest arc joining Y to the tree. She can then select the next vertex to add by choosing the smallest value of $\min(Y)$.

When a new vertex, X, is added to the tree, she considers each vertex Y not contained in the new tree. She compares the weight of the arc XY with the current value of $\min(Y)$. If $\min(Y)$ is greater than the weight of XY, she replaces $\min(Y)$ with the new, shorter value.

She is then able to select the smallest of the updated values of $\min(Y)$ to determine which vertex to add next.

Show that this implementation of Prim's algorithm has order n^2.

Summary of key points

1. A **minimum spanning tree** (MST) is a spanning tree such that the total length of its arcs (edges) is as small as possible.

2. **Kruskal's algorithm** can be used to find a minimum spanning tree.
   - Sort the arcs into ascending order of weight and use the arc of least weight to start the tree. Then add arcs in order of ascending weight, unless an arc would form a cycle, in which case reject it.

3. **Prim's algorithm** can be used to find a minimum spanning tree.
   - Choose any vertex to start the tree. Then select an arc of least weight that joins a vertex already in the tree to a vertex not yet in the tree. Repeat this until all vertices are connected.

4. Prim's algorithm can be applied to a **distance matrix**.
   - Choose any vertex to start the tree. Delete the row in the matrix for the chosen vertex and number the column in the matrix for the chosen vertex. Ring the lowest undeleted entry in the numbered columns, which becomes the next arc. Repeat this until all rows are deleted.

5. **Dijkstra's algorithm** can be used to find the shortest path between two vertices in a network.
   - Label the start vertex with final value 0.
   - Record a working value at every vertex, Y, that is directly connected to the vertex that has just received its final label, X:

     final label at X + weight of arc XY
   - Select the smallest working value. This is the final label for that vertex.
   - Repeat until the destination vertex receives its final label.
   - Find the shortest path by tracing back from destination to start. Include an arc AB on the route if B is already on the route and

     final label B − final label A = weight of arc AB

   Each vertex is replaced by a box:

Vertex	Order of labelling	Final label
Working values		

6. Dijkstra's algorithm finds the shortest route between the start vertex and each intermediate vertex completed on the way to the destination vertex.

7. It is possible to use Dijkstra's algorithm on networks with directed arcs, such as a route with one-way streets.

8. **Floyd's algorithm** can be used to find the shortest path between every pair of vertices in a network.
   - Floyd's algorithm applied to a network with n vertices produces two tables as a result of n iterations. One table shows the shortest distances and the other contains information about the corresponding paths taken.

Route inspection

4

Objectives

After completing this unit you should be able to:

* Use the orders of nodes to determine whether a graph is Eulerian, semi-Eulerian or neither → **pages 86 – 89**
* Use the route inspection (Chinese postman) algorithm to find the shortest route in a network → **pages 89 – 94**
* Use the route inspection algorithm in networks with more than four odd nodes → **pages 94 – 98**

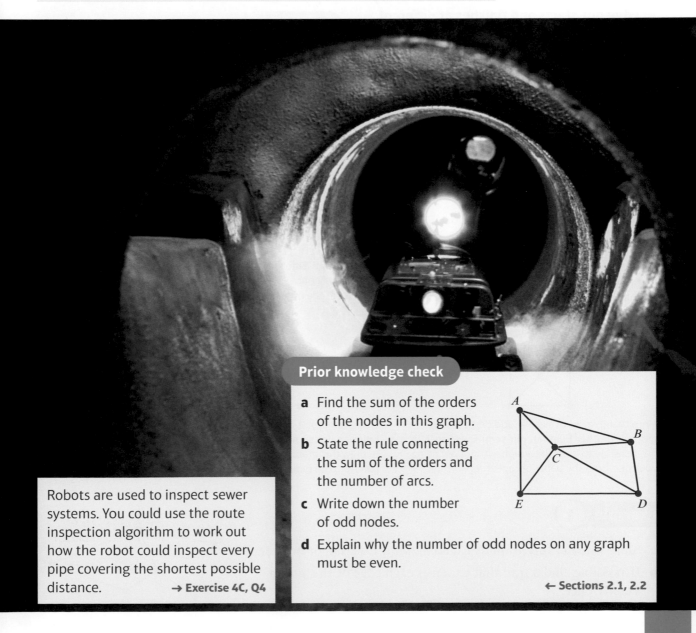

Robots are used to inspect sewer systems. You could use the route inspection algorithm to work out how the robot could inspect every pipe covering the shortest possible distance. → **Exercise 4C, Q4**

Prior knowledge check

a Find the sum of the orders of the nodes in this graph.

b State the rule connecting the sum of the orders and the number of arcs.

c Write down the number of odd nodes.

d Explain why the number of odd nodes on any graph must be even.

← **Sections 2.1, 2.2**

4.1 Eulerian graphs

You can determine whether a graph is Eulerian, semi-Eulerian, or neither, by considering the degrees of the vertices.

- **An Eulerian graph or network is one which contains a trail that includes every edge and starts and finishes at the same vertex. This trail is called an Eulerian circuit. Any connected graph whose vertices are all even is Eulerian.**

- **A semi-Eulerian graph or network is one which contains a trail that includes every edge but starts and finishes at different vertices. Any connected graph with exactly two odd vertices is semi-Eulerian.**

Links The **degree** of a vertex is the number of arcs that are incident to it. Vertices with even degree are called **even vertices**, and those with odd degree are called **odd vertices**.

← Section 2.2

Notation An Eulerian circuit is sometimes called an **Eulerian cycle**. However, it is not a true cycle since, in general, vertices will be visited more than once.

Watch out The trail must start at one odd vertex and end at the other odd vertex.

Eulerian and semi-Eulerian graphs can be drawn without removing your pen from the paper, and without repeating any arcs.

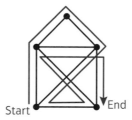

All vertices have even degree, so this graph is **Eulerian**. An Eulerian cycle covering every edge exactly once can start and end at any vertex.

Exactly two vertices have odd degree, so this graph is **semi-Eulerian**. A trail covering every edge exactly once must start at one odd vertex and end at the other.

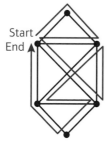

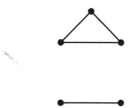

More than two vertices have odd degree, so this graph is **neither Eulerian nor semi-Eulerian**. It is not possible to draw this graph without removing your pen from the paper or repeating an edge.

This graph is not connected so it is **neither Eulerian nor semi-Eulerian**. It is not possible to draw this graph without removing your pen from the paper or repeating an edge.

Example 1

a State whether this graph is Eulerian, semi-Eulerian or neither.

b If possible, find a trail that traverses each edge of this graph exactly once.

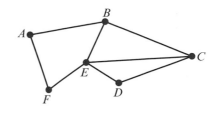

a	Vertex	A	B	C	D	E	F
	Degree	2	3	3	2	4	2

Exactly 2 odd vertices so graph is semi-Eulerian.

Use the degree of each vertex to work out whether the graph is Eulerian, semi-Eulerian or neither.

b $BAFEDCEBC$

The trail must start and end at the odd vertices

Example 2

A connected graph has 5 nodes. The orders of the nodes are 4, 6, 3, p and 2.

a Explain why the graph cannot be Eulerian.

b Explain why the graph must be semi-Eulerian.

c Give the number of edges contained in the network in terms of p.

d Draw a graph with two nodes which is neither Eulerian nor semi-Eulerian.

a The graph has an odd node. It cannot be Eulerian since every node of an Eulerian graph must have even order.

b The number of odd nodes must be even, so p must be odd. The graph will have exactly two odd nodes so is semi-Eulerian.

c Sum of the orders of the nodes = 2 × number of edges.

$$\text{Number of edges} = \frac{4 + 6 + 3 + p + 2}{2}$$

$$= \frac{p + 15}{2}$$

d $A \bullet$

$\bullet B$

For any graph, the hand-shaking lemma states that:

Sum of orders of nodes = 2 × number of edges

This means that the sum of the orders of the nodes must be even, and there must be an even number of odd vertices.   ← **Section 2.2**

Watch out Eulerian and semi-Eulerian graphs must be **connected**. The graph shown has two vertices and no edges. It is not connected so it is neither Eulerian nor semi-Eulerian.

Exercise 4A

1 List the valency of each vertex and hence determine if each of the graphs below is

i Eulerian **ii** semi-Eulerian **iii** neither.

For those that are Eulerian or semi-Eulerian, find a route that traverses each edge exactly once.

a

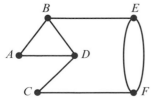

b

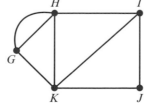

c

2 a Show that each of the graphs below is Eulerian.

b In each case, find a route that starts and finishes at A and traverses each arc exactly once.

i

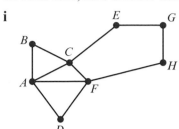

ii

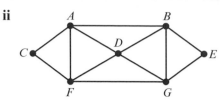

3 a Show that each of these graphs is semi-Eulerian.

b In each case, find a route, starting and finishing at different vertices, that traverses each edge exactly once.

i

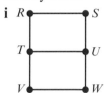

ii

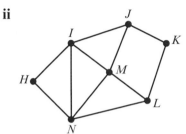

(E/P) 4 a Explain why it is impossible to draw a graph with 6 vertices such that the orders of the nodes are 4, 5, 6, 7, 8 and 9. **(2 marks)**

b A connected graph has 5 nodes and 9 arcs. The orders of the vertices are $2x + 1$, $2x$, $4x - 1$, $4x$ and $6x$.

  i Find the value of x.

  ii State whether the graph is Eulerian, semi-Eulerian or neither, giving a reason for your answer. **(4 marks)**

c Draw a semi-Eulerian graph with 6 vertices. **(2 marks)**

5 Here is the adjacency matrix for a graph.

a State, with reasons, whether the graph is connected.

b State, with reasons, whether the graph is Eulerian, semi-Eulerian or neither.

c Draw the graph.

	A	B	C	D	E
A	0	1	1	0	0
B	1	0	1	0	0
C	1	1	0	0	0
D	0	0	0	0	1
E	0	0	0	1	0

(P) 6 Here is the adjacency matrix of a connected graph.

Determine whether the graph is Eulerian, semi-Eulerian or neither, giving reasons for your answer.

	A	B	C	D	E
A	0	1	0	0	1
B	1	0	0	1	0
C	0	0	0	1	1
D	0	1	1	0	2
E	1	0	1	2	0

E/P **7 a** Determine under what conditions K_n is Eulerian. **(3 marks)**

b Give an example of a complete, semi-Eulerian graph. **(1 mark)**

E/P **8** A student makes the following claim:

> Any graph which contains a Hamiltonian cycle must be Eulerian.

Give a counter-example to show that this claim is incorrect. **(2 marks)**

Challenge

The diagram represents the city of Königsberg (now Kaliningrad, Russia). The Pregel river runs through the city and creates two large islands in the centre. The two islands (C and D) were linked to each other and the mainland (A and B) by seven bridges.

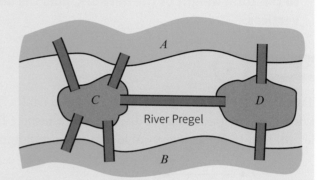

The citizens of Königsberg wished to find a route which allowed them to cross all the bridges in the city exactly once.

a By constructing a graph to model this situation, prove that it is impossible to find such a route.

Johannes works at A, Gregor works at B and Peter works at D. There is a hotel at C.

b Johannes builds an eighth bridge so that he can start at A and finish at his home at C, crossing each bridge once. However, he does not want Gregor to be able to find a similar route from B to C. Where should Johannes build his eighth bridge?

c Gregor decides to build a ninth bridge so that he can start at B and finish at his home at C, crossing each bridge once. He does not want Johannes to be able to find a similar route from A to C. Where should Gregor build his ninth bridge?

d Peter decides to build a tenth bridge, so that every person in the city can cross all the bridges in turn and return to their starting point. Where should Peter build the tenth bridge?

4.2 Using the route inspection algorithm

The **route inspection algorithm** or Chinese postman algorithm, can be used to find the shortest route in a network that traverses every arc at least once and returns to its starting point.

You can solve this problem easily if the network is Eulerian.

- **If all the vertices in the network have even degree, then the length of the shortest route will be equal to the total weight of the network.**

Example 3

The diagram shows a network of trails on a BMX course. The number on each edge is the length, in metres, of the trail. Hannah wants to start at A, traverse every trail on the network, then return to A. Find the length of the shortest path she could take.

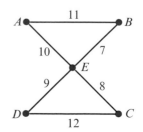

> A, B, C, D and E have degrees 2, 2, 2, 2 and 4 respectively.
> All vertices are even so the network is Eulerian.
> Shortest route = 11 + 7 + 8 + 12 + 9 + 10 = 57 m

Any Eulerian cycle starting at A would traverse each edge exactly once then return to A, so the total length is equal to the total weight of the network. A possible route is $ABECDEA$.

If the network is semi-Eulerian, you need to traverse some edges more than once.

- **If a network has exactly two odd vertices, then the length of the shortest route will be equal to the total weight of the network, plus the length of the shortest path between the two odd vertices.**

Example 4

The diagram shows a network of undersea cables. The number on each arc is the length, in kilometres, of that cable. A diver needs to inspect the cables, and needs to find a route of minimum length that traverses every cable and starts and finishes at S.

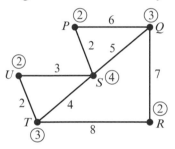

(Total weight of network is 37)

Problem-solving

The numbers in circles show the degree of each vertex. You can label graphs like this in your exam to speed up your working.

a State the cables that will need to be traversed twice.

b State the minimum length of the diver's route.

> **a** There are two odd vertices, Q and T.
> The shortest path connecting Q and T is QST, which has length 4 + 5 = 9.

There are two odd vertices, so this network is semi-Eulerian.

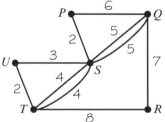

Problem-solving

You can redraw the network and **add extra arcs** to represent the shortest path from Q to T. You have added 1 extra arc to each odd vertex, and either 0 or 2 extra arcs to the other vertices, so this new network is Eulerian.

> The cables that need to be traversed twice are ST and QS.
>
> **b** Total length of route = Total weight + Length of path from Q to T
> $= 37 + 9 = 46$ km

You are adding the lengths of the arcs that need to be traversed twice. This is the total weight of the redrawn graph.

If the network has more than two odd vertices, then you need to consider all the possible pairings of the odd vertices. You need to include additional arcs to join the odd vertices. Select the arcs with the smallest total additional weight and add this to the total weight of the network. This method produces an algorithm that can be used with any connected network.

■ **Here is the route inspection algorithm.**

1 Identify any vertices with odd degree

2 Consider all possible complete pairings of these vertices

3 Select the complete pairing that has the least sum

4 Add a repeat of the arcs indicated by this pairing to the network

Example 5

A route is needed that starts at A, traverses every arc in this network, and returns to A. Find a possible route and state its length.

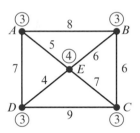

(Total weight of network is 52)

A, B, C and D have odd degree.
You can pair these four nodes in three ways:
A with B and C with D
A with C and B with D
A with D and B with C.
These give the following lengths.
$AB + CD = 8 + 9 = 17$
$AC + BD = 12 + 10 = 22$
$AD + BC = 7 + 6 = 13$ ← least sum
Add AD and BC to the network.

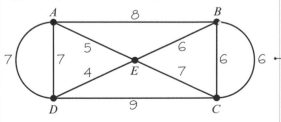

A possible route is
$A\ B\ C\ E\ B\ C\ D\ E\ A\ D\ A$
Total length = 52 + 13 = 65

Consider all the possible pairings of the four odd nodes. The order is not important so there are three possible pairings.

Watch out You must consider all three pairings of the four odd nodes.

Work out the shortest path between each pair. Add these together to find the total additional length that would be needed with each pairing.

Pairing A with D and B with C gives the least sum, so you need to traverse these arcs twice. Redraw the network with these arcs duplicated.

Example 6

The diagram represents a network of roads. The numbers on the arcs represent the lengths of each road in miles. The roads need to be resurfaced. The council wants to find the shortest possible route which traverses every road and starts and ends at A.

a Determine the length of this route. You must show your method and working, and state clearly which roads must be traversed twice.

The council decides that, in order to save money, the route can start and end at different vertices.

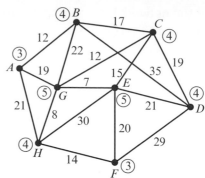

(Total weight of network is 301)

b State which two vertices should be chosen to minimise the length of the new route. Give a reason for your answer, and state the length of the new route.

---

a The odd vertices are A, E, F and G.

The minimum path lengths for each possible pairing are:

$AE + FG = 26 + 22 = 48$

$AF + EG = 35 + 7 = 42$

$AG + EF = 19 + 20 = 39 \leftarrow$ least sum

The most efficient pairing is A with G and E with F.

So repeat arcs AG and EF.

Shortest route = $301 + 39 = 340$ miles

b Least length of all pairs of odd vertices is EG with length 7.

So start at A and finish at F.

Then only arc EG must be repeated.

New route has total length $301 + 7 = 308$ miles.

Label the degree of each vertex and identify the odd vertices.

Find the shortest path between each possible pairing of odd vertices. You can do this by inspection, but check carefully. The shortest path between:

A and E is AGE of length 26

F and G is FHG of length 22

A and F is AHF of length 35

Problem-solving

You must select two odd vertices. Choose them to minimise the length of the path between the remaining odd vertices. Watch out: the single pair with the least distance might not be one of the pairs you identified in part **a**.

Exercise 4B

Answer templates for questions marked * are available at www.pearsonschools.co.uk/d1maths

1* For each network, find a route of minimal length which traverses every arc and starts and ends at vertex A. In each case, state the length of your route and the arcs that must be repeated.

a

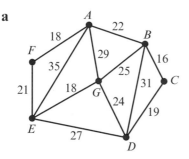

(Total weight of network is 285)

b

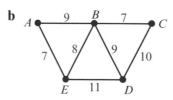

(Total weight of network is 61)

c

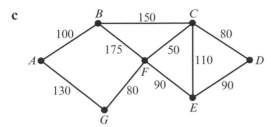

(Total weight of network is 1055)

d

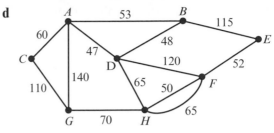

(Total weight of network is 995)

2* Each of the diagrams below shows a network of roads that need to be inspected. In each case, find the length of the shortest route that traverses each arc at least once and returns to the start vertex. List any roads that will need to be traversed twice.

a

A — 110 — B — 100 — C with 75, 125, 130, 80, 110, F — 85 — E — 93 — D

(Total weight of network is 908)

b

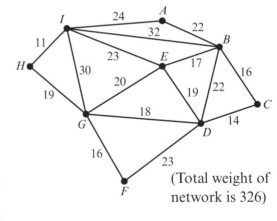

(Total weight of network is 938)

(E) **3*** The diagram represents a network of cables joining computer servers in a university. The number on each arc represents the time, in milliseconds, taken for a data packet to travel along each cable. An analyst wishes to test the network by sending a data packet on the shortest possible route that traverses every cable at least once, and which starts and finishes at server *A*.

a Work out which cables will need to be traversed twice, making your working clear. **(4 marks)**

b State the total time taken for the data packet to complete this route. **(1 mark)**

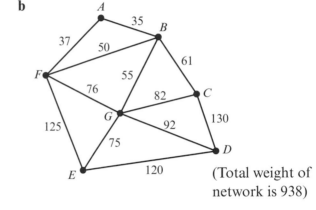

(Total weight of network is 326)

Watch out The shortest route between adjacent odd vertices may not be the arc that directly links them.

(E/P) **4*** The diagram shows the paths in a park. The number on each arc gives the length, in metres, of that path. The vertices show the park entrances, *A*, *B*, *C*, *D*, *E* and *F*.

A gardener needs to inspect each path for weeds.

She will walk along each path once and wishes to minimise her route.

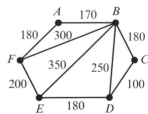

(Total weight of network is 1910)

a Use the route inspection algorithm to find a minimum route, starting and finishing at entrance A. State the length of your route. **(4 marks)**

Given that it is now permitted to start and finish at two different entrances,

b find the start and finish points that would give the shortest route, and state the length of the route. **(4 marks)**

(E/P) **5*** The diagram represents a system of roads.

The number on each arc gives the distance, in kilometres, of that road.

The town council needs to renew the road markings.

Cherry will be renewing the kerbside markings and Mac will renew the centre road markings.

Cherry needs to travel along each road twice, once on each side of the road.

a Explain how this differs from the standard route inspection problem and find the length of Cherry's route. **(4 marks)**

Mac must travel along each road once.

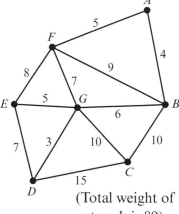

(Total weight of network is 89)

b Use the route inspection algorithm to find a minimal route. You should state the roads he will traverse twice and the length of his route. **(4 marks)**

Road EG is being resurfaced soon and it is decided not to renew its road markings until after the resurfacing.

Given that EG may be omitted from his route,

c find the length of Mac's minimal route. **(3 marks)**

4.3 Networks with more than four odd nodes

If the network has more than four odd nodes then additional information will be provided that will restrict the number of pairings that will need to be considered. With four odd nodes, there are three possible pairings. With six odd nodes there are 15 possible pairings. Beyond six odd nodes, the number of possible pairings increases rapidly.

Example 7

A night watchman has to patrol a network of paths as shown in the diagram. The number shown on each arc represents the time taken, in minutes, to walk between the labelled points.

a Use the route inspection algorithm, starting at A and finishing at C, to find the minimum time taken to traverse each arc at least once.

b State a possible route.

An extra path is added joining B and F directly. After the addition of this path, the minimum time needed to traverse all the paths, starting at A and finishing at C, is reduced by twice the length of time needed to traverse this path.

c Calculate the time needed to traverse the new path, BF.

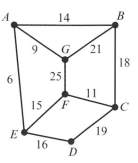

(Total weight of network is 154)

A

a There are 6 odd nodes: A, B, C, E, F, G.

Identify the odd nodes.

Since the required route must start at A and finish at C, these nodes remain odd. You now choose pairings of the remaining odd nodes B, E, F, G.

The constraint regarding the start and finish positions effectively simplifies the problem to a consideration of 4 odd nodes.

By inspection, these path lengths are:

$BE + FG = 20 + 25 = 45$

$BF + EG = 29 + 15 = 44$

$BG + EF = 21 + 15 = 36$

Repeating the paths BG and EF minimises the total time required.

When using the route inspection algorithm you should always state which arcs need to be traversed twice.

The minimum total time is $154 + 36 = 190$ minutes.

b The modified network is now:

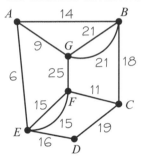

Draw the modified network. Add extra arcs from B to G and from E to F to show that these are the repeated arcs. The nodes at B, G, E and F are now even, so the network is semi-Eulerian.

A possible route is $ABGAEFGBCFEDC$.

c With the extra path, the network now looks like this.

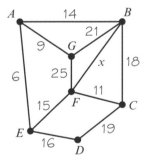

Problem-solving

Draw the network with the new path. You don't know its weight, so label it x. You can find the new shortest route and use this to write an equation involving x.

A, C, E and G are the only odd vertices. Nodes A and C remain odd, so find the shortest way to join E and G:

$EA + AG = 15$

You need to repeat arcs EA and AG, with total weight 15.

Minimum time to traverse all paths =

$154 + x + 15$

So $154 + x + 15 = 190 - 2x$

$3x = 21$

$x = 7$

You know that the total length of the route is reduced by twice the length of BF, or $2x$, so the new route must have total weight $190 - 2x$. Write an equation and solve it to find x.

The new path BF takes 7 minutes to traverse.

Example 8

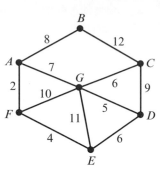

The network represents the roads that must be travelled by a police patrol car. The length of each road in kilometres is shown on each arc.

The patrol starts at A and can finish at either of two police stations, at G or D.

a Find the length of the shortest route such that every road is traversed at least once.

b State which police station the patrol car finishes at, and which roads must be traversed twice.

(Total weight of network is 80)

a The odd nodes are A, C, D, E, F, G.
Starting at A and finishing at G leaves C, D, E, and F to be paired.
By inspection, the path lengths are:
$CD + EF = 9 + 4 = 13$
$CE + DF = 15 + 10 = 25$
$CF + DE = 16 + 6 = 22$
Starting at A and finishing at D, leaves C, E, F and G to be paired.
$CE + FG = 15 + 9 = 24$
$CF + EG = 16 + 11 = 27$
$CG + EF = 6 + 4 = 10$
The minimum distance to be added is 10.
So the shortest route has length
$80 + 10 = 90$ km.

b The patrol car finishes at station D, and roads CG and EF must be traversed twice.

> **Problem-solving**
>
> The situations where the patrol car finishes at G or D need to be considered separately.

Compare all the possible pairings for *either* situation, before choosing the pairing that adds the least weight.

Exercise 4C

Answer templates for questions marked * are available at www.pearsonschools.co.uk/d1maths

1* Find the length of the shortest route in this network that starts at B, traverses every edge at least once and finishes at G. You should show your method clearly, and state any edges that must be traversed twice.

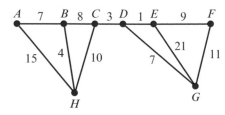

(Total weight of network is 96)

A **2*** The diagram shows a network of cross-country skiing trails
E/P at a resort. The number on each arc shows the length of each
trail in kilometres.

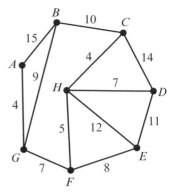

Each week, the trails are inspected by snowmobile.
The inspector wants to find a route of minimum length
that starts at *C* and traverses each trail at least once,
before finishing at either *E* or *G*.

a Use a suitable algorithm to find the trails that must
be traversed twice. **(5 marks)**

b State the length of the route. **(1 mark)**

(Total weight of
network is 106)

A new trail, *BD*, is added. The inspector is still required to start
at *C* and end at either *E* or *G*. Given that the new trail increases
the total length of the inspector's minimum route by half the length of *BD*,

c find the length of *BD*. **(4 marks)**

3* The diagram shows a network of roads. The weights of the arcs
E/P represent the lengths of the roads in metres. A traffic warden
wishes to traverse every road at least once, by the shortest route,
starting and finishing at *A*. She knows that the road joining
vertices *E* and *F* has double yellow lines so she wants to traverse
it twice.

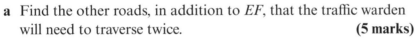

a Find the other roads, in addition to *EF*, that the traffic warden
will need to traverse twice. **(5 marks)**

b Write down the total minimum length of her route. **(1 mark)**

(Total weight of
network is 2220)

On a certain day road *BG* is closed, and she does not need to include it in her route.

c Find the new minimum length of her route. **(3 marks)**

E/P **4*** Each arc in this network represents a sewage pipe. The numbers on the arcs represent the lengths
of the pipes in metres.

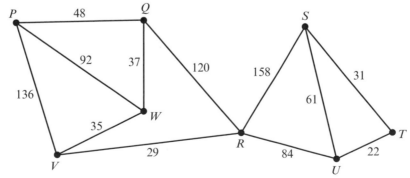

(Total weight of network is 853)

a Use Dijkstra's algorithm to find the shortest path from *P* to *S*. State the path and its
length. **(5 marks)**

A The network is to be inspected using a robot that travels along the sewage pipes. The robot must start and end at vertex P, and must traverse each pipe at least once.

b Explain why the shortest path between any pair of odd vertices in the network lies within the path found in part **a**. **(2 marks)**

c Find the length of the shortest route the robot can take, showing your working clearly. You should state which pipes the robot must traverse twice. **(4 marks)**

Mixed exercise 4

Answer templates for questions marked * are available at www.pearsonschools.co.uk/d1maths

1 For each of these graphs, state, with a reason, whether the graph is Eulerian, semi-Eulerian or neither:

a

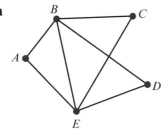

b

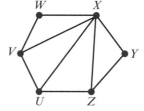

(E/P) **2** Draw a graph that satisfies the following conditions:

- The graph has exactly 6 nodes
- All nodes have even degree
- The graph is not Eulerian **(2 marks)**

(E/P) **3** A connected graph has exactly 4 vertices and 35 edges. The orders of the vertices are $3^{2x} - 700$, $3^{x+1} - 60$, $20 - x$ and x.

a Calculate x. **(4 marks)**

b State whether the graph is Eulerian, semi-Eulerian or neither. You must justify your answer. **(2 marks)**

(E) **4*** The network of paths in a garden is shown below. The numbers on the paths give their lengths in metres. The gardener wishes to inspect each of the paths to check for broken paving slabs so that they can be repaired before the garden is opened to the public. The gardener has to walk along each of the paths at least once.

a Write down the degree (valency) of each of the ten vertices. **(1 mark)**

b Hence find a route of minimum length. You should clearly state, with reasons, which paths, if any, will be covered twice. **(4 marks)**

c State the total length of your shortest route. **(1 mark)**

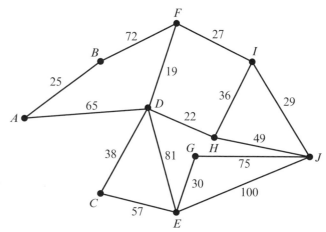

(Total weight of network is 725)

E/P **5*** The network represents the roads on a housing development. The weight on each arc represents the length of that road in metres. A postman parks his van at *P*. He wants to traverse each road at least once before returning to his van.

Given that the postman wants to find a route of minimum total length,

a determine which roads must be traversed twice, showing your working clearly **(5 marks)**

b state the total length of the postman's route. **(1 mark)**

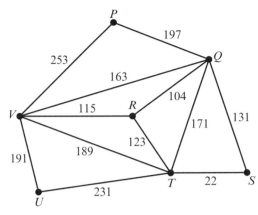

(Total weight of network is 1890)

On a particular day, there is no mail to deliver for the road shown by arc *RT*.

c Explain what difference, if any, this makes to the total length of the postman's route. **(3 marks)**

E/P **6*** The diagram shows the lengths of different tracks in a miniature railway. The number on each track gives its length in metres.

a Use Dijkstra's algorithm to find the shortest path from *G* to *D*. You should state the path and its length. **(5 marks)**

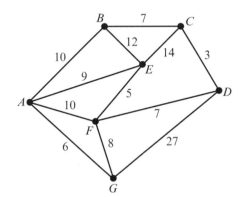

(Total weight of network is 118)

The tracks in the miniature railway are to be tested. A route of minimum length is needed which traverses each track at least once, starting and finishing at *A*.

b Use a suitable algorithm to determine the length of a suitable route. You should show all your working, stating which tracks must be traversed more than once. **(5 marks)**

A new track *CG* of length *x* m is added to the network. After adding the new track, the length of the minimum route traversing each track at least once and starting and finishing at *A* is unchanged.

c Find the value of *x*. **(3 marks)**

E/P **7*** The network shows the major roads that are to be gritted by a council in bad weather. The number on each arc is the length of the road in kilometres.

a List the degree of each of the vertices. **(1 mark)**

b Starting and finishing at *A*, use an algorithm to find a route of minimum length that covers each road at least once. You should clearly state, with reasons, which (if any) roads will be traversed twice. **(4 marks)**

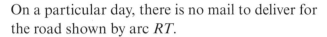

(Total weight of network is 51.4)

c Obtain the total length of your shortest route. **(1 mark)**

There is a minor road *BD* (not shown) between *B* and *D* of length 6.4 km. It is not a major road so it does not need gritting urgently.

d Decide whether or not it is sensible to include *BD* as a part of the main gritting route, giving your reasons. (You may ignore the cost of the grit.) **(3 marks)**

(E/P) **8*** The network opposite represents the streets in a village. The number on each arc represents the length of the street in metres.

The junctions have been labelled *A, B, C, D, E, F, G* and *H*.

An aerial photographer has taken photographs of the houses in the village. A salesman visits each house to see if the occupants would like to buy a photograph of their house. He needs to travel along each street at least once. He parks his car at *A* and starts and finishes there. He wishes to minimise the total distance he has to walk.

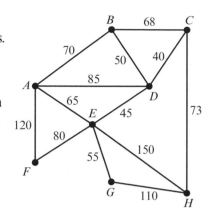

(Total weight of network is 1011)

a Use a suitable algorithm to find a route that the salesman could take, stating the total distance he has to walk. **(5 marks)**

A friend offers to drive the salesman to *B* at the start of the day and collect him from *C* later in the day.

b Explaining your reasoning, carefully determine whether this would increase or decrease the total distance the salesman has to walk. **(3 marks)**

(E/P) **9*a** Describe an algorithm that is used to solve the route inspection (Chinese postman) problem. **(2 marks)**

b Apply the algorithm and find a route, starting and finishing at *A*, that solves the route inspection problem for the network shown. **(4 marks)**

c State the total length of your route. **(1 mark)**

The situation is now altered so that, instead of starting and finishing at *A*, the route starts at one vertex and finishes at another vertex.

(Total weight of network is 249)

d i State the starting vertex and the finishing vertex which minimises the total length of the route. Give a reason for your selections.

ii State the length of your route. **(3 marks)**

e Explain why, in any network, there is always an even number of vertices of odd degree. **(2 marks)**

10* The network represents a system of oil pipelines. The number on each arc represents the length of each pipeline in miles. All of the pipelines are to be inspected each month by aeroplane. The inspection route must start at B and can finish at landing strips at either D or F.

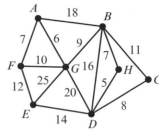

a Determine which landing strip should be chosen in order to minimise the length of the inspection route. **(4 marks)**

b Find the length of the minimum inspection route. **(1 mark)** (Total weight of network is 168)

In one month, pipelines FG, GE and EF are out of operation, and do not need to be inspected.

c Determine how this affects the minimum length of the inspection route. **(3 marks)**

Challenge

***Each arc in this network represents a section of an obstacle course. Contestants must start and finish at A, and every section must be traversed at least once. Jess has worked out the time needed, in minutes, for each section and these are shown as weights on the arcs.

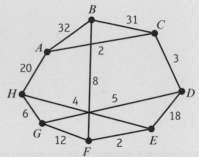

a Use Dijkstra's algorithm to find the shortest time needed to travel from A to B and list the vertices visited in order.

b Deduce which sections Jess needs to repeat to complete the course in the minimum time, and calculate this time.

(Total weight of network is 143)

Summary of key points

1 An **Eulerian graph** or network is one which contains a trail that includes every edge and starts and finishes at the same vertex. This trail is called an Eulerian circuit. Any connected graph whose vertices are all even is Eulerian.

2 A **semi-Eulerian graph** or network is one which contains a trail that includes every edge but starts and finishes at different vertices. Any connected graph with exactly two odd vertices is semi-Eulerian.

3 The **route inspection algorithm** can be used to find the shortest route in a network that traverses every arc at least once and returns to its starting point.
   • If all the vertices in the network have even degree, then the length of the shortest route will be equal to the total weight of the network.
   • If a network has exactly two odd vertices, then the length of the shortest route will be equal to the total weight of the network, plus the length of the shortest path between the two odd vertices.
   • If the network has more than two odd vertices, then you need to consider all the possible pairings of the odd vertices.

4 Here is the route inspection algorithm.
   • Identify any vertices with odd degree
   • Consider all possible complete pairings of these vertices
   • Select the complete pairing that has the least sum
   • Add a repeat of the arcs indicated by this pairing to the network

5 The travelling salesman problem

Objectives

After completing this chapter you should be able to:

● Explain the differences between the classical and practical problems
→ **pages 103–107**

● Use a minimum spanning tree method to find an upper bound
→ **pages 107–113**

● Use a minimum spanning tree method to find a lower bound
→ **pages 114–118**

● Use the nearest neighbour algorithm to find an upper bound
→ **pages 118–122**

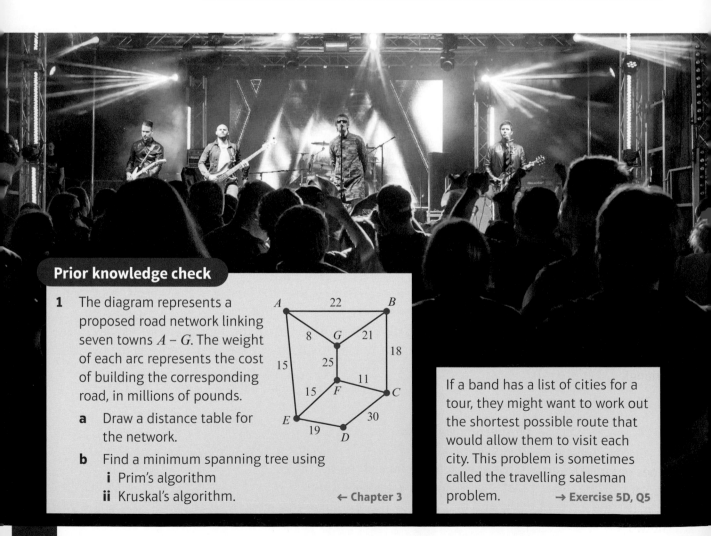

Prior knowledge check

1 The diagram represents a proposed road network linking seven towns $A – G$. The weight of each arc represents the cost of building the corresponding road, in millions of pounds.

   a Draw a distance table for the network.

   b Find a minimum spanning tree using
   i Prim's algorithm
   ii Kruskal's algorithm.   ← **Chapter 3**

If a band has a list of cities for a tour, they might want to work out the shortest possible route that would allow them to visit each city. This problem is sometimes called the travelling salesman problem.   → **Exercise 5D, Q5**

5.1 The classical and practical travelling salesman problems

A There are two variations of the travelling salesmen problem. Before investigating the difference between the **classical** and **practical** travelling salesman problems, there is some essential terminology you need to learn.

- **A walk in a network is a finite sequence of edges such that the end vertex of one edge is the start vertex of the next.**

- **A walk which visits every vertex, returning to its starting vertex, is called a tour.**

Links A **tour** could visit some vertices more than once. If a tour visits every vertex **exactly once** then it is a **Hamiltonian cycle**.

← Section 2.2

The travelling salesman problem involves finding a tour of minimum total weight. There is no efficient algorithm for solving this problem. In real-life situations, it is much more useful to make use of an heuristic algorithm. This is an algorithm which will find a good solution which is not necessarily the optimal solution.

In practice we can find an **upper bound** and **lower bound** for the solution and use these to 'trap' the optimal solution. If our upper and lower bounds are close, then a solution between the two may be acceptable.

You will therefore find upper bounds and select the **smallest**, and then find lower bounds and select the **largest**, trying to 'trap' the optimal solution in as narrow an interval as possible.

Upper bounds
Better upper bounds
Best upper bound
Optimal solution
Best lower bound
Better lower bounds
Lower bounds

Hint If you know that your shortest route is between 123 and 145 miles, say, and you find a route that is 123 miles long, then you know you have found the optimal route. If you find a route that is 130 miles long you may decide that it is 'optimal enough' and use it.

You need to know the difference between the two variations of the travelling salesman problem.

- **In the classical problem, each vertex must be visited exactly once before returning to the start.**

- **In the practical problem, each vertex must be visited at least once before returning to the start.**

If you convert the network into **a complete network of least distances**, the classical and practical travelling salesman problems are equivalent.

To create a complete network of least distances you ensure that the **triangle inequality** holds for all triangles in the network.

- **The triangle inequality states**

    the longest side of any triangle $\leqslant$ the sum of the two shorter sides.

If you have a network where the triangle inequality does not hold in one or more triangles, you simply replace the longest arc in those triangles by the sum of the two smaller ones, thereby creating a network which shows the shortest distances.

Watch out This inequality is satisfied if the longest side is equal to the sum of the two shorter sides.

A The distance matrix for a complete network of least distances is called a **table of least distances**. It shows the shortest path between any two points in the network.

← Section 3.5

Hint In simple cases you can find the table of least distances by inspection. You might also be asked to use Floyd's algorithm to find a complete table of least distances.

Example **1**

Create a table of least distances for the network opposite.

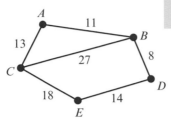

The top row shows the shortest routes starting at A.

	A	B	C	D	E
A	–	11	13	19	31
B		–			
C			–		
D				–	
E					–

You can use this to fill in the first column.

	A	B	C	D	E
A	–	11	13	19	31
B	11	–			
C	13		–		
D	19			–	
E	31				–

Move on to the remaining routes starting from B.

	A	B	C	D	E
A	–	11	13	19	31
B	11	–	24	8	22
C	13	24	–		
D	19	8		–	
E	31	22			–

Once again you use the table's symmetry to complete the second column.
Considering the remaining routes starting at C, you get

	A	B	C	D	E
A	–	11	13	19	31
B	11	–	24	8	22
C	13	24	–	32	18
D	19	8	32	–	
E	31	22	18		–

Watch out The shortest distance between a pair of vertices may not always be the direct route so check carefully for non-direct shorter routes.

Look at routes from A.
There are direct routes from AB and AC and you can see these are the shortest routes.
Using inspection you complete the table to show the shortest routes AD and AE.
You need to check AE carefully. Using ACE the route is 31. Using $ABDE$ the route is 33, so you record 31 as the value.
This enables you to complete the top row of the table.

Since this is not a directed network, the shortest distance from A to D is the same as the shortest distance from D to A.

The direct route BC on the network is given as 27, but if you use BAC as your route you get 24, so 24 is the least distance from B to C.

You complete BD and BE by observation, using BDE as the shortest route from B to E.

Starting at C you have two routes to find: CD and CE. The direct arc CE is the shortest route.
For CD you need to check CED (length 32) and $CABD$ (length 32), so you can record 32. (You do not need to check CBD since you found that the direct route CB was longer than the route CAB earlier.)

A

Finally you complete the last entries, giving the completed table of least distances.

	A	B	C	D	E
A	–	11	13	19	31
B	11	–	24	8	22
C	13	24	–	32	18
D	19	8	32	–	14
E	31	22	18	14	–

This table corresponds to the following **completed network of least distances**. This is different to the original network.

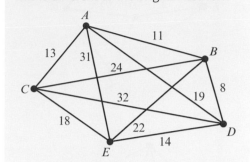

Example 2

The network on the right shows the lengths, in km, of the cross-country skiing trails connecting five towns. Due to a steep incline, the trail between B and E can only be travelled in one direction.

Create a complete table of least distances for this network.

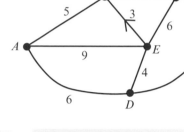

	A	B	C	D	E
A	–	5	7	6	9
B	5	–	2	10	8
C	7	2	–	8	6
D	6	7	8	–	4
E	8	3	5	4	–

The shortest route from E to B is the direct path of length 3 km. However, you cannot travel directly from B to E so the shortest route from B to E is the path BCE of total length 8 km.

Watch out This is a **directed network** so the table of least distances will not necessarily be symmetrical about its leading diagonal. Remember that the **row heading** tells you the start vertex and the **column heading** tells you the end vertex.

Exercise 5A

Answer templates for questions marked * are available at www.pearsonschools.co.uk/d1maths

1* For each network, complete the table of least distances.

a

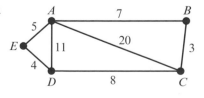

Table of least distances

	A	B	C	D	E
A	–	7			5
B	7	–	3		
C		3	–	8	12
D			8	–	4
E	5		12	4	–

A

b

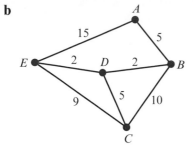

Table of least distances

	A	B	C	D	E
A	–	5		7	
B	5	–		2	
C			–	5	
D	7	2	5	–	2
E				2	–

c

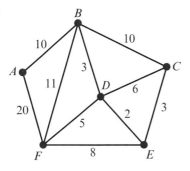

Table of least distances

	A	B	C	D	E	F
A	–	10		13	15	
B	10	–		3		
C			–		3	
D	13	3		–	2	5
E	15		3	2	–	
F				5		–

d

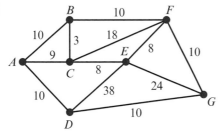

Table of least distances

	A	B	C	D	E	F	G
A	–	10	9	10	17		
B	10	–	3	20		10	20
C	9	3	–	19	8		
D	10	20	19	–		20	10
E	17		8		–	8	18
F		10		20	8	–	10
G		20		10	18	10	–

e

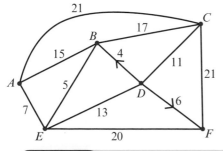

Watch out This is a directed network. The completed table of least distances will not be symmetrical about its leading diagonal.

	A	B	C	D	E	F
A	–	12	21		7	26
B	12	–	17		5	
C		15	–			17
D		4		–		
E	7	5	22		–	
F	27		21		20	–

A **2 a** Describe the difference between the practical and the classical travelling salesman problems.

b For the network given on the right find, by inspection, the solution of:

 i the practical travelling salesman problem
 ii the classical travelling salesman problem.

State the route used in each case, starting from A.

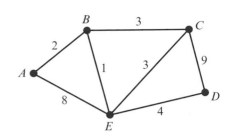

P **3 a** Find, by inspection, the solution of the practical travelling salesman problem for this network, starting from P. State the route used.

b Explain why the choice of start point does not affect the minimum distance travelled.

c Explain why it isn't possible to find a solution to the classical travelling salesman problem for this network.

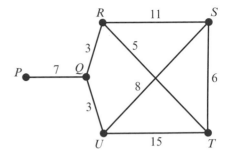

P **4** Construct a complete network in which the solutions to the classical and practical travelling salesman problems are different.

Challenge

1 Explain why a solution to the classical travelling salesman problem must exist for any complete network.

2 The following algorithm is suggested for finding a solution to the classical travelling salesman problem on a complete network:

- Find all possible Hamiltonian cycles in the network and compute the weight of each
- Select the cycle with the least weight

Determine the order of this algorithm in terms of the number of vertices, n, and comment on its usefulness for large values of n.

5.2 Using a minimum spanning tree method to find an upper bound

- **You can use a minimum spanning tree method to find an upper bound for the practical travelling salesman problem in an undirected network by following these steps:**

1 Find the minimum spanning tree for the network (using Prim's algorithm or Kruskal's algorithm). This guarantees that each vertex is included.

2 Double this minimum connector (in effect you keep on retracing your steps) so that completing the cycle is guaranteed.

3 Finally, seek 'shortcuts'. (Make use of some of the non-included arcs that enable you to bypass a repeat of some of the minimum spanning tree.)

Note You are seeking a **minimum** route. There is therefore a logic in trying to use, as a starting point, the **minimum** spanning tree, which you know how to find.

A The **initial upper bound** found by doubling each arc on a minimum spanning tree is shown in this diagram.

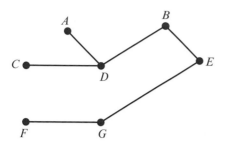

Minimum spanning tree

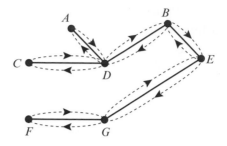

Initial upper bound

This visits each vertex and returns to the starting vertex.

- **An initial upper bound is found by finding the weight of the minimum spanning tree for the network and doubling it.**

Example 3

a Use Kruskal's algorithm to find a minimum spanning tree for the network on the right.

b Hence find an initial upper bound for the travelling salesman problem.

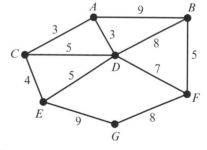

a Using Kruskal's algorithm:

Putting the arcs in order: *AC*, *AD*, *CE*, *BF*, *CD*, *DE*, *DF*,
BD, *FG*, *AB*, *EG*

Include *AC*, *AD*, *CE*, *BF*, reject *CD*, reject *DE*,
include *DF*, reject *BD*, include *FG*. Tree complete.

Arcs that have the same weight can be placed in any order.

← Section 3.1

This gives the following minimum spanning tree

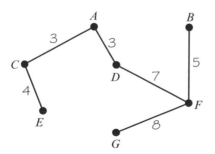

If you are not directed to a particular algorithm, you could use either Prim's algorithm or Kruskal's algorithm.

The weight of the minimum spanning tree
= 3 + 3 + 4 + 5 + 7 + 8 = 30

b The initial upper bound is 2 × 30 = 60.

Example 4

A The table of least distances for a network is given below. Find an initial upper bound for the travelling salesman problem for this network.

	A	B	C	D	E
A	–	11	13	19	31
B	11	–	24	8	22
C	13	24	–	32	18
D	19	8	32	–	14
E	31	22	18	14	–

The network associated with this table of least distances is shown in Example 1 on page 104.

Starting at *A*:

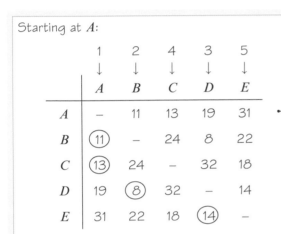

Order of arc inclusion: *AB*, *BD*, *AC*, *DE*

When using a table it is easier to use Prim's algorithm.   **← Section 3.3**

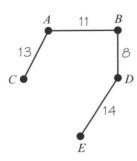

The weight of the minimum spanning tree
= 11 + 8 + 13 + 14 = 46
The initial upper bound is 2 × 46 = 92.

The initial upper bound found in the previous two examples is not very good, since you repeat each arc in the minimum spanning tree.

- **You can improve the initial upper bound by looking for shortcuts.**

Example 5

Starting from the initial upper bound found in Example 4, use a shortcut to reduce the upper bound to below 70.

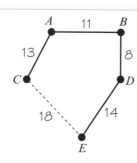

Use CE(18) instead of repeating CA(13), AB(11), BD(8) and DE(14), this saves $13 + 11 + 8 + 14 - 18 = 28$
This gives the following route:

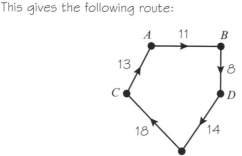

Improved upper bound is now $11 + 8 + 14 + 18 + 13 = 64$

Example 6

The table shows the distances, in miles, between eight cities. A politician has to visit each city, starting and finishing at A. She wishes to minimise the total distance travelled.

	A	B	C	D	E	F	G	H
A	–	47	84	382	120	172	299	144
B	47	–	121	402	155	193	319	165
C	84	121	–	456	200	246	373	218
D	382	402	456	–	413	220	155	289
E	120	155	200	413	–	204	286	131
F	172	193	246	220	204	–	144	70
G	299	319	373	155	286	144	–	160
H	144	165	218	289	131	70	160	–

a Find a minimum spanning tree for this network.

b Hence find an upper bound for this problem.

c Use shortcuts to reduce this upper bound to a value below 1300 miles.

A

a Using Prim's algorithm

	1 ↓ A	2 ↓ B	3 ↓ C	8 ↓ D	4 ↓ E	6 ↓ F	7 ↓ G	5 ↓ H
A	–	47	84	382	120	172	299	144
B	(47)	–	121	402	155	193	319	165
C	(84)	121	–	456	200	246	373	218
D	382	402	456	–	413	220	(155)	289
E	(120)	155	200	413	–	204	286	131
F	172	193	246	220	204	–	144	(70)
G	299	319	373	155	286	(144)	–	160
H	144	165	218	289	(131)	70	160	–

Order of arc selection: AB, AC, AE, EH, HF, FG, GD

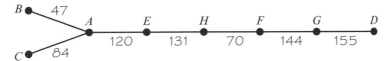

b The initial upper bound is $2 \times 751 = 1502$ miles

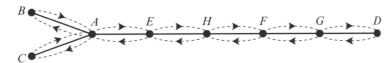

c Looking at the tree, likely shortcuts are AD and BC.

AD saves $120 + 131 + 70 + 144 + 155 - 382 = 238$ miles

BC saves $47 + 84 - 121 = 10$ miles

This leaves the following tour

Tour $ABCAEHFGDA$

Tour length = $47 + 121 + 84 + 120 + 131 + 70 + 144 + 155 + 382$
$\qquad\qquad = 1254$ miles

Problem-solving

You can check your answer by comparing your initial upper bound, the savings from your short cuts, and your improved upper bound:
$1502 - 238 - 10 = 1254$

A Selecting shortcuts

In Example 6 there are many other shortcuts that could be tried.

For example, here are three others.

- BD saves 265 miles, so this shortcut alone would have been sufficient.

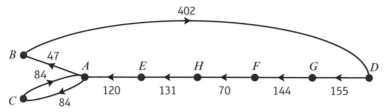

Tour $ABDGFHEACA$, length 1237 miles

- $CB + AF + FD$ saves 238 miles.

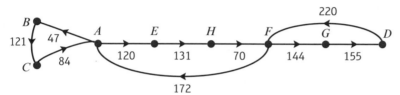

Tour $ABCAEHFGDFA$, length 1264 miles

- CD alone saves 248 miles.

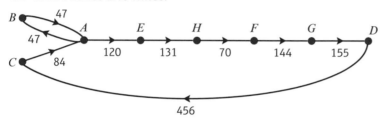

Tour $ABAEHFGDCA$, length 1254 miles

Of these, the best upper bound is 1237 miles since this is the smallest.

- **Aim to make the upper bound as low as possible to reduce the interval in which the optimal solution is contained.**

Exercise 5B

1

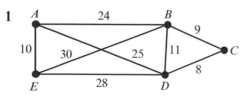

a Find a minimum spanning tree for the network above and hence find an initial upper bound for the travelling salesman problem.

b Use a shortcut to find a better upper bound.

c State the route given by your improved upper bound and state its length.

A 2 A council employee needs to service five sets of traffic lights located at A, B, C, D and E.

The table shows the distance, in miles, between the lights. She will start and finish at A and
wishes to minimise her total travelling distance.

	A	B	C	D	E
A	–	13	11	19	14
B	13	–	12	7	16
C	11	12	–	11	8
D	19	7	11	–	14
E	14	16	8	14	–

a Find a minimum spanning tree for the network. **(4 marks)**

b Hence find an initial upper bound for the length of the employee's route. **(1 mark)**

c Use shortcuts to reduce the upper bound to a value below 65 miles. **(3 marks)**

d State the route given by your improved upper bound and state its length. **(2 marks)**

E 3 a Use Kruskal's algorithm to find a minimum
spanning tree for the network shown on the
right, and hence find an initial upper bound for
the travelling salesman problem. **(5 marks)**

b Use shortcuts to reduce the upper bound to
below 240. **(3 marks)**

c State the route given by your improved upper
bound and state its length. **(2 marks)**

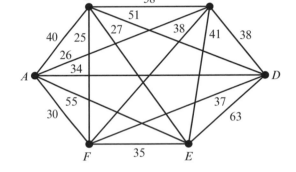

> **Hint** Start with the edge of minimum
> weight, BF ← **Section 3.1**

E 4 The table shows the times, in minutes, taken to travel between a surgery S and five farms V, W,
X, Y and Z. A vet needs to visit animals at each of the farms and wishes to minimise the total
travel time. He will start and finish at the surgery, S.

	S	V	W	X	Y	Z
S	–	75	30	55	70	70
V	75	–	55	30	40	15
W	30	55	–	65	45	55
X	55	30	65	–	15	10
Y	70	40	45	15	–	20
Z	70	15	55	10	20	–

a Use Prim's algorithm, starting at S, to find a minimum spanning tree for the network above
and hence find an initial upper bound for the travelling salesman problem. **(4 marks)**

b Use the method of shortcuts to reduce the upper bound to below 200 minutes. **(3 marks)**

c State the route given by your improved upper bound and state the total time taken on this
route. **(2 marks)**

5.3 Using a minimum spanning tree method to find a lower bound

A
- **You can use a minimum spanning tree method to find a lower bound for the classical problem by following these steps.**

Note This algorithm only produces a lower bound for the **classical** problem. If you are solving the **practical** problem you need to apply the algorithm to a completed network of least distances.

1. Remove each vertex in turn, together with its arcs.
2. Find the **residual minimum spanning tree** (RMST) and its length.
3. Add to the RMST the 'cost' of reconnecting the deleted vertex **by the two shortest, distinct, arcs** and note the totals.
4. The greatest of these totals is used for the lower bound.
5. Make the lower bound as high as possible to reduce the interval in which the optimal solution is contained.
6. You have found an optimal solution if the lower bound gives a Hamiltonian cycle, or if the lower bound has the same value as the upper bound.

Notation

The **residual spanning tree** is the minimum spanning tree for the resulting network after removing a vertex.

Watch out This algorithm will not, in general, generate a Hamiltonian cycle. As such, the lower bound is generally not a solution to the original problem, as it would be necessary to repeat arcs to create a tour.

Example 7

	A	B	C	D	E
A	–	11	13	19	31
B	11	–	24	8	22
C	13	24	–	32	18
D	19	8	32	–	14
E	31	22	18	14	–

The table of least distances for a network is shown.

a By deleting vertex A, find a lower bound to the travelling salesman problem for this network.

b State whether this lower bound represents an optimal solution. Give a reason for your answer.

---

a When A is deleted, the table for the residual network becomes

	1	4	2	3
	↓	↓	↓	↓
	B	C	D	E
B	–	24	8	22
C	24	–	32	18
D	8	32	–	14
E	22	18	14	–

Using Prim's algorithm starting at B, the order of arc selection is BD, DE and EC

This is the same network that you created a table of least distances for in Example 1 on page 104, and you found upper bounds for in Examples 4 and 5 on pages 109 and 110.

The residual minimum spanning tree is

Weight of residual minimum spanning tree = 8 + 14 + 18 = 40

The two least arcs from A are AB (11) and AC (13)

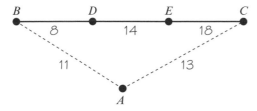

Lower bound = weight of RMST + weights of two least arcs from A

$$= 40 + 11 + 13$$
$$= 64$$

b The lower bound represents a Hamiltonian cycle, so it must be an optimal solution.

Problem-solving

The spanning tree can be drawn in a straight line, and the two least arcs join A to the start and end of this line, so the solution represents a Hamiltonian cycle. You could also compare the lower bound with a known upper bound. In Example 5 you found an improved upper bound of 64 for this network. Since upper bound = lower bound the solution must be optimal.

Example 8

	A	B	C	D	E	F	G	H
A	–	47	84	382	120	172	299	144
B	47	–	121	402	155	193	319	165
C	84	121	–	456	200	246	373	218
D	382	402	456	–	413	220	155	289
E	120	155	200	413	–	204	286	131
F	172	193	246	220	204	–	144	70
G	299	319	373	155	286	144	–	160
H	144	165	218	289	131	70	160	–

You found an upper bound for this network in Example 6 on page 110.

a By deleting vertices A then G, find two lower bounds to the travelling salesman problem for the network above.

b Select the better lower bound of the two found in part **a**, giving a reason for your answer.

c Taking your answer to **b** and using the better upper bound, 1237 miles, found in Example 6, write down the smallest interval that must contain the length of the optimal route.

A

a i Deleting A and using Prim's algorithm starting at B

	1 ↓ B	2 ↓ C	7 ↓ D	3 ↓ E	5 ↓ F	6 ↓ G	4 ↓ H
B	–	121	402	155	193	319	165
C	(121)	–	456	200	246	373	218
D	402	456	–	413	220	(155)	289
E	(155)	200	413	–	204	286	131
F	193	246	220	204	–	144	(70)
G	319	373	155	286	(144)	–	160
H	165	218	289	(131)	70	160	–

Order of arc selection: BC, BE, EH, HF, FG, GD

The residual minimum spanning tree is:

C •———121———• B ———155——— E ———131——— H ———70——— F ———144——— G ———155——— D •

Weight of RMST = 776 miles

Two least arcs from A are AB (47) and AC (84)

C 121 B 155 E 131 H 70 F 144 G 155 D

84 ⋅⋅⋅⋅ ⋅⋅⋅ 47

A

Lower bound by deleting A = 776 + 47 + 84 = 907 miles

> You need to make your method for finding the RMST clear. The order of arc selection is sufficient to demonstrate that you have applied Prim's algorithm correctly.

> This is not a Hamiltonian cycle. In general you do not get a solution to the travelling salesman problem when finding a lower bound.

ii Deleting G and using Prim's algorithm starting at A

	1 ↓ A	2 ↓ B	3 ↓ C	7 ↓ D	4 ↓ E	6 ↓ F	5 ↓ H
A	–	47	84	382	120	172	144
B	(47)	–	121	402	155	193	165
C	(84)	121	–	456	200	245	218
D	382	402	456	–	413	(220)	289
E	(120)	155	200	413	–	204	131
F	172	193	246	220	204	–	(70)
H	144	165	218	289	(131)	70	–

Order of arc selection: AB, AC, AE, EH, HF and FD

The residual minimum spanning tree is:

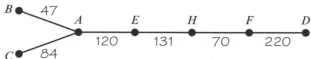

Weight of RMST = 672 miles

A

Two least arcs are **GF** (144) and **GD** (155)

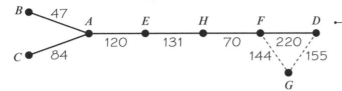

Once again we do not get a
Hamiltonian cycle when we
'reconnect' *G*, so this lower
bound does not represent an
optimal solution.

Lower bound by deleting **G** = 672 + 144 + 155 = 971 miles

b The better lower bound is the higher one, 971 miles, the
one obtained by deleting **G**.

This will reduce the size of the interval containing the
optimal solution.

c The better lower bound is 971 miles, the better upper
bound is 1237 miles.

　　971 miles ≤ optimal solution ≤ 1237 miles

Exercise 5C

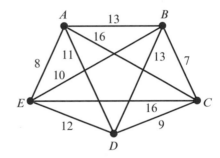

1 The network shows the least distances, in miles,
between 5 towns.

　a By deleting vertex *A*, find a lower bound to the
　　travelling salesman problem for the network opposite.

　b State, with reasons, whether your answer represents
　　an optimal solution.

E/P **2**

	A	B	C	D	E
A	–	13	11	19	14
B	13	–	12	7	16
C	11	12	–	11	8
D	19	7	11	–	14
E	14	16	8	14	–

A council employee needs to service five sets of traffic
lights located at *A*, *B*, *C*, *D* and *E*.
The table shows the least distances, in miles between
the lights. She will start and finish at *A* and
wishes to minimise her total travelling distance.

　a By deleting vertices *A* then *B* find two lower
　　bounds for the employee's route.　　　**(4 marks)**

　b Select the better lower bound, giving a reason for
　　your answer.　　　　　　　　　　　　**(1 mark)**

E/P **3** The network shows the least distances, in km,
between 6 locations on an orienteering course.

　a By deleting vertices *A* then *B*, find two lower bounds
　　for the travelling salesman problem.　　**(4 marks)**

　b Select the better lower bound, giving a reason
　　for your answer.　　　　　　　　　　**(1 mark)**

An upper bound for the solution is given as 190 km.

　c Write down the smallest interval that you can be
　　confident contains the optimal length
　　of the route.　　　　　　　　**(2 marks)**

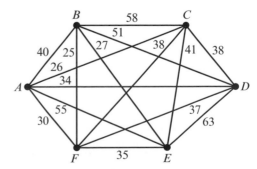

Hint　This upper bound was found as a solution to Q3
in Exercise 5B.

A 4 The table shows the time, in minutes, taken to travel between a surgery S and five farms **E/P** V, W, X, Y and Z. A vet needs to visit animals at each of the farms and wishes to minimise the total travel time. He will start and finish at the surgery, S.

	S	V	W	X	Y	Z
S	–	75	30	55	70	70
V	75	–	55	30	40	15
W	30	55	–	65	45	55
X	55	30	65	–	15	10
Y	70	40	45	15	–	20
Z	70	15	55	10	20	–

a By deleting vertices S then V, find two lower bounds for the vet's route. **(4 marks)**

b Select the better lower bound, giving a reason for your answer. **(1 mark)**

An upper bound for the solution is given as 190 minutes.

c Write down the smallest interval that you can be confident contains the optimal solution. **(2 marks)**

Hint This upper bound was found as a solution to Q4 in Exercise 5B.

5.4 Using the nearest neighbour algorithm to find an upper bound

The minimum spanning tree method for finding an upper bound given in section 5.2 can be hard to use for large networks, and cannot be applied to directed networks.

- **You can use the nearest neighbour algorithm to find an upper bound for the travelling salesman problem in a completed network of least distances. Follow these steps.**

1 Select each vertex in turn as a starting point.

2 Go to the nearest vertex which has not yet been visited.

3 Repeat step **2** until all vertices have been visited and then return directly to the start vertex.

4 Once all vertices have been used as the starting vertex, select the tour with the smallest length as the upper bound.

The upper bound given by the nearest neighbour algorithm always represents a possible tour. However, in general, it will not represent the optimal solution.

Note You will usually be directed to start with specific vertices. You will only have to check all the vertices if you are specifically instructed to do so.

Watch out Do not confuse the nearest neighbour algorithm with Prim's algorithm. In Prim's algorithm you look for the vertex nearest *any* of the vertices in your growing tree. In the nearest neighbour algorithm, you look for the vertex nearest to the *last vertex chosen*.

Example 9

A This is the network from Example 2. The completed table of least distances for this network is given to the right.

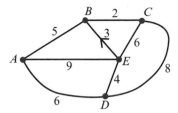

	A	B	C	D	E
A	–	5	7	6	9
B	5	–	2	10	8
C	7	2	–	8	6
D	6	7	8	–	4
E	8	3	5	4	–

a Use the nearest neighbour algorithm on the completed table of least distances, starting at C, to find an upper bound for the travelling salesman problem on this network.

b State the corresponding route in the original network.

a

	A	B	C	D	E
A	–	5	7	6	9
B	5	–	2	10	8
C	7	②	–	8	6
D	6	7	8	–	4
E	8	3	5	4	–

Route: $C \to B$
Length: 2

Start at C. Look across the C row. The smallest number is 2, so the nearest vertex is B. You have visited C so delete the C column.

	A	B	C	D	E
A	–	5	7	6	9
B	⑤	–	2	10	8
C	7	②	–	8	6
D	6	7	8	–	4
E	8	3	5	4	–

Route: $C \to B \to A$
Length: 2 + 5

Now look across the B row. The smallest number is 5, so the nearest remaining vertex is A. Delete the B column.

	A	B	C	D	E
A	–	5	7	⑥	9
B	⑤	–	2	10	8
C	7	②	–	8	6
D	6	7	8	–	4
E	8	3	5	4	–

Route: $C \to B \to A \to D$
Length: 2 + 5 + 6

Now look across the A row. The smallest number is 6, so the nearest remaining vertex is D. Delete the A column.

	A	B	C	D	E
A	–	5	7	⑥	9
B	⑤	–	2	10	8
C	7	②	–	8	6
D	6	7	8	–	④
E	8	3	5	4	–

Route: $C \to B \to A \to D \to E$
Length: 2 + 5 + 6 + 4

Now look across the D row. The only number left is 4, so the nearest remaining vertex is E. Delete the D column.

	A	B	C	D	E
A	–	5	7	⑥	9
B	⑤	–	2	10	8
C	7	②	–	8	6
D	6	7	8	–	④
E	8	3	⑤	4	–

Route: $C \to B \to A \to D \to E \to C$
Length: 2 + 5 + 6 + 4 + 5 = 22 km

You have visited every vertex so return to C. The total length of your route is 22 km. You do not need to show the table at each step of the algorithm. You only need to show the vertices visited in order (including returning to C), and the total length of your tour.

An upper bound for the travelling salesman problem in this network is 22 km.

b $C \to B \to A \to D \to E \to B \to C$

Watch out The algorithm found a Hamiltonian cycle in the completed network of least distances. However, in the original network, the least distance route from E to C is via B. So this route corresponds to $CBADEBC$ in the original network. This is not a Hamiltonian cycle in the original network as it visits vertex B twice, but it is a tour that satisfies the practical travelling salesman problem.

Example 10

	A	B	C	D	E	F	G	H
A	–	47	84	382	120	172	299	144
B	47	–	121	402	155	193	319	165
C	84	121	–	456	200	246	373	218
D	382	402	456	–	413	220	155	289
E	120	155	200	413	–	204	286	131
F	172	193	246	220	204	–	144	70
G	299	319	373	155	286	144	–	160
H	144	165	218	289	131	70	160	–

This is the table of distances from Example 8. In that example you found that the optimal solution lay in the following interval

971 miles ≤ optimal solution ≤ 1237 miles.

a Use the nearest neighbour algorithm, using A then B as starting vertices, to find upper bounds for the travelling salesman problem.

b Review the interval containing the optimal solution and amend it if necessary, giving a reason for your answer.

a Nearest neighbour tour starting at A

A		B		C		E		H		F		G		D		A
	47		121		200		131		70		144		155		382	

= 1250 miles

Start at A.
Look across the A row. The smallest number is 47, AB. Delete column A.
Look across the B row. The smallest number is 121, BC. Delete column B.
Look across the C row. The smallest number is 200, CE. Delete column C.
Look across the E row. The smallest number is 131, EH. Delete column E.
Look across the H row. The smallest number is 70, HF. Delete column H.
Look across the F row. The smallest number is 144, FG. Delete column F.
Look across the G row. The smallest number is 155, DG. Delete column G.
You have now visited each vertex, so you return directly from D to A, 382.

Nearest neighbour tour starting at B

B		A		C		E		H		F		G		D		B
	47		84		200		131		70		144		155		402	

= 1233 miles

Start at B.
Look across the B row. The smallest number is 47, BA. Delete column B.
Look across the A row. The smallest number is 84, AC. Delete column A.
Look across the C row. The smallest number is 200, CE. Delete column C.
Look across the E row. The smallest number is 131, EH. Delete column E.
Look across the H row. The smallest number is 70, HF. Delete column H.
Look across the F row. The smallest number is 144, FG. Delete column F.
Look across the G row. You have now visited each vertex but D, so you must choose GD, 155.
You have now visited each vertex, so you return directly from D to B, 402.

b The interval is now

971 miles ≤ optimal solution ≤ 1233 miles.

We replace the upper bound of 1237 miles with the better upper bound of 1233 miles, since it is lower. This reduces the interval containing the optimal solution.

Exercise 5D

A **1** (This is the same problem as described in Exercise 5C question **2**)

A council employee needs to service five sets of traffic lights located at *A*, *B*, *C*, *D* and *E*. The table shows the least distances, in miles, between the lights. She wishes to minimise her total travelling distance.

	A	*B*	*C*	*D*	*E*
A	–	13	11	19	14
B	13	–	12	7	16
C	11	12	–	11	8
D	19	7	11	–	14
E	14	16	8	14	–

> **Hint** When there are two equal least values in a row you can choose either.

a Starting at *D*, find a nearest neighbour route to give an upper bound for the council employee's route.

b Show that there are two nearest neighbour routes starting from *E*.

c Select the value that should be given as the upper bound. Give a reason for your answer.

2 (This is the same problem as described in Exercise 5C question **4**)

The table shows the time, in minutes, taken to travel between a surgery *S* and five farms *V*, *W*, *X*, *Y* and *Z*. A vet needs to visit animals at each of the farms and wishes to minimise the total travel time.

	S	*V*	*W*	*X*	*Y*	*Z*
S	–	75	30	55	70	70
V	75	–	55	30	40	15
W	30	55	–	65	45	55
X	55	30	65	–	15	10
Y	70	40	45	15	–	20
Z	70	15	55	10	20	–

a Starting at *Z*, find a nearest neighbour route.

b Find two further nearest neighbour routes starting at *X* then *V*.

c Select the value that should be given as the upper bound. Give a reason for your answer.

E/P **3** The nearest neighbour algorithm of selecting a route from a given vertex has quadratic order.

> **Problem-solving**
> The application to a *single* vertex has quadratic order.

A computer program finds all the possible nearest neighbour routes and selects the route of least weight. With 12 towns, the computer program takes 0.27 seconds to complete the task.

Estimate the length of time the computer program will take to compute the least length of the nearest neighbour route for a network of 20 towns. **(3 marks)**

4 A printing company prints six magazines R, S, T, U, V and W, each week. The printing equipment needs to be set up differently for each magazine and the table shows the time, in minutes, needed to set up the equipment from one magazine to another.

a If the magazines were printed in the order $RSTUVWR$, how long would it take in total to set up the equipment? **(2 marks)**

b Show that there are two nearest neighbour routes starting from U. **(2 marks)**

c Show that there are three nearest neighbour routes starting from V. **(3 marks)**

d Select the value that should be given as the upper bound. Give a reason for your answer. **(1 mark)**

	R	S	T	U	V	W
R	–	150	210	150	120	240
S	150	–	210	120	210	240
T	210	210	–	120	150	180
U	150	120	120	–	180	270
V	120	210	150	180	–	300
W	240	240	180	270	300	–

5 The table shows the least driving distances between Aberdeen (A), Berwick-upon-Tweed (B), Carlisle (C), Dundee (D), Edinburgh (E), Fort William (F) and Glasgow (G). The distances are given in miles. The least distance between Aberdeen and Dundee is x miles, where $x < 126$.

A band is planning a tour. The band wants to plan an itinerary that will visit each town at least once. The band would like to minimise the total distance travelled.

The band finds the nearest neighbour routes from Berwick-upon-Tweed and Aberdeen, and finds that the sum of the lengths of these routes is 1419 miles.

a Find x, showing your working clearly. **(4 marks)**

	A	B	C	D	E	F	G
A	–	192	216	x	129	157	146
B	192	–	92	123	64	196	110
C	216	92	–	168	92	211	95
D	x	123	168	–	56	126	81
E	129	64	92	56	–	132	43
F	157	196	211	126	132	–	116
G	146	110	95	81	43	116	–

b Hence find an upper bound for the optimal length of the band's route. **(1 mark)**

6 The network shows the direct paths connecting five holes on a golf course. The paths connecting A to C and D to E are one-way paths. The weight on each arc is the time taken in minutes to travel along the path.

The partially complete table of shortest times is shown below.

a Complete the table by adding the shortest times needed to travel from A to E and from E to A. In each case state the route taken. **(1 mark)**

	A	B	C	D	E
A	–	5	3	7	
B	6	–	2	7	8
C	8	2	–	5	6
D	7	7	5	–	4
E		8	6	11	–

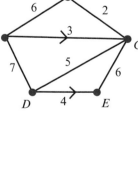

b Use the nearest neighbour algorithm starting at A to find a Hamiltonian cycle in your completed network of shortest times, and state the total time taken for this cycle. **(2 marks)**

A **c** Interpret this cycle in terms of the actual holes visited and state whether it represents a
Hamiltonian cycle in the original network. **(1 mark)**

d Prove that this cycle is not an optimal solution to the
travelling salesman problem in this network. **(1 mark)**

> **Problem-solving**
>
> You can prove the solution is not optimal by finding a shorter cycle.

Mixed exercise 5

1 a Use an appropriate algorithm to find a
minimum connector for the network above.
You must make your method clear.

b Hence find an initial upper bound
for the travelling salesman problem.

c Use the method of shortcuts to
find an upper bound below 6100.

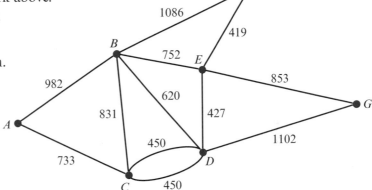

E/P **2** The network on the right shows a number of
hostels in a national park and the possible paths
joining them. The numbers on the edges give the
lengths, in km, of the paths.

a Draw a complete table of least distances for
the network. (You may do this by inspection.
The application of an algorithm is not
required.) **(2 marks)**

b Use the nearest neighbour algorithm on the
complete network to obtain an upper bound
to the length of a tour in this network which
starts and finishes at *A* and visits each hostel
exactly once. **(3 marks)**

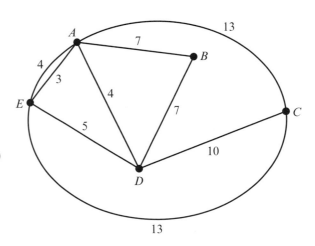

c Interpret your result in part **b** in terms of the original network. **(2 marks)**

A computer program takes 0.85 seconds to apply the nearest neighbour algorithm to the above
network. A different national park has 12 hostels joined by paths.

d Given that the nearest neighbour algorithm has cubic order, estimate the time needed for the
same computer program to apply the nearest neighbour algorithm to this network of paths
and hostels. **(2 marks)**

e Explain why your answer to part **d** is only an estimate. **(1 mark)**

A **3** The table of least distances below was formed from the
E network, N, on the right.

	S	A	B	C	D	E	F
S	–	8	7	2	14	19	5
A	8	–	2	7	19	17	3
B	7	2	–	5	17	19	5
C	2	7	5	–	12	21	7
D	14	19	17	12	–	13	19
E	19	17	19	21	13	–	14
F	5	3	5	7	19	14	–

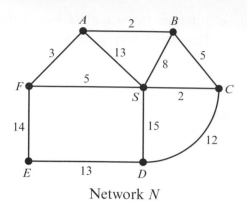

Network N

The table shows the distances, in km, between the central sorting office at S and six post offices
A, B, C, D, E and F.

A postal worker will leave the sorting office, go to each post office to collect mail and return to
the sorting office. He wishes to minimise his route.

a Use Prim's algorithm, starting at S, to obtain two minimum spanning trees. State the order in
which you select the arcs. **(4 marks)**

b Hence find an initial upper bound for the postal worker's route. **(1 mark)**

c Starting from this upper bound, use shortcuts to reduce the upper bound to a value below
60 km. You must state the shortcuts you use. **(2 marks)**

d Starting at C, and then at D, find two nearest neighbour routes stating their lengths. **(4 marks)**

e Select the better upper bound from your answers to parts **c** and **d**. Give a reason for your
answer. **(1 mark)**

f Interpret your answer to **e** in terms of the original network, N, of roads. **(1 mark)**

g Using the table of least distances, and by deleting C, find a lower bound for the postal
worker's route. **(3 marks)**

E **4 a** Explain the difference between the classical and practical travelling salesman problems.
(1 mark)

The table shows the travel time, in minutes, between seven town halls P, Q, R, S, T, U and V.
Kim works at P and must visit each of the other town halls to deliver leaflets. She wishes to
minimise her route.

b Find a minimum connector for the
network. You must make your
method clear by listing the arcs in
order of selection. **(3 marks)**

c Use the minimum connector and
shortcuts to find an upper bound
below 220. You must list the shortcuts
you use and your final route. **(2 marks)**

d Starting at P, find a nearest neighbour
route and state its length. **(3 marks)**

	P	Q	R	S	T	U	V
P	–	19	30	45	38	33	29
Q	19	–	28	27	50	23	55
R	30	28	–	51	29	49	50
S	45	27	51	–	77	21	71
T	38	50	29	77	–	69	37
U	33	23	49	21	69	–	56
V	29	55	50	71	37	56	–

A

e Find a lower bound for the length of the route by deleting P. **(3 marks)**

f Looking at your answers to **c**, **d** and **e**, use inequalities to write down the smallest interval containing the optimal solution. **(1 mark)**

E/P **5** A computer supplier has outlets in seven cities A, B, C, D, E, F and G. The table shows the distances, in km, between each of these seven cities. John lives in city A and has to visit each of these cities to advise on displays. He wishes to plan a route starting and finishing at A, visiting each city and covering a minimum distance.

a Obtain a minimum spanning tree for this network explaining briefly how you applied the algorithm that you used. (Start with A and state the order in which you selected the arcs used in your tree.) **(3 marks)**

b Hence determine an initial upper bound for the length of the route travelled by John. **(1 mark)**

	A	B	C	D	E	F	G
A	–	103	89	42	54	143	153
B	103	–	60	98	56	99	59
C	89	60	–	65	38	58	77
D	42	98	65	–	45	111	139
E	54	56	38	45	–	95	100
F	143	99	58	111	95	–	75
G	153	59	77	139	100	75	–

c Explain why the upper bound found in this way is unlikely to give the minimum route length. **(2 marks)**

d Starting from your initial upper bound and using an appropriate method, find an upper bound for the length of the route which is less than 430 km. **(2 marks)**

e By deleting city A, determine a lower bound for the length of John's route. **(3 marks)**

f Explain under what circumstances a lower bound obtained by this method might be an optimum solution. **(2 marks)**

E **6** A sales representative, Sheila, has to visit clients in six cities, London, Cambridge, Oxford, Birmingham, Nottingham and Exeter. The table shows the distances, in miles, between these six cities. Sheila lives in London and plans a route starting and finishing in London. She wishes to visit each city and drive the minimum distance.

	L	C	O	B	N	E
London (L)	–	80	56	120	131	200
Cambridge (C)	80	–	100	98	87	250
Oxford (O)	56	100	–	68	103	154
Birmingham (B)	120	98	68	–	54	161
Nottingham (N)	131	87	103	54	–	209
Exeter (E)	200	250	154	161	209	–

a Starting from London, use Prim's algorithm to obtain a minimum spanning tree. Show your working. State the order in which you selected the arcs and draw the tree. **(3 marks)**

b **i** Hence determine an initial upper bound for the length of the route planned by Sheila.

ii Starting from your initial upper bound and using shortcuts, obtain a route which is less than 660 miles. **(3 marks)**

c By deleting Exeter from the table determine a lower bound for the length of Sheila's route. **(3 marks)**

A **7** The table shows the least distances, in miles, by road between seven towns labelled A, B, C, D, E, F and G. The least distance between F and D is x miles.

E/P

	A	B	C	D	E	F	G
A	–	16	21	17	12	15	19
B	16	–	24	18	30	26	20
C	21	24	–	31	22	35	23
D	17	18	31	–	28	x	33
E	12	30	22	28	–	27	28
F	15	26	35	x	27	–	30
G	19	20	23	33	28	30	–

Tom is putting up posters for a fund-raising event, and wants to plan a route that will visit each town.

He applies the nearest neighbour algorithm, starting and finishing at A, and obtains a total distance of 140 miles.

a Find the value of x. **(3 marks)**

b Find the length of the nearest neighbour route starting and finishing at B. **(2 marks)**

c Starting by deleting G and its arcs, find a lower bound for the length of Tom's route. **(3 marks)**

d Write down the smallest interval that must contain the optimal length of Tom's route. **(1 mark)**

E/P **8** The network shows the lengths in miles of the direct roads connecting 6 villages. The roads between B and D and between D and F are one-way roads.

A table of shortest distances for this network is shown below.

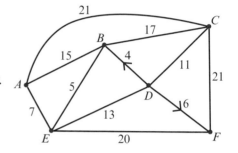

	A	B	C	D	E	F
A	–	12	21	20	7	26
B	12	–	17	18	5	24
C	21	15	–	11	20	17
D	16	4	11	–	9	6
E	7	5	22	13	–	19
F	27	25	21	32	20	–

A delivery driver wishes to find the shortest cycle which visits each village at least once.

a Use the nearest neighbour algorithm starting at D, and then at E, to find two Hamiltonian cycles in the completed network of shortest distances. Give the total length of each cycle. **(4 marks)**

b State, with a reason, which of these two cycles provides the best upper bound for the length of the delivery driver's route. **(1 mark)**

c Interpret this cycle in terms of the actual villages visited. **(1 mark)**

Challenge

Show that the nearest neighbour algorithm has cubic order. Clearly explain each step of your working.

Summary of key points

A

1 A **walk** in a network is a finite sequence of edges such that the end vertex of one edge is the start vertex of the next.

A walk which visits every vertex, returning to its starting vertex, is called a **tour**.

2 There are two variations of the travelling salesman problem. In the **classical problem**, each vertex must be visited exactly once before returning to the start. In the **practical problem**, each vertex must be visited at least once before returning to the start.

3 The **triangle inequality** states

the longest side of any triangle ⩽ the sum of the two shorter sides.

4 You can use a **minimum spanning tree method** to find an upper bound for the practical travelling salesman problem in an undirected network.

- Find the minimum spanning tree for the network (using Prim's algorithm or Krunkal's algorithm).
- An **initial upper bound** for the practical travelling salesman problem is found by finding the weight of the minimum spanning tree for the network and doubling it.
- You can improve the initial upper bound by looking for **shortcuts**. Aim to make the upper bound as low as possible to reduce the interval in which the optimal solution is contained.

5 You can use a **minimum spanning tree method** to find a lower bound from a table of least distances for the classical problem.

- Remove each vertex in turn, together with its arcs.
- Find the residual minimum spanning tree (RMST) and its length.
- Add to the RMST the 'cost' of reconnecting the deleted vertex **by the two shortest, distinct, arcs** and note the totals.
- The greatest of these totals is used for the lower bound.
- Make the lower bound as high as possible to reduce the interval in which the optimal solution is contained.
- You have found an optimal solution if the lower bound gives a Hamiltonian cycle, or the lower bound has the same value as the upper bound.

6 You can use the **nearest neighbour algorithm** to find an upper bound.

- Select each vertex in turn as a starting point.
- Go to the nearest vertex which has not yet been visited.
- Repeat step **2** until all vertices have been visited and then return directly to the start vertex.
- Once all vertices have been used as the starting vertex, select the tour with the smallest length as the upper bound.

Review exercise

1

Answer templates for questions marked * are available on www.pearsonschools.co.uk/d1maths

(E/P) 1* An algorithm is described by the flow chart below.

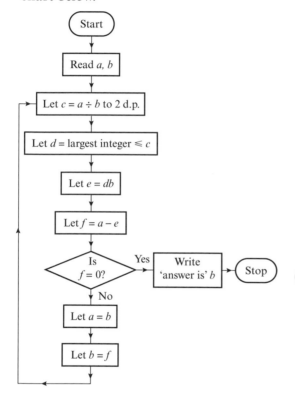

a Given that $a = 645$ and $b = 255$, draw a table to show the results obtained at each step when the algorithm is applied. **(6)**

b Explain how your solution to part **a** would be different if you had been given that $a = 255$ and $b = 645$. **(2)**

c State what the algorithm achieves. **(2)**

← Sections 1.1, 1.2

(E/P) 2 Nine pieces of wood are required to build a small cabinet. The lengths, in cm, of the pieces of wood are listed below.

  10 15 55 40 75 25 55 60 55

Planks can be purchased in one-metre lengths.

a Calculate a lower bound for the number of planks that will be needed to make the cabinet. **(2)**

b Use the first-fit bin packing algorithm to determine how many planks are needed. **(3)**

c Use the full-bin algorithm to determine how many planks are needed. **(2)**

d Explain why it is not possible to make the cabinet using fewer planks than the number found in part **c**. **(2)**

← Section 1.5

(E) 3   55 80 25 84 25 34 17 75 3 5

a The list of numbers above is to be sorted into descending order. Perform a bubble sort to obtain the sorted list, giving the state of the list after each complete pass. **(4)**

The numbers in the list represent masses, in grams, of objects which are to be packed into bins that hold up to 100 g.

b Determine the least number of bins needed. **(1)**

c Use the first-fit decreasing algorithm to fit the objects into bins which hold up to 100 g. **(2)**

← Sections 1.3, 1.5

(E) 4   45 56 37 79 46 18 90 81 51

a Using the quick sort algorithm, perform one complete iteration towards sorting these numbers into ascending order. **(3)**

b Using the bubble sort algorithm, perform one complete pass towards sorting the original list into descending order. **(3)**

Bubble sort is an algorithm of quadratic order.

c A computer applies a bubble sort to a list of 500 numbers in 0.016 seconds. Estimate the time required for the computer to apply the bubble sort to a list of 3000 numbers. **(2)**

← **Sections 1.3, 1.4, 1.6**

(E/P) **5** The following list gives the names of some students who have represented Britain in the International Mathematics Olympiad.

Roper (R), Palmer (P), Boase (B), Young (Y), Thomas (T), Kenney (K), Morris (M), Halliwell (H), Wicker (W), Garesalingam (G).

a Use the quick sort algorithm to sort the names above into alphabetical order. **(4)**

b A computer applies the quick sort algorithm with order $n \log n$.

The computer takes 0.024 seconds to sort 1200 names into alphabetical order. Estimate the time the computer would need to order 2000 names. **(3)**

← **Sections 1.4, 1.6**

(E/P) **6** A simple graph has 6 vertices.

a Explain why if the graph has two vertices of order 5, it cannot have any vertices of order 1. **(2)**

b Draw a simple graph with exactly 6 vertices with degrees 4, 4, 3, 3, 1 and 1. **(2)**

← **Section 2.2**

(E) **7 a** Draw the graph K_5. **(1)**

b State the number of edges in a minimum spanning tree for the graph K_5. **(1)**

c State the number of edges in a Hamiltonian cycle for the graph K_5. **(1)**

← **Section 2.3**

 8 a Define a Hamiltonian cycle. **(1)**

A Hamiltonian cycle for this graph begins *AB...*

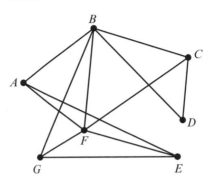

b Complete the Hamiltonian cycle. **(4)**

c Use the planarity algorithm to determine whether the graph is planar. **(2)**

A new edge *DE* is added to the graph.

d Use the planarity algorithm to show that this new graph is not planar. **(3)**

← **Sections 2.2, 2.4, 2.5**

(E) **9** The diagram shows 7 locations *A*, *B*, *C*, *D*, *E*, *F* and *G* which are to be connected by pipelines. The arcs show the possible routes. The number on each arc gives the cost, in thousands of pounds, of laying that particular section.

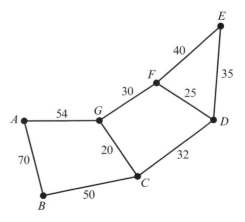

a Use Kruskal's algorithm to obtain a minimum spanning tree for the network, giving the order in which you selected the arcs. **(6)**

b Draw your minimum spanning tree and find the least cost of pipelines. **(3)**

← **Sections 2.4, 3.1**

(E/P) 10 a The table shows the distances, in metres, between six nodes A, B, C, D, E and F of a network.

	A	B	C	D	E	F
A	–	10	12	13	20	9
B	10	–	7	15	11	7
C	12	7	–	11	18	3
D	13	15	11	–	27	8
E	20	11	18	27	–	18
F	9	7	3	8	18	–

i Use Prim's algorithm, starting at A, to solve the minimum connector problem for this table of distances. Explain your method and indicate the order in which you selected the edges. **(6)**

ii Draw your minimum spanning tree and find its total length. **(3)**

iii State whether your minimum spanning tree is unique. Justify your answer. **(2)**

b A connected network N has seven vertices.

i State the number of edges in a minimum spanning tree for N. **(1)**

A minimum spanning tree for a connected network has n edges.

ii State the number of vertices in the network. **(1)**

← **Section 2.4, 3.3**

(E) 11* The matrix represents a network of roads between six villages A, B, C, D, E and F. The value in each cell represents the distance, in km, along these roads.

	A	B	C	D	E	F
A	–	7	3	–	8	11
B	7	–	4	2	–	7
C	3	4	–	5	9	–
D	–	2	5	–	6	3
E	8	–	9	6	–	–
F	11	7	–	3	–	–

a Show this information on a diagram. **(3)**

b Use Kruskal's algorithm to determine the minimum spanning tree. State the order in which you include the arcs and the length of the minimum spanning tree. Draw the minimum spanning tree. **(4)**

c Starting at D, use Prim's algorithm on the matrix given to find the minimum spanning tree. State the order in which you include the arcs. **(4)**

← **Sections 2.4, 3.1, 3.3**

(E) 12 a Describe two differences between Prim's algorithm and Kruskal's algorithm. **(2)**

b Find a minimum spanning tree for the network below using:

i Prim's algorithm, starting with vertex G **(4)**

ii Kruskal's algorithm. **(4)**

In each case write down the order in which you made your selection of arcs.

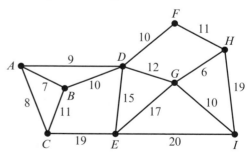

c State the weight of a minimum spanning tree. **(1)**

← **Sections 3.1, 3.2, 3.3**

(E) 13*

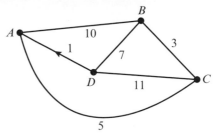

a Use Dijkstra's algorithm to find the shortest route from S to T in this network. Show all necessary working by drawing a diagram. State your shortest route and its length. **(6)**

b Explain how you determined the shortest route from your labelling. **(2)**

c It is now necessary to go from S to T via H. Obtain the shortest route and its length. **(2)**

← Section 3.4

(E) 14* The diagram shows a network of roads. Erica wishes to travel from A to L as quickly as possible. The number on each edge gives the time, in minutes, to travel along that road.

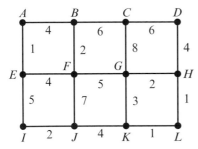

a Use Dijkstra's algorithm to find the quickest route from A to L. Complete all the boxes on the answer sheet and explain clearly how you determined the quickest route from your labelling.

b Show that there is another route which also takes the minimum time.

← Section 3.4

(A)
(E) **15** The diagram shows the direct distances in miles, by road, between 4 villages A, B, C and D. The road connecting A and D is one-way.

(A)

a Use Floyd's algorithm to produce a table of least distances. **(7)**

Show the distance table and the route table after each iteration.

b State the shortest distance from A to D. **(1)**

c Show how to use the route table to find the shortest route from A to D. **(2)**

Ibrahim lives in village B. He needs to visit shops in each of the other three villages before returning to B, and wants to find the shortest possible route.

d Starting at B, Ibrahim applies the nearest neighbour algorithm to the final table of least distances.

i State the cycle obtained in the completed network of least distances.

ii State the total length of this cycle.

iii State the actual villages visited and their order for this cycle.

iv Prove that Ibrahim's cycle is not optimal. **(4)**

← Sections 3.5, 5.4

(E/P) 16 Floyd's algorithm is applied to a network. The final distance and route tables are shown below.

Distance table

	A	B	C	D
A	–	5	y	z
B	5	–	3	6
C	x	13	–	7
D	10	6	7	–

Route table

	A	B	C	D
A	A	B	B	B
B	A	B	C	D
C	D	D	C	D
D	A	B	C	D

Deduce the values of x, y and z. You must show enough working to justify your answers. **(5)**

← Section 3.5

17

(E/P)

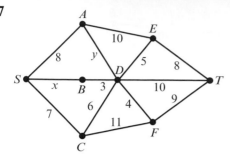

A weighted network is shown above.

Given that the shortest path from S to T is 17 and that $x \geqslant 0$, $y \geqslant 0$,

a i explain why A and C cannot lie on the shortest path **(2)**

  ii find the value of x. **(1)**

b Given that $x = 12$ and $y \geqslant 0$, find the possible range of values for the length of the shortest path. **(3)**

c Give an example of a practical problem that could be solved by drawing a network and finding the shortest path through it. **(2)**

← Sections 4.1, 4.2

(E/P) **18**

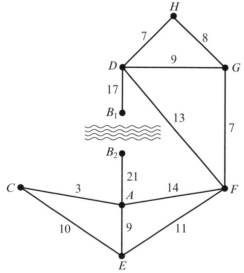

(Total weight of network is 129)

The diagram shows a network of roads connecting villages. The length of each road, in km, is shown. Village B has only a small footbridge over the river which runs through the village. It can be accessed by two roads, from A and D.

The driver of a snowplough, based at F, is planning a route to enable her to clear all the roads of snow. The route should be of minimum length. Each road can be cleared by driving along it once. The snowplough cannot cross the footbridge.

Showing all your working and using an appropriate algorithm,

a find the route the driver should follow, starting and ending at F, to clear all the roads of snow. Give the length of this route. **(6)**

The local authority decides to build a road bridge over the river at B. The snowplough will be able to cross the road bridge.

b Reapply the algorithm to find the minimum distance the snowplough will have to travel (ignore the length of the new bridge). **(4)**

← Section 4.2

(E/P) **19**

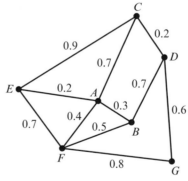

(Total weight of network is 6)

A local council is responsible for maintaining pavements in a district. The roads for which it is responsible are represented by arcs in the diagram. The road junctions are labelled A, B, C, ..., G. The number on each arc represents the length of that road in km.

The council has received a number of complaints about the condition of the pavements. In order to inspect the pavements, a council employee needs to walk along each road twice (once on each side of the road) starting and ending at the council offices at C. The length of the route is to be minimal. Ignore the widths of the roads.

a Explain how this situation differs from the standard route inspection problem. **(1)**

b Find a route of minimum length and state its length. **(5)**

← Section 4.2

E **20***

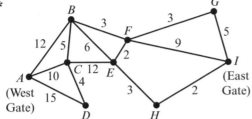

(Total weight of network is 91)

The diagram shows the network of paths in a country park. The number on each path gives its length in km. The vertices A and I represent the two gates in the park and the vertices B, C, D, E, F, G and H represent places of interest.

a Use Dijkstra's algorithm to find the shortest route from A to I. Show all necessary working and state your shortest route and its length. **(6)**

The park warden wishes to inspect each of the paths to check for frost damage. She has to cycle along each path at least once, starting and finishing at A.

b i Use an appropriate algorithm to find which paths will be covered twice and state these paths. **(2)**

ii Find a route of minimum length. **(3)**

iii Find the total length of this shortest route. **(2)**

← Sections 3.4, 4.2

A **E/P** **21** The diagram shows the road network of a small housing development. The number on each arc represents the length of the road in metres. A security guard has to patrol the length of each road at least once, starting at T and finishing at P.

A The total length of the route should be as small as possible.

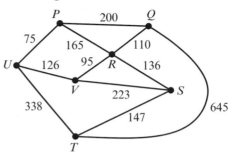

The total length of all the roads is 2260 m. Using the route inspection algorithm,

a determine which roads must be traversed more than once **(6)**

b calculate the length of the shortest route. **(2)**

← Section 4.3

E/P **22**

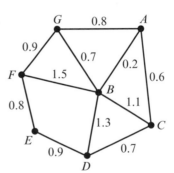

(Total weight of network is 9.5)

An engineer needs to check the state of a number of roads to see whether they need resurfacing. The roads that need to be checked are represented by the arcs in the diagram. The number on each arc represents the length of that road in km. To check all the roads, he needs to travel along each road at least once. He wishes to minimise the total distance travelled.

a Use the route inspection algorithm, starting at A and finishing at G, to find the minimum time taken to traverse each arc at least once. **(4)**

b State a possible route. **(4)**

The engineer believes that he can reduce the distance travelled by starting at D and finishing G.

A   **c** State whether the engineer is correct in his belief. If so, calculate how much shorter his new route is. If not, explain why not. **(4)**

← Section 4.3

E **23 a** Explain the difference between the classical and practical travelling salesperson problems. **(2)**

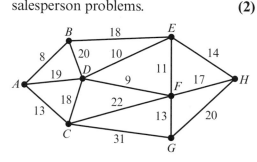

The network above shows the distances, in kilometres, between eight McBurger restaurants. An inspector from head office wishes to visit each restaurant. His route should start and finish at A, visit each restaurant at least once and cover a minimum distance.

b Obtain a minimum spanning tree for the network using Kruskal's algorithm. You should draw your tree and state the order in which the arcs were added. **(4)**

c Use your answer to part **b** to determine an initial upper bound for the length of the route. **(1)**

d Starting from your initial upper bound and using an appropriate method, find an upper bound which is less than 135 km. State your tour. **(4)**

← Sections 5.1, 5.2

E/P **24***

A   The diagram shows a network of roads connecting six villages A, B, C, D, E and F. The lengths of the roads are given in km.

a Complete a table of least distances for this network, by inspection. **(2)**

The table can now be taken to represent a complete network.

b Use the nearest neighbour algorithm, starting at A, on your completed table in part **a**. Obtain an upper bound to the length of a tour in this complete network, which starts and finishes at A and visits every village exactly once. **(4)**

c Interpret your answer in part **b** in terms of the original network of roads connecting the six villages. **(1)**

d By choosing a different vertex as your starting point, use the nearest neighbour algorithm to obtain a shorter tour than that found in part **b**. State the tour and its length. **(4)**

← Sections 5.1, 5.4

E **25** The table shows the least distances, in km, between five towns, A, B, C, D and E.

	A	B	C	D	E
A	–	153	98	124	115
B	153	–	74	131	149
C	98	74	–	82	103
D	124	131	82	–	134
E	115	149	103	134	–

Nassim wishes to find an interval which contains the solution to the travelling salesman problem for this network.

a Making your method clear, find an initial upper bound starting at A and using:

i the minimum spanning tree method **(4)**

ii the nearest neighbour algorithm. **(4)**

b By deleting E, find a lower bound. **(4)**

A

c Using your answers to parts **a** and **b**, state the smallest interval that Nassim could correctly write down. **(2)**

← **Sections 5.2, 5.3, 5.4**

E **26** The diagram shows six towns A, B, C, D, E and F and the roads joining them. The number on each arc gives the length of that road in miles.

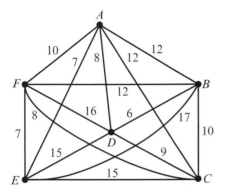

a By deleting vertex A, obtain a lower bound for the solution to the travelling salesman problem. **(4)**

The nearest neighbour algorithm for finding a possible salesman tour is as follows:

Step 1: Let V be the current vertex.

Step 2: Find the nearest unvisited vertex to the current vertex, move directly to that vertex and call it the current vertex.

Step 3: Repeat step 2 until all vertices have been visited and then return directly to the start vertex.

b **i** Use this algorithm to find a tour starting at vertex A. State clearly the tour and give its length. **(4)**

ii Starting at an appropriate vertex, use the algorithm to find a tour of shorter length. **(4)**

← **Section 5.4**

E **27** The table shows the distances, in km, between six towns A, B, C, D, E and F.

A

	A	B	C	D	E	F
A	–	85	110	175	108	100
B	85	–	38	175	160	93
C	110	38	–	148	156	73
D	175	175	148	–	110	84
E	108	160	156	110	–	92
F	100	93	73	84	92	–

a Starting from A, use Prim's algorithm to find a minimum connector and draw the minimum spanning tree. You must make your method clear by stating the order in which the arcs are selected. **(6)**

b **i** Using your answer to part **a** obtain an initial upper bound for the solution of the travelling salesperson problem. **(1)**

ii Use a short cut to reduce the upper bound to a value less than 680. **(2)**

c Starting by deleting F, find a lower bound for the solution of the travelling salesperson problem. **(4)**

← **Sections 3.3, 5.2, 5.3**

E **28** The table shows the distances, in km, between six towns A, B, C, D, E and F.

	A	B	C	D	E	F
A	–	113	53	54	87	68
B	113	–	87	123	38	100
C	53	87	–	106	58	103
D	54	123	106	–	140	48
E	87	38	58	140	–	105
F	68	100	103	48	105	–

a Starting from A, use Prim's algorithm to find a minimum connector and draw the minimum spanning tree. You must make your method clear by stating the order in which the arcs are selected. **(4)**

b **i** Hence form an initial upper bound for the solution to the travelling salesman problem. **(1)**

ii Use a short cut to reduce the upper bound to a value below 360. **(2)**

A

c By deleting A, find a lower bound for the solution to the travelling salesman problem. **(4)**

d Use your answers to parts **b** and **c** to make a comment on the value of the optimal solution. **(1)**

e Draw a diagram to show your best route. **(2)**

← Sections 3.3, 5.2, 5.3

E **29** A retailer has shops in seven cities A, B, C, D, E, F and G. The table below shows the distances, in km, between each of these seven cities. Susie lives in city A and has to visit each of the shops. She wishes to plan a route starting and finishing at A and covering a minimum distance.

	A	B	C	D	E	F	G
A	–	55	125	160	135	65	95
B	55	–	82	135	140	100	83
C	125	82	–	85	120	140	76
D	160	135	85	–	65	132	63
E	135	140	120	65	–	90	55
F	65	100	140	132	90	–	75
G	95	83	76	63	55	75	–

a Starting at A, use an algorithm to find a minimum spanning tree for this network. State the order in which you added vertices to the tree and draw your final tree. Explain briefly how you applied the algorithm. **(6)**

b Hence determine an initial upper bound for the length of Susie's route. **(1)**

c Starting from your initial upper bound, obtain an upper bound for the route which is less than 635 km. State the route which has a length equal to your new upper bound and cities which are visited more than once. **(2)**

d Obtain the minimum spanning tree for the reduced graph produced by deleting the vertex G and all edges joined to it. Draw the tree. **(4)**

A

e Hence obtain a lower bound for the length of Susie's route. **(2)**

f Using your solution to part **d**, obtain a route of length less than 500 km which visits each vertex exactly once. **(2)**

← Sections 5.2, 5.3, 5.4

Challenge

1 a Explain why any simple graph with two or more vertices must contain at least two vertices with the same degree.

b The diagram shows an edge-colouring of K_6, where every edge has been coloured either red or blue.

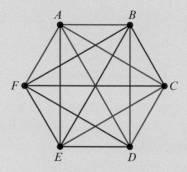

For the above edge-colouring, it is possible to find sets of three vertices which are all linked by edges of the same colour. Here are two examples:

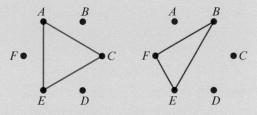

i Find two additional sets of three vertices in the above graph which are all linked by edges of the same colour.

ii Prove that for any edge-colouring of the K_6 with two colours, it will be possible to find a set of three vertices which are all linked by edges of the same colour.

iii By constructing a suitable graph, show that this result is not true for K_5.

← Section 3.3

2* Peter wishes to minimise the time spent driving from his home at H, to a campsite at G. The network above shows a number of towns and the time, in minutes, taken to drive between them. The volume of traffic on the roads into G is variable, and so the length of time taken to drive along these roads is expressed in terms of x, where $x \geqslant 0$.

   a Use Dijkstra's algorithm to find two routes from H to G (one via A and one via B) that minimise the travelling time from H to G. State the length of each route in terms of x.

   b Find the range of values of x for which Peter should follow the route via A.

   ← **Section 3.4**

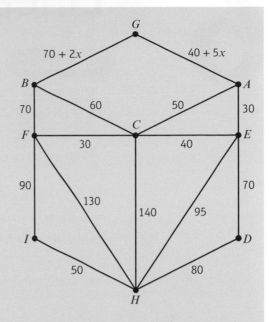

3 The route inspection problem is solved for this network giving a route of length 100.

   a Find an expression for the total weight of the network.

   b Prove by contradiction that $BA + AC \geqslant BC$.

   c Find the value of x showing all your working.

   ← **Section 4.2**

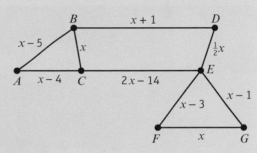

4 The table shows the distances, in miles, between some cities. A politician has to visit each city, starting and finishing at A. She wishes to minimise her total travelling distance.

   a Show that she will definitely be able to complete her trip in under 1400 miles, and state a suitable route.

   b Show that her total trip must be at least 971 miles in length, fully justifying why your answer represents a lower bound.

   ← **Sections 5.1, 5.2, 5.3**

	A	B	C	D	E	F	G	H
A	–	47	84	382	120	172	299	144
B	47	–	121	402	155	193	319	165
C	84	121	–	456	200	246	373	218
D	382	402	456	–	413	220	155	289
E	120	155	200	413	–	204	286	131
F	172	193	246	220	204	–	144	70
G	299	319	373	155	286	144	–	160
H	144	165	218	289	131	70	160	–

6 Linear programming

Objectives

After completing this chapter you should be able to:

● Formulate a problem as a linear programming problem
→ pages 139–145

● Illustrate a two-variable linear programming problem graphically
→ pages 145–148

● Locate the optimal point in a feasible region using the objective line (ruler) method
→ pages 148–155

● Use the vertex testing method to locate the optimal point
→ pages 156–162

● Determine solutions that need integer values
→ pages 162–166

Prior knowledge check

1 Use three inequalities to describe the shaded region of this graph.

← Pure Year 1, Chapter 3

2 Represent these inequalities on the same set of axes.

$2x + 3y \leqslant 12$, $x > 1$, $x \leqslant 4$, $y > 2$

Shade the region that satisfies all four inequalities.

← Pure Year 1, Chapter 3

Linear programming is a method of solving problems involving inequalities and more than one variable. It is widely used in finance, planning, control, design, manufacturing and resourcing, and can help to maximise profits or minimise costs. → Exercise 6A, Q1

6.1 Linear programming problems

You need to be able to formulate a problem as a linear programming problem.

- The **decision variables** in a linear programming problem are the numbers of each of the things that can be varied. The variables, which are often called x, y, z, etc., will be the 'letters' in the inequalities and **objective function**.

  For example, the cost of one teddy bear, the number of teddy bears made, etc.

- The objective is the aim of the problem. It may be to maximise profit or to minimise cost.

  There are two parts: a word 'maximise' or 'minimise', and an algebraic expression called the **objective function**, which is usually written as an equation in terms of the decision variables, for example, $P = 3x + 2y$, $C = 4x + 7y + 3z$

- The **constraints** are the things that will prevent you making, or using, an infinite number of each of the variables. **Each constraint will give rise to one inequality**.

- If you find values for the decision variables that satisfy each constraint you have a **feasible solution**.

- In a graphical linear programming problem, the region that contains *all* the feasible solutions is called the **feasible region**.

  Examples of constraints are the quantity of raw materials available, the time available, the fact that you cannot have a negative quantity, etc.

- The **optimal solution** is the feasible solution that meets the objective. **There might be more than one optimal solution**.

- **To formulate a problem as a linear programming problem:**

  1 **Define the decision variables (x, y, z, etc.).**

  2 **State the objective (maximise or minimise, together with an objective function).**

  3 **Write the constraints as inequalities.**

Example 1

Mrs Cook is making cakes to sell for charity. She makes two types of cake, fruit and chocolate.

Amongst other ingredients, each fruit cake requires 1 egg, 250 g of flour and 200 g of sugar. Each chocolate cake requires 2 eggs, 250 g of flour and 300 g of sugar.

Mrs Cook has 36 eggs, 7 kg of flour and 6 kg of sugar.

She will sell the fruit cakes for £3.50 and the chocolate cakes for £5. She wishes to maximise the money she makes from these sales.

You may assume she sells all the cakes that she makes.

This is an assumption that is made in linear programming problems and is not usually stated.

Formulate this as a linear programming problem.

Type of cake	Eggs	Flour	Sugar	Price
Fruit	1	250 g	200 g	£3.50
Chocolate	2	250 g	300 g	£5.00
Total available	36	7000 g	6000 g	

It is sometimes useful to summarise the information in a table.

First define the decision variables.

You need to make the units agree.

Let f be the number of fruit cakes made.

Let c be the number of chocolate cakes made.

The decision variables always start 'Let x be the number of …' etc.

Next, state the objective.

Maximise because Mrs Cook wishes to maximise her income.

Choose a letter to represent Mrs Cook's income.

There are two parts to the objective: a word maximise or minimise and an objective function.

maximise $P = 3.5f + 5c$

Mrs Cook will get £3.50 for each fruit cake she sells. If she sells f of them she will make £3.5f.

Each chocolate cake sold raises £5. If c cakes are sold, Mrs Cook will make £5c.

You don't need to include units in the objective function, but make sure that all the units agree.

Finally, identify the constraints.

These will be inequalities.

eggs: $f + 2c \leqslant 36$

flour: $250f + 250c \leqslant 7000$

This simplifies to
$$f + c \leqslant 28$$

Mrs Cook needs 1 egg for each fruit cake and 2 to make each chocolate cake. There is a maximum of 36 eggs that can be used.

Each fruit cake requires 250 g and each chocolate cake 250 g. There are up to 7000 g available.

sugar: $200f + 300c \leqslant 6000$

This simplifies to
$$2f + 3c \leqslant 60$$

Each fruit cake requires 200 g and each chocolate cake 300 g. There are 6000 g available.

non-negativity: $f \geqslant 0 \ c \geqslant 0$

These are often written together as
$$f, c \geqslant 0$$

A formal summary of the problem is

You cannot have negative values for f or c. (A negative cake is not possible!)

Let f be the number of fruit cakes made.

Let c be the number of chocolate cakes made.

maximise $\quad P = 3.5f + 5c$

subject to: $\quad f + 2c \leqslant 36$

$\qquad\qquad\quad f + c \leqslant 28$

$\qquad\qquad\quad 2f + 3c \leqslant 60$

$\qquad\qquad\qquad f, c \geqslant 0$

Note This is the formal template for presenting a linear programming problem. It is a good idea to follow this model when formulating a linear programming problem.

Example 2

A company buys two types of diary to send to its customers, a desk top diary and a pocket diary.

They will need to place a minimum order of 200 desk top and 80 pocket diaries.

They will need at least twice as many pocket diaries as desk top diaries.

They will need a total of at least 400 diaries.

Each desk top diary costs £6 and each pocket diary costs £3.

The company wishes to minimise the cost of buying the diaries.

Formulate this as a linear programming problem.

Summarise the information in a table.

Type of diary	Minimum order	Cost
Desk top	200	£6
Pocket	80	£3

Also require:
- twice as many pocket as desk top
- total of at least 400.

Define the decision variables.

Let x be the number of desk top diaries bought.

Let y be the number of pocket diaries bought.

State the objective.

minimise $\quad C = 6x + 3y$

State the constraints.

minimum order: $x \geqslant 200$

$\qquad y \geqslant 80$

at least twice as many pocket as desk top:

$\qquad 2x \leqslant y$

a total of at least 400 diaries:

$\qquad x + y \geqslant 400$

non-negativity:

Since we have already stated that $x \geqslant 200$ and $y \geqslant 80$, we do not need to state that $x \geqslant 0$ and $y \geqslant 0$.

So the non-negativity constraint is already satisfied by the previous constraints.

Here is the summary.

Let x be the number of desk top diaries bought.

Let y be the number of pocket diaries bought.

minimise $\qquad c = 6x + 3y$

subject to: $\quad x \geqslant 200$

$\qquad\qquad y \geqslant 80$

$\qquad\qquad 2x \leqslant y$

$\qquad\qquad x + y \geqslant 400$

Each desk top diary costs £6 and each pocket diary costs £3. The company wants to minimise the cost.

The company must order at least 200 desk top diaries and at least 80 pocket diaries.

This type of comparative constraint can be tricky. It sometimes helps to see it as two steps: first getting the algebra correct and then getting the inequality correct.

Problem-solving

To get the algebra correct, change the statement to read 'exactly twice as many pocket as desk top'. This tells you which one needs to be doubled. The number of pocket diaries (y) is double the number of desk top (x). So we get $2x = y$. To get the direction of the inequality you can consider which one of these can be made larger whilst still satisfying the given condition. Here the number of pocket diaries can increase so $2x \leqslant y$.

The total number of diaries is simply $x + y$.

Example (3)

A company produces two types of syrup, *A* and *B*. The syrups are a blend of sugar, fruit and juice.

Syrup *A* contains 30% sugar, 50% fruit and 20% juice.

Syrup *B* contains 20% sugar, 35% fruit and 45% juice.

Each litre of syrup *A* costs 50p and each litre of syrup *B* costs 40p.

There is a maximum daily production of 40 000 litres of syrup *A* and 45 000 litres of syrup *B*.

A confectionery manufacturer places an order for 60 000 litres of syrup but requires

- below 25% sugar
- at least 40% fruit
- no more than 35% juice.

The company will blend syrups *A* and *B* to meet the confectionery manufacturer's requirements.

The company wishes to minimise its costs.

Letting *x* be the number of litres of syrup *A* used, and *y* be the number of litres of syrup *B* used, formulate this as a linear programming problem.

Summarise the information.

Syrup	Sugar %	Fruit %	Juice %	Maximum amount (ℓ)	Cost per litre
A (*x*)	30	50	20	40 000	50p
B (*y*)	20	35	45	45 000	40p
Required in final blend	< 25	⩾ 40	⩽ 35	60 000	

Define the decision variables.

This has already been done in the question.

State the objective.

minimise $C = 0.5x + 0.4y$

> Each litre of syrup *A* costs 50p and each litre of syrup *B* costs 40p. The company wants to minimise its costs.

State the constraints.

30% of syrup *A* is sugar.

Note the strict inequality. It must be *below* 25%.

sugar: $0.3x + 0.2y < 0.25 (x + y)$

20% of syrup *B* is sugar.   less than …   … 25% of the combined blend must be sugar.

This simplifies to $x < y$

> $0.3x + 0.2y < 0.25x + 0.25y$
> $0.05x < 0.05y$

Each litre of syrup *A* contains 50% fruit.

fruit: $0.5x + 0.35y \geqslant 0.4 (x + y)$

Each litre of syrup *B* contains 35% fruit.   at least …   … 40% fruit in the combined blend.

This simplifies to $2x \geqslant y$

juice: $0.2x + 0.45y \leqslant 0.35\,(x + y)$

which simplifies to $2y \leqslant 3x$

amount required: $x + y \geqslant 60\,000$

$x \leqslant 40\,000$

$y \leqslant 45\,000$

non-negativity: $x, y \geqslant 0$

$x \leqslant 40\,000$

$y \leqslant 45\,000$

$0.5x + 0.35y \geqslant 0.4x + 0.4y$

$0.1x \geqslant 0.05y$

$0.2x + 0.45y \leqslant 0.35x + 0.35y$

$0.1y \leqslant 0.15x$

$10y \leqslant 15x$

Here is the summary

minimise $C = 0.5x + 0.4y$

subject to $x < y$

$2x \geqslant y$

$3x \geqslant 2y$

$x + y \geqslant 60\,000$

$x \leqslant 40\,000$

$y \leqslant 45\,000$

$x, y \geqslant 0$

Exercise 6A

(E/P) 1 A chocolate manufacturer is producing two hand-made assortments, gold and silver, to commemorate 50 years in business.

It will take 30 minutes to make all the chocolates for one box of gold assortment and 20 minutes to make the chocolates for one box of silver assortment.

It will take 12 minutes to wrap and pack the chocolates in one box of gold assortment and 15 minutes for one box of silver assortment.

The manufacturer needs to make at least twice as many silver as gold assortments.

The gold assortment will be sold at a profit of 80p, and the silver at a profit of 60p.

There are 300 hours available to make the chocolates and 200 hours to wrap them. The profit is to be maximised.

Watch out Remember to define your variables clearly, and include a non-negativity constraint.

Letting the number of boxes of gold assortment be x and the number of boxes of silver assortment be y, formulate this as a linear programming problem. **(5 marks)**

(E/P) 2 A floral display is required for the opening of a new building. The display must be at least 30 m long and is to be made up of two types of planted displays, type A and type B.

Type A is 1 m in length and costs £6 and

Type B is 1.5 m in length and costs £10

The client wants at least twice as many type A as type B, and at least 6 of type B.

The cost is to be minimised.

Letting x be the number of type A used and y be the number of type B used, formulate this as a linear programming problem. **(5 marks)**

(E/P) 3 A toy company makes two types of board game, Cludopoly and Trivscrab. As well as the board, each game requires playing pieces and cards.

The company uses two machines, one to produce the pieces and one to produce the cards. Both machines can only be operated for up to ten hours per day.

The first machine takes 5 minutes to produce a set of pieces for Cludopoly and 8 minutes to produce a set of pieces for Trivscrab.

The second machine takes 8 minutes to produce a set of cards for Cludopoly and 4 minutes to produce a set of cards for Trivscrab.

The company knows it will sell at most three times as many games of Cludopoly as Trivscrab.

The profit made on each game of Cludopoly is £1.50 and £2.50 on each game of Trivscrab.

The company wishes to maximise its daily profit.

Let x be the number of games of Cludopoly and y the number of games of Trivscrab.

Formulate this problem as a linear programming problem. **(5 marks)**

(E/P) 4 A librarian needs to purchase bookcases for a new library. She has a budget of £3000 and 240 m² of available floor space. There are two types of bookcase, type 1 and type 2, that she is permitted to buy.

Type 1 costs £150, needs 15 m² of floor space and has 40 m of shelving.

Type 2 costs £250, needs 12 m² of floor space and has 60 m of shelving.

She must buy at least 8 type 1 bookcases and wants at most $\frac{1}{3}$ of all the bookcases to be type 2.

She wishes to maximise the total amount of shelving.

Letting x and y be the number of type 1 and type 2 bookcases bought respectively, formulate this as a linear programming problem. **(5 marks)**

(E/P) 5 A garden supplies company produces two different plant feeds, one for indoor plants and one for outdoor plants.

In addition to other ingredients, the plant feeds are made by combining three different natural ingredients A, B and C.

Each kilogram of indoor feed requires 10 g of A, 20 g of B and 20 g of C.

Each kilogram of outdoor feed requires 20 g of A, 10 g of B and 20 g of C.

The company has 5 kg of A, 5 kg of B and 6 kg of C available each week to use to make these feeds.

The company will sell at most three times as much outdoor as indoor feed, and will sell at least 50 kg of indoor feed.

The profit made on each kilogram of indoor and outdoor feed is £7 and £6 respectively. The company wishes to maximise its weekly profit.

Formulate this as a linear programming problem, defining your decision variables. **(6 marks)**

(E) 6 Sam makes three types of fruit smoothies, A, B and C. As well as other ingredients all three smoothies contain oranges, raspberries, kiwi fruit and apples, but in different proportions. Sam has 50 oranges, 1000 raspberries, 100 kiwi fruit and 60 apples. The table below shows the number of these 4 fruits used to make each smoothie and the profit made per smoothie. Sam wishes to maximise the profit.

Smoothie	Oranges	Raspberries	Kiwi fruit	Apples	Profit
A	1	10	2	2	60p
B	$\frac{1}{2}$	40	3	$\frac{1}{2}$	65p
C	2	15	1	2	55p
Total available	50	1000	100	60	

Letting x be the number of A smoothies, y the number of B smoothies and z the number of C smoothies, formulate this as a linear programming problem. **(5 marks)**

(E/P) **7** A dairy manufacturer has two factories, R and S. Each factory can process milk and yoghurt.

Factory R can process 1000 litres of milk and 200 litres of yoghurt per hour.

Factory S can process 800 litres of milk and 300 litres of yoghurt per hour.

It costs £300 per hour to operate factory R and £400 per hour to operate factory S. In order to safeguard jobs it has been agreed that each factory will operate for at least $\frac{1}{3}$ of the total, combined, operating time.

The manufacturer needs to process 20 000 litres of milk and 6000 litres of yoghurt. He wishes to distribute this between the 2 factories in such a way as to minimise operating costs. Formulate this as a linear programming problem in x and y, defining your decision variables. **(6 marks)**

6.2 Graphical methods

You can illustrate a two-variable linear programming problem graphically. You use the x- and y-axes to represent the two variables, and shade any areas of the graph that **fail to satisfy** any of the inequalities in the linear programming problem.

- **The region of a graph that satisfies all the constraints of a linear programming problem is called the feasible region.**

Watch out By convention, you leave the feasible region **unshaded**, and shade all other regions of the graph.

Example 4

A linear programming problem is given as:

minimise $C = 0.5x + 0.4y$

subject to

$$x < y$$
$$2x \geqslant y$$
$$3x \geqslant 2y$$
$$x + y \geqslant 60\,000$$
$$x, y \geqslant 0$$

On a graph, represent the feasible region for this problem, and label it R.

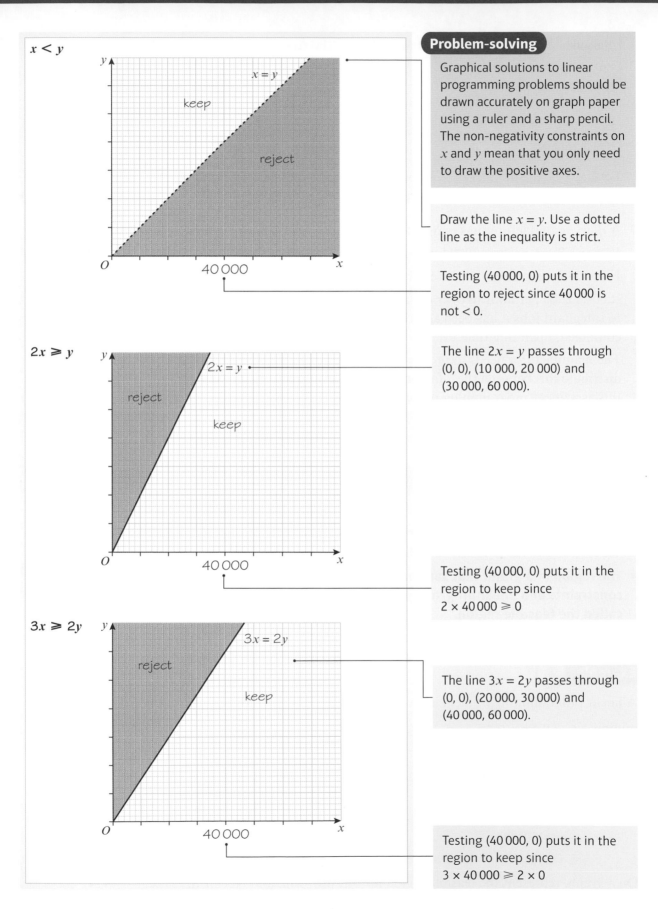

$x < y$

$x = y$

keep

reject

O 40 000 x

$2x \geqslant y$

$2x = y$

reject

keep

O 40 000 x

$3x \geqslant 2y$

$3x = 2y$

reject

keep

O 40 000 x

Problem-solving

Graphical solutions to linear programming problems should be drawn accurately on graph paper using a ruler and a sharp pencil. The non-negativity constraints on x and y mean that you only need to draw the positive axes.

Draw the line $x = y$. Use a dotted line as the inequality is strict.

Testing (40 000, 0) puts it in the region to reject since 40 000 is not < 0.

The line $2x = y$ passes through (0, 0), (10 000, 20 000) and (30 000, 60 000).

Testing (40 000, 0) puts it in the region to keep since $2 \times 40\,000 \geqslant 0$

The line $3x = 2y$ passes through (0, 0), (20 000, 30 000) and (40 000, 60 000).

Testing (40 000, 0) puts it in the region to keep since $3 \times 40\,000 \geqslant 2 \times 0$

$x + y \geqslant 60\,000$

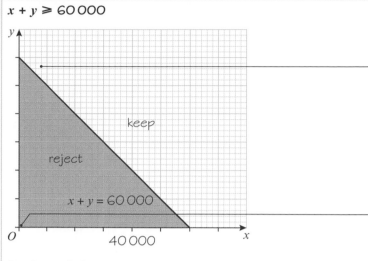

The line $x + y = 60\,000$ passes through $(0, 60\,000)$ and $(60\,000, 0)$.

Testing $(0, 0)$ puts it in the region to *reject* since $0 + 0$ is not $\geqslant 60\,000$.

Combine all these on one diagram.

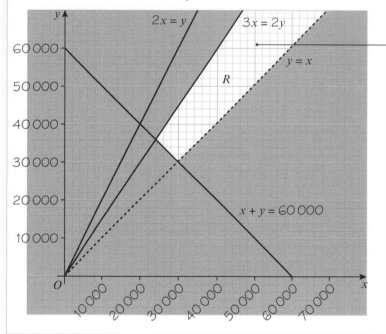

Label the feasible region R.

The inequalities $2x \geqslant y$, $x \geqslant 0$ and $x \geqslant 0$ are not actually necessary to define the feasible region.

The question only asked for the feasible region, so you do not need to consider the objective function.

Online Explore graphical solutions to linear programming problems using GeoGebra.

Exercise 6B

1 Represent each set of inequalities on a graph. Shade any regions of the graph that fail to satisfy all the inequalities, and label the feasible region R.

a $2x + 3y > 18$
$y > x$
$y \leqslant 5$
$x, y \geqslant 0$

b $2x \geqslant 3y$
$3x + 4y \leqslant 24$
$x \geqslant 3$
$y \geqslant 1$

c $x + y \leqslant 20$
$5x + 6y \geqslant 60$
$2x \geqslant y$
$y \leqslant 10$

d $2x - 3 < y$
$y > 3$
$y > 6 - 2x$
$x \geqslant 0$

2 A linear programming problem is given as:

maximise $P = 3.5f + 5c$

subject to: $f + 2c \leqslant 36$

$\qquad f + c \leqslant 28$

$\qquad 2f + 3c \leqslant 60$

$\qquad f, c \geqslant 0$

On a graph, represent the feasible region for this problem, and label it R.

3 A linear programming problem is given as:

minimise $C = 6x + 3y$

subject to $\quad x \geqslant 200$

$\qquad y \geqslant 80$

$\qquad 2x \leqslant y$

$\qquad x + y \geqslant 400$

On a graph, represent the feasible region for this problem, and label it R.

(P) **4** A company manufactures two types of mp3 player, type A and type B. The company decides that each month:

- at least 200 type A mp3 players should be produced

- the number of type A mp3 players should be between 10% and 40% of the total number of mp3 players produced

- a maximum of 3000 mp3 players can be produced.

The company makes a profit of £75 on each type A mp3 player produced and a profit of £55 on each type B mp3 player produced. The firm wishes to maximise its monthly profit.

Show the feasible region for this linear programming problem on a suitable graph.

(E) **5** This graph is being used to solve a linear programming problem. Three of the constraints have been drawn on the graph and the rejected regions shaded.

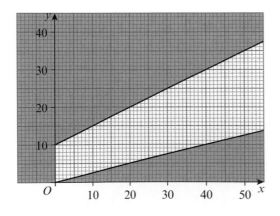

a Write down the constraints shown on the graph. **(3 marks)**

Two further constraints are:

$x + y \geqslant 20$

$3x + 4y \leqslant 120$

b Copy the graph and add two lines and shading to represent these constraints. Hence determine the feasible region and label it R. **(2 marks)**

6.3 Locating the optimal point

Every point in the feasible region satisfies all the constraints, so is a potential solution to the linear programming problem.

- **To solve a linear programming problem, you need to find the point in the feasible region which maximises or minimises the objective function.**

There are two methods for finding the optimal solution. The first method is called the **objective line method** or **ruler method**.

Example 5

Nigel is making ice cream for sale at a charity fair. He makes two flavours of ice cream: vanilla and chocolate. Let the number of litres of vanilla ice cream made be x and the number of litres of chocolate ice cream made be y. Nigel decides to use linear programming to determine the number of litres of each type of ice cream he should make. The constraints and the feasible region, R, are illustrated in the diagram below.

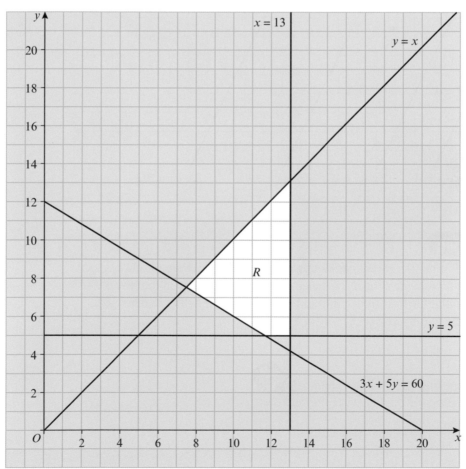

Determine the optimal solution for this problem, given that the objective is to

a maximise the profit on sales, $P = 2x + y$,

b minimise the production costs, $C = 5x + 2y$.

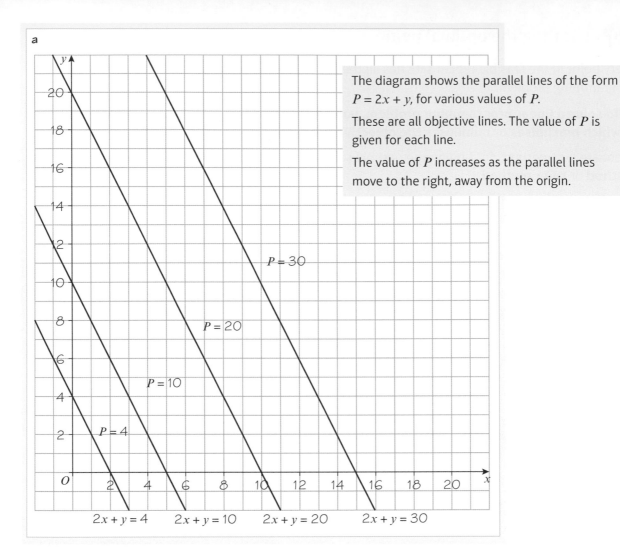

a

The diagram shows the parallel lines of the form $P = 2x + y$, for various values of P.

These are all objective lines. The value of P is given for each line.

The value of P increases as the parallel lines move to the right, away from the origin.

$P = 30$

$P = 20$

$P = 10$

$P = 4$

$2x + y = 4$ $2x + y = 10$ $2x + y = 20$ $2x + y = 30$

Problem-solving

Imagine a ruler sliding over the feasible region on page 149 so that it is always parallel to the profit lines above. The maximum value of P will be in the feasible region at the point furthest from the origin (the last point the ruler touches as it slides out of the feasible region).

Online   Explore how the optimal solution can be found using the objective line method with GeoGebra.

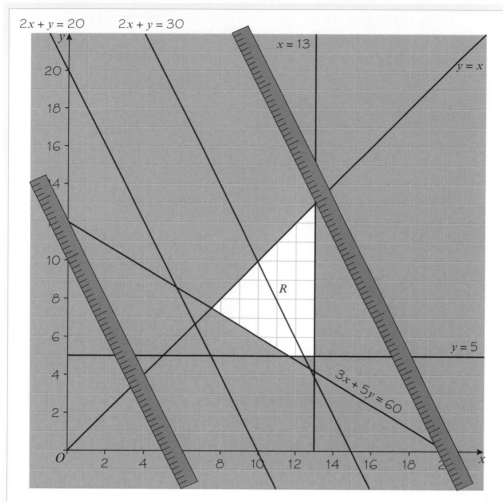

From the diagram, the optimal point is (13, 13), giving an optimal value for P of
$2 \times 13 + 13 = 39$.

So Nigel should make 13 litres of vanilla ice cream, and 13 litres of chocolate ice cream, and makes a profit of £39.

Watch out You should always make sure you give your answer in the context of the original question. Don't just state the coordinates of the optimal point.

b

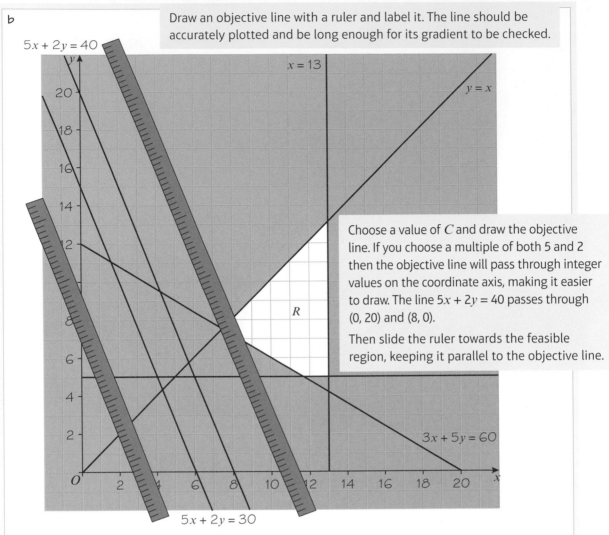

Draw an objective line with a ruler and label it. The line should be accurately plotted and be long enough for its gradient to be checked.

Choose a value of C and draw the objective line. If you choose a multiple of both 5 and 2 then the objective line will pass through integer values on the coordinate axis, making it easier to draw. The line $5x + 2y = 40$ passes through $(0, 20)$ and $(8, 0)$.

Then slide the ruler towards the feasible region, keeping it parallel to the objective line.

The minimum value will occur at the first point covered by the objective line as it moves into the feasible region. In this case the optimal point is found where the line $y = x$ meets the line $3x + 5y = 60$. Solving these equations simultaneously gives $x = 7.5$ and $y = 7.5$.
So at this point $C = 5x + 2y = 7 \times 7.5 = 52.5$.
So Nigel should make 7.5 litres of vanilla ice cream and 7.5 litres of chocolate ice cream, with production costs of £52.50.

- **For a maximum point, look for the last point covered by an objective line as it leaves the feasible region.**
- **For a minimum point, look for the first point covered by an objective line as it enters the feasible region.**

It is very important, when using the objective line method, that the ruler is kept parallel to an objective line. To do this you need two straight edges: a ruler and either a second ruler or a set square.

Place the ruler along an objective line, then place the set square (or second ruler) at the base of the ruler.

Hold the set square firmly and slide the ruler along the edge of the set square.

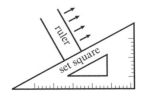

Example 6

Using the feasible region determined by the inequalities

$$x \geqslant 2 \qquad 4x + 3y \leqslant 12 \qquad 2y \leqslant x \qquad x, y \geqslant 0$$

find the optimal point and the optimal value when the objective is to:

a maximise $P = 2x + y$ **b** maximise $P = x + 2y$

a

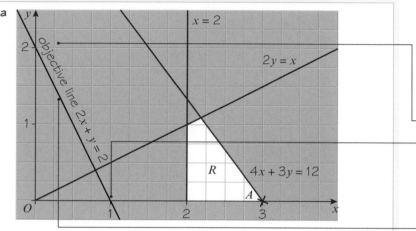

An objective line of the form $ab = ax + by$ always passes through $(b, 0)$ and $(0, a)$.

The objective line must be labelled.

Draw an objective line.

The diagram shows the feasible region and the objective line $2x + y = 2$, which passes through (1, 0) and (0, 2).
The final point is point A, where $x = 3$, $y = 0$.
Optimal point is (3, 0). Optimal value is $P = 2 \times 3 + 0 = 6$.

b

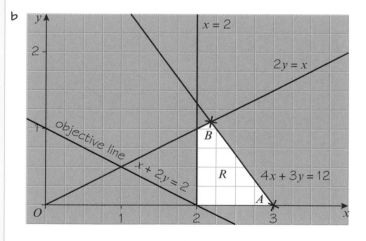

The diagram shows the feasible region and the objective line $x + 2y = 2$, which passes through (2, 0) and (0, 1).
The final point is point B.
B is at the intersection of $2y = x$ and $4x + 3y = 12$.
Solving these equations simultaneously gives $y = 1\frac{1}{11}$ and $x = 2\frac{2}{11}$
Optimal point is $\left(2\frac{2}{11}, 1\frac{1}{11}\right)$. Optimal value is $P = 2\frac{2}{11} + 2\frac{2}{11} = 4\frac{4}{11}$

Online Explore how the optimal solution can be found using the objective line method with GeoGebra.

153

Example 7

In a linear programming problem the constraints are given by

$$3x + y \geqslant 90$$
$$2x + 7y \geqslant 140$$
$$x + y \geqslant 50$$
$$x, y \geqslant 0$$

a Minimise $C = 3x + 2y$. **b** Minimise $C = 3x + 7y$.

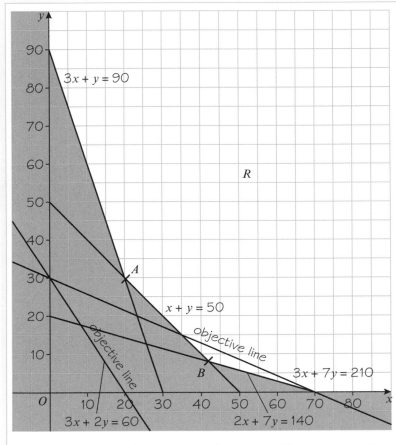

Online Explore how the optimal solution can be found using the objective line method with GeoGebra.

a The objective line has equation $3x + 2y = 60$ and passes through $(20, 0)$ and $(0, 30)$.

The first point in the feasible region as the objective line moves away from the origin is A.

A lies at the intersection of

$$3x + y = 90$$
and $$x + y = 50$$

Solving simultaneously gives

$$x = 20, y = 30$$

which gives $C = 3 \times 20 + 2 \times 30 = 120$

b The objective line has equation $3x + 7y = 210$ and passes through $(70, 0)$ and $(0, 30)$.

The first point in the feasible region is **B**.

B lies at the intersection of
$$2x + 7y = 140$$
and $x + y = 50$

Solving simultaneously gives
$$x = 42, y = 8$$
which gives $C = 3 \times 42 + 7 \times 8 = 182$

Example 8

Using the same feasible region as in Example 7, find an optimal solution given that the objective is to minimise $C = x + y$.

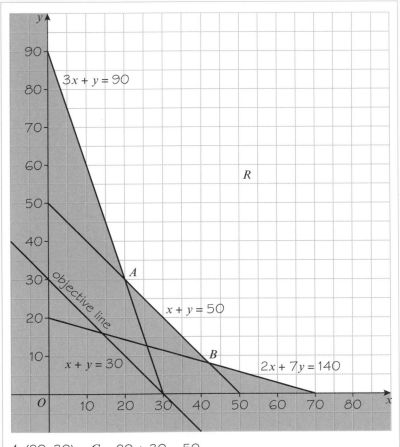

Problem-solving

As the objective line slides into the feasible region, it lies along the line segment AB. This means that all points along this part of the line are optimal solutions.

Watch out The point $(10, 40)$ also lies on this line, but it is not a solution, since it does not lie in the feasible region.

A $(20, 30)$ $C = 20 + 30 = 50$
B $(42, 8)$ $C = 42 + 8 = 50$
    $(25, 25)$ $C = 25 + 25 = 50$
    $(40, 10)$ $C = 40 + 10 = 50$

The optimal solutions are the points on the line $x + y = 50$ for which $20 \leqslant x \leqslant 42$

Online Explore how the optimal solution can be found using the objective line method with GeoGebra.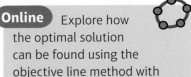

You may have noticed that the optimal point occurs at one (or more) of the vertices of the feasible region. You can use this fact to find optimal points using the **vertex testing method**.

- **To find an optimal point using the vertex method:**
  1. **First find the coordinates of each vertex of the feasible region.**
  2. **Evaluate the objective function at each of these points.**
  3. **Select the vertex that gives the optimal value of the objective function.**

Example 9

Use the vertex testing method to solve the following linear programming problem:

minimise $\quad\quad\quad x + 3y$

subject to $\quad\quad\quad y \leqslant x$

$\quad\quad\quad\quad\quad 3x + 5y \geqslant 60$

$\quad\quad\quad\quad\quad\quad\quad y \geqslant 5$

$\quad\quad\quad\quad\quad\quad\quad x \leqslant 13$

$\quad\quad\quad\quad\quad x, y \geqslant 0$

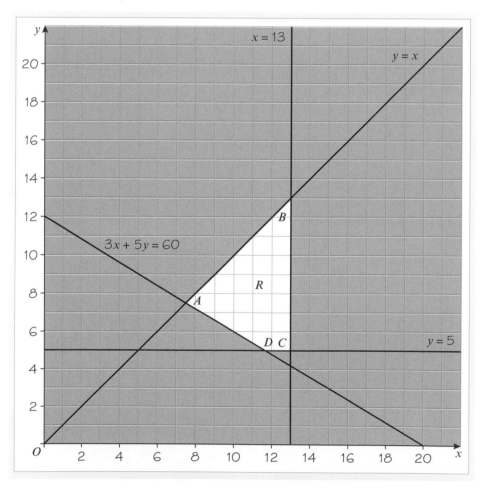

This is the feasible region determined in Example 5.

The feasible region has 4 vertices, A, B, C and D.

Vertex	Coordinates	Value of $x + 3y$
A	$x = 7.5$, $y = 7.5$	$7.5 + 3 \times 7.5 = 30$
B	$x = 13$, $y = 13$	$13 + 3 \times 13 = 52$
C	$x = 13$, $y = 5$	$13 + 3 \times 5 = 28$
D	$x = 11\frac{2}{3}$, $y = 5$	$11\frac{2}{3} + 3 \times 5 = 26\frac{2}{3}$

The minimum value occurs at $D\left(11\frac{2}{3}, 5\right)$ and is $26\frac{2}{3}$

When using this method all the vertices of the feasible region should be tested, even if they are obviously not optimal.

Online Explore how the optimal solution can be found using vertex testing with GeoGebra.

Example **10**

A feasible region is defined by the following constraints

$$9x + 11y \leqslant 99$$
$$4x + y \leqslant 28$$
$$5x + 3y \geqslant 30$$
$$x + 2y \geqslant 8$$
$$y \leqslant x$$

Find the optimal point and optimal value given that the objective is:

a maximise $x + y$ **b** minimise $3x + 4y$

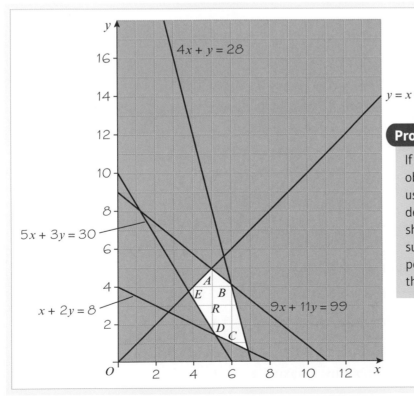

Problem-solving

If the points of intersection are not obvious integer values, you should use simultaneous equations to determine their coordinates. You should always draw a graph to make sure you don't waste time finding points of intersection that lie outside the feasible region.

The feasible region has 5 vertices, A, B, C, D and E. Use simultaneous equations to find the coordinates of A, B, C, D and E.

Vertex	Coordinates	Value of $x + y$	Value of $3x + 4y$
A	$x = 4\frac{19}{20}, \ y = 4\frac{19}{20}$	$4\frac{19}{20} + 4\frac{19}{20} = 9\frac{9}{10}$	$3 \times 4\frac{19}{20} + 4 \times 4\frac{19}{20} = 34\frac{13}{20}$
B	$x = 5\frac{34}{35}, \ y = 4\frac{4}{35}$	$5\frac{34}{35} + 4\frac{4}{35} = 10\frac{3}{35}$	$3 \times 5\frac{34}{35} + 4 \times 4\frac{4}{35} = 34\frac{13}{35}$
C	$x = 6\frac{6}{7}, \ \ y = \frac{4}{7}$	$6\frac{6}{7} + \frac{4}{7} = 7\frac{3}{7}$	$3 \times 6\frac{6}{7} + 4 \times \frac{4}{7} = 22\frac{6}{7}$
D	$x = 5\frac{1}{7}, \ \ y = 1\frac{3}{7}$	$5\frac{1}{7} + 1\frac{3}{7} = 6\frac{4}{7}$	$3 \times 5\frac{1}{7} + 4 \times 1\frac{3}{7} = 21\frac{1}{7}$
E	$x = 3\frac{3}{4}, \ \ y = 3\frac{3}{4}$	$3\frac{3}{4} + 3\frac{3}{4} = 7\frac{1}{2}$	$3 \times 3\frac{3}{4} + 4 \times 3\frac{3}{4} = 26\frac{1}{4}$

a Maximum value of $x + y$ is at $B\left(5\frac{34}{35}, 4\frac{4}{35}\right)$ and has value of $10\frac{3}{35}$

b Minimum value of $3x + 4y$ is at $D\left(5\frac{1}{7}, 1\frac{3}{7}\right)$ and has value of $21\frac{1}{7}$

Online

Explore how the optimal solution can be found using vertex testing with GeoGebra.

Exercise 6C

1 The diagram shows a feasible region, R.

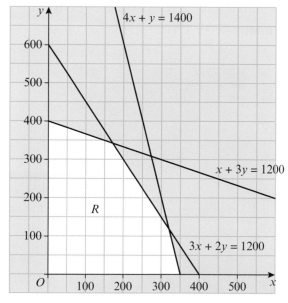

Find the optimal point and the optimal value, using:

a the objective line method, with the objective 'maximise $M = 2x + y$'

b the objective line method, with the objective 'maximise $N = x + 4y$'

c the vertex testing method, with the objective 'maximise $P = x + y$'

d the vertex testing method, with the objective 'maximise $Q = 6x + y$'

2 The diagram shows a feasible region, R.

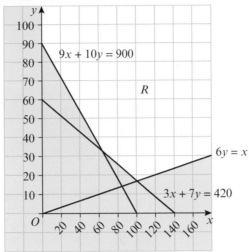

Find the optimal point and the optimal value, using:

a the vertex testing method, with the objective 'minimise $E = 2x + y$'

b the vertex testing method, with the objective 'minimise $F = x + 4y$'

c the objective line method, with the objective 'minimise $G = 3x + 4y$'

d the objective line method, with the objective 'minimise $H = x + 6y$'

3 The diagram shows a feasible region, R.

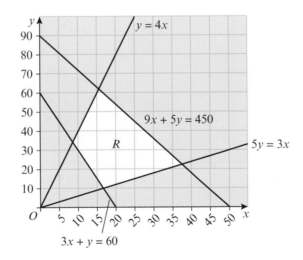

Find the optimal point and the optimal value, using:

a the vertex testing method, with the objective 'minimise $J = x + 4y$'

b the vertex testing method, with the objective 'maximise $K = x + y$'

c the objective line method, with the objective 'minimise $L = 6x + y$'

d the objective line method, with the objective 'maximise $M = 2x + y$'

4 The diagram shows a feasible region, R, which is defined by:

$$3x + y \geqslant 12$$
$$y \leqslant 2x$$
$$3y \geqslant x$$
$$6x + 5y \leqslant 120$$

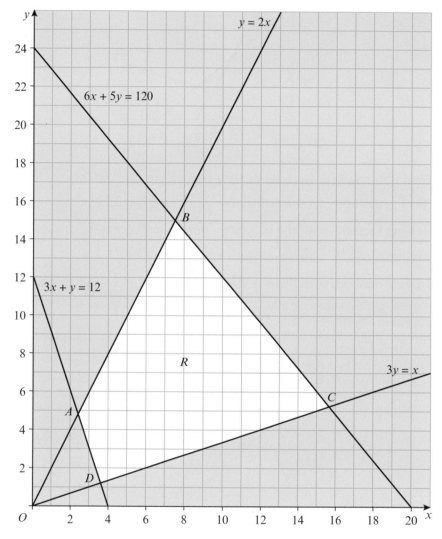

Determine which vertex, A, B, C or D, is the optimal point for each of the following objectives.

a maximise x

b minimise x

c maximise y

d minimise y

e maximise $6x + y$

f minimise $6x + y$

g maximise $2x + 5y$

h minimise $2x + 5y$

i maximise $3x + 2y$

j minimise $3x + 2y$

5 Solve the linear programming problem posed in Exercise 6A, question **5** (page 144).

6 Solve the linear programming problem posed in Exercise 6A, question **7** (page 145).

(E/P) **7** A linear programming problem is given as

minimise $\quad C = 3x + 2y$

subject to $\quad 2x + y \geqslant 160$

$\qquad\qquad x + y \geqslant 120$

$\qquad\qquad x + 3y \geqslant 180$ and $x \geqslant 0, y \geqslant 0$

a Draw a graph to illustrate the feasible, R. **(3 marks)**
(Use the values $0 \leqslant x \leqslant 200$ and $0 \leqslant y \leqslant 160$ for your axes.)

b Use the vertex testing method, on the four vertices, to identify the optimal point and optimal value. **(2 marks)**

Given that the objective changes to

minimise $C_1 = 2x + 3y$,

c draw a suitable objective line and use it to identify the optimal point and optimal value. **(2 marks)**

Given that the objective changes to

minimise $C_2 = x + y$,

d explain why there is more than one solution to the problem. **(2 marks)**

(E) **8** A feasible region, R, is defined by

$\qquad\qquad y \leqslant 5x$

$\qquad 14x + 9y \leqslant 630$

$\qquad\qquad 2y \geqslant x$

$\qquad\quad 4x + y \geqslant 60$

a Draw a graph to illustrate the feasible region, R.
(Use the values $0 \leqslant x \leqslant 45$ and $0 \leqslant y \leqslant 70$ for your axes.) **(3 marks)**

b Use the objective line method to determine the optimal point and the optimal value given the objective

i minimise $P = x + 3y$ **(4 marks)**

ii maximise $Q = 6x + y$. **(4 marks)**

In each case you must draw, and label, an objective line, and find *exact* values.

(E) **9** A feasible region, R, is defined by

$\qquad\qquad y \leqslant 10x$

$\qquad\qquad x \leqslant 120$

$\qquad 2y - x \geqslant 100$

$\qquad 2x + y \leqslant 400$

a Draw a graph to illustrate R. **(3 marks)**

Given that the objective function is $z = 5x + y$,

b determine the optimal value of z, if z is to be

   i maximised **(3 marks)**

   ii minimised. **(3 marks)**

c Determine the maximum value of $x + 2y$. Give your answer as an exact value. **(3 marks)**

Challenge

Here is an example of a non-linear programming problem:

maximise $P = 3x + y$

subject to $x \geqslant 1$

            $5y - 4x \geqslant 0$

            $x^2 + y \leqslant 10$

a Sketch the feasible region for this problem.

b By considering the gradient of the objective line, find the maximum value of P within the feasible region.

6.4 Solutions with integer values

There is an additional constraint in some problems – that the solution has integer values. For example, questions **1**, **2**, **3** and **4** in Exercise 6A all require integer solutions, since it is not possible to sell a fraction of a box of chocolates, etc. In such cases the optimal point in the feasible region may not be an acceptable solution, so you have to find the optimal integer solution.

■ **If a linear programming problem requires integer solutions, you need to consider points with integer coordinates near the optimal vertex.**

Example **11**

Given that x and y must be integers, solve the following linear programming problem.

Maximise $P = x + 2y$

subject to $3x + 4y \leqslant 36$

            $13x + 9y \leqslant 117$

             $5y - 4x \leqslant 10$

             $6x + 5y \geqslant 30$

                 $y \geqslant 2$

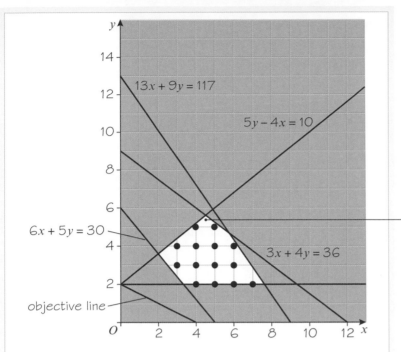

Draw a graph of the feasible region. Show the integer solutions using dots.

Using the objective line method, the optimal point is at the intersection of
$5y - 4x = 10$ and
$3x + 4y = 36$, but this does not have integer solutions.

There are two methods that can be used to locate the optimal integer value.

Method 1

If it is possible to plot the integer value solutions accurately, simply select the last integer solution covered by the objective line as it leaves the feasible region, moving away from the origin.

In this case, it is $(5, 5)$ giving $P = 15$.

Watch out This method is only possible if the feasible region is sufficiently clear to identify the integer solutions accurately. This may depend on the scales used on the axes.

Method 2

Locate the optimal (non-integer) solution, then test the integer solutions that are close to it. Evaluate the objective function and, most importantly, check whether the integer solutions lie in the feasible region.

The optimal solution is $\left(4\frac{16}{31}, 5\frac{19}{31}\right)$ so test $(4, 5)$, $(4, 6)$, $(5, 5)$ and $(5, 6)$.

Watch out It could be that the integer solution lies some distance from the optimal solution.

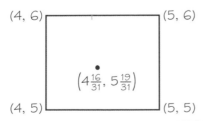

Online Explore how the optimal solution can be found using the objective line method with GeoGebra.

163

(4, 5) obeys all the constraints, so it lies in the feasible region.

$P = 4 + 2 \times 5 = 14$

(5, 5) obeys all the constraints, so it lies in the feasible region.

$P = 5 + 2 \times 5 = 15$

(4, 6) does not obey $5y - 4x \leqslant 10$,

since $5 \times 6 - 4 \times 4 = 14$ and is not $\leqslant 10$, so is outside the feasible region.

(5, 6) does not obey $3x + 4y \leqslant 36$,

since $3 \times 5 + 4 \times 6 = 39$ and is not $\leqslant 36$, so is outside the feasible region.

The optimal integer solution is (5, 5) and $P = 15$.

> Looking at the graph may suggest which constraints may be broken for each point, so these can be checked first.

Example 12

Minimise $x + y$

subject to $3x + 5y \geqslant 1500$

$5x + 2y \geqslant 1000$

$x, y \geqslant 0$

given that x and y must be integers.

> You need to decide whether integer solutions were required from the content of the question.
>
> For example, you cannot have fractions of flower displays!   ← Exercise 6A, Question 2

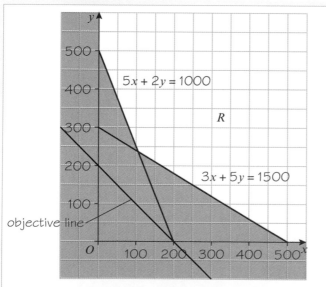

Using the objective line method, the optimal point is at the intersection of $5x + 2y = 1000$

and $3x + 5y = 1500$,

which is $\left(105\frac{5}{19}, 236\frac{16}{19}\right)$.

> **Watch out**   The scale on this graph is such that you cannot plot all integer solutions in the feasible region, so you test integer solutions close to the optimal vertex.

You cannot use the objective line to help any more.
Investigate integer solutions close to this point.

Using a table may help to organise your working.

Point	$5x + 2y \geqslant 1000$	$3x + 5y \geqslant 1500$	In R?	$x + y$
(105, 236)	$997 \not\geqslant 1000$	—	No	
(105, 237)	$999 \not\geqslant 1000$	—	No	
(106, 236)	$1002 \geqslant 1000$	$1498 \not\geqslant 1500$	No	
(106, 237)	$1004 \geqslant 1000$	$1503 \geqslant 1500$	Yes	343

As one constraint has failed, you do not need to check others.

The optimal integer solution lies at (106, 237) and has a value of 343.

You only need to evaluate the objective function for points in the feasible region.

Exercise 6D

1 Solve the following linear programming problems, given that integer values are required for the decision variables.

a Maximise $3x + 2y$

    subject to $x + 5y \geqslant 10$

               $3x + 4y \leqslant 24$

               $4x + 3y \leqslant 24$

                   $x, y \geqslant 0$

b Minimise $2x + y$

    subject to $5x + 6y \geqslant 60$

               $4x + y \geqslant 28$

                   $x, y \geqslant 0$

c Maximise $5x + 2y$

    subject to $2y \geqslant x$

               $5x + 4y \leqslant 800$

                  $y \leqslant 4x$

                  $x, y \geqslant 0$

d Maximise $2x + y$

    subject to $3x + 5y \leqslant 1500$

               $3x + 16y \geqslant 2400$

                   $y \leqslant x$

                   $x, y \geqslant 0$

(E/P) 2 The graph opposite is being used to solve a linear programming problem, subject to the following constraints:

$$y \leqslant 2x + 10$$

$$y + 2x \leqslant 60$$

$$x \geqslant 10$$

$$y \geqslant 0$$

The feasible region is labelled R.

Given that x and y must both be integers, use the vertex testing method to maximise the objective function $P = 5x + 3y$, and write down the corresponding values of x and y. **(7 marks)**

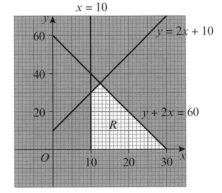

(E/P) 3 A chocolate manufacturer is producing two hand-made assortments, gold and silver, to commemorate 50 years in business.

It will take 30 minutes to make all the chocolates for one box of gold assortment and 20 minutes to make the chocolates for one box of silver assortment.

It will take 12 minutes to wrap and pack the chocolates in one box of gold assortment and 15 minutes for one box of silver assortment.

The manufacturer needs to make at least twice as many silver as gold assortments.

The gold assortment will be sold at a profit of 80p, and the silver at a profit of 60p.

There are 300 hours available to make the chocolates and 200 hours to wrap them.

Maximise the profit, P.   **(9 marks)**

(E/P) **4** A floral display is required for the opening of a new building. The display must be at least 30 m long and is to be made up of two types of planted displays, type A and type B.

Type A is 1 m in length and costs £6.

Type B is 1.5 m in length and costs £10.

The client wants at least twice as many type A as type B, and at least 6 of type B.

  a Write each of the constraints as an inequality.   **(3 marks)**

  b Represent the inequalities on a diagram and identify the feasible region.   **(3 marks)**

  c Find the minimum cost and justify your answer.   **(3 marks)**

(E/P) **5** A toy company makes two types of board game, Cludopoly and Trivscrab. As well as the board each game requires playing pieces and cards.

The company uses two machines, one to produce the pieces and one to produce the cards. Both machines can only be operated for up to ten hours per day.

The first machine takes 5 minutes to produce a set of pieces for Cludopoly and 8 minutes to produce a set of pieces for Trivscrab.

The second machine takes 8 minutes to produce a set of cards for Cludopoly and 4 minutes to produce a set of cards for Trivscrab.

The company knows it will sell at most three times as many games of Cludopoly as Trivscrab.

The profit made on each game of Cludopoly is £1.50 and £2.00 on each game of Trivscrab.

  a Write each of the constraints as an inequality.   **(3 marks)**

  b Represent the inequalities on a diagram and identify the feasible region.   **(3 marks)**

  c Find the maximum profit and justify your answer.   **(3 marks)**

(E/P) **6** A librarian needs to purchase bookcases for a new library. She has a budget of £3000 and 240 m² of available floor space. There are two types of bookcase, type 1 and type 2, that she is permitted to buy.

Type 1 costs £150, needs 15 m² of floor space and has 40 m of shelving.

Type 2 costs £250, needs 12 m² of floor space and has 60 m of shelving.

She must buy at least 8 type 1 bookcases and wants at most $\frac{1}{3}$ of all the bookcases to be type 2.

She wishes to maximise the total amount of shelving.

  a Formulate the above information as a linear programming problem.   **(5 marks)**

  b Use a suitable method to solve the problem, fully justifying your answer.   **(5 marks)**

Mixed exercise 6

(E/P) **1** Mr Baker is making cakes and fruit loaves for sale at a charity cake stall. Each cake requires 200 g of flour and 125 g of fruit. Each fruit loaf requires 200 g of flour and 50 g of fruit. He has 2800 g of flour and 1000 g of fruit available.

Let the number of cakes that he makes be x and the number of fruit loaves he makes be y.

a Show that these constraints can be modelled by the inequalities

$x + y \leqslant 14$ and $5x + 2y \leqslant 40$. **(4 marks)**

Each cake takes 50 minutes to cook and each fruit loaf takes 30 minutes to cook. There are 8 hours of cooking time available.

b Obtain a further inequality, other than $x \geqslant 0$ and $y \geqslant 0$, which models this time constraint.
(2 marks)

c On graph paper illustrate these three inequalities, indicating clearly the feasible region.
(3 marks)

d It is decided to sell the cakes for £3.50 each and the fruit loaves for £1.50 each. Assuming that Mr Baker sells all that he makes, write down an expression for the amount of money P, in pounds, raised by the sale of Mr Baker's products. **(1 mark)**

e Explaining your method clearly, determine how many cakes and how many fruit loaves Mr Baker should make in order to maximise P. **(2 marks)**

f Write down the greatest value of P. **(1 mark)**

(E/P) **2** A junior librarian is setting up a section of a library to loan CDs and DVDs. He has a budget of £420 to spend on storage units to display these items.

Let x be the number of CD storage units and y the number of DVD storage units he plans to buy.

Each type of storage unit occupies 0.08 m³, and there is a total area of 6.4 m³ available for the display.

a Show that this information can be modelled by the inequality **(2 marks)**

$x + y \leqslant 80$

The CD storage units cost £6 each and the DVD storage units cost £4.80 each.

b Write down a second inequality, other than $x \geqslant 0$ and $y \geqslant 0$, to model this constraint. **(2 marks)**

The CD storage unit displays 30 CDs and the DVD storage unit displays 20 DVDs. The chief librarian advises the junior librarian that he should plan to display at least half as many DVDs as CDs.

c Show that this implies that $3x \leqslant 4y$. **(2 marks)**

d On graph paper, display your three inequalities, indicating clearly the feasible region.

The librarian wishes to maximise the total number of items, T, on display. Given that **(3 marks)**

$T = 30x + 20y$

e determine how many CD storage units and how many DVD storage units he should buy, briefly explaining your method. **(2 marks)**

(E/P) 3 The headteacher of a school needs to hire coaches to transport all the year 7, 8 and 9 pupils to take part in the recording of a children's television programme. There are 408 pupils to be taken and 24 adults will accompany them on the coaches. The headteacher can hire either 54 seater (large) or 24 seater (small) coaches. She needs at least two adults per coach. The bus company has only seven large coaches but an ample supply of small coaches.

Let x and y be the number of large and small coaches hired respectively.

a Show that the situation can be modelled by the three inequalities:

  i $9x + 4y \geqslant 72$ **ii** $x + y \leqslant 12$ **iii** $x \leqslant 7$ **(3 marks)**

b On graph paper display the three inequalities, indicating clearly the feasible region. **(3 marks)**

A large coach costs £336 and a small coach costs £252 to hire.

c Write down an expression, in terms of x and y, for the total cost of hiring the coaches. **(1 mark)**

d Explain how you would locate the best option for the headteacher, given that she wishes to minimise the total cost. **(2 marks)**

e Find the number of large and small coaches that the headteacher should hire in order to minimise the total cost and calculate this minimum total cost. **(2 marks)**

(P) 4 The graph below was drawn to solve a linear programming problem. The feasible region, R, includes the points on its boundary.

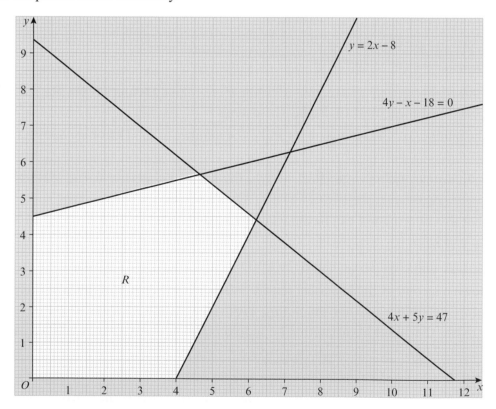

a Write down the inequalities that define the region R.

The objective function, P, is given by $P = 3x + 2y$.

b Find the value of x and the value of y that lead to the maximum value of P. Make your method clear.

   c i Give an example of a practical linear programming problem in which it would be necessary for the variables to have integer values.

     ii Given that the solution must have integer values of x and y, find the values that lead to the maximum value of P.

(E/P) 5 A company produces plates and mugs for local souvenir shops. The plates and mugs are manufactured in a two-stage process. Each day there are 300 minutes available for the completion of the first stage and 400 minutes available for the completion of the second stage. In addition the mugs require some hand painting. There are 150 minutes available each day for hand painting.

Product	Stage 1	Stage 2	Hand painting
Plate	$2\frac{1}{2}$	5	—
Mug	3	2	2

The above table shows the production time, in minutes, required for the plates and the mugs.

All plates and mugs made are sold. The profit on each plate sold is £2 and the profit on each mug sold is £4. The company wishes to determine how many plates and mugs to make so as to maximise its profits each day.

Let x be the number of plates made and y the number of mugs made each day.

   a Write down the three constraints, other than $x \geqslant 0$, $y \geqslant 0$, satisfied by x and y. **(5 marks)**

   b Write down the objective function to be maximised. **(1 mark)**

   c Using the graphical method, solve the resulting linear programming problem. Determine the optimal number of plates and mugs to be made each day and the resulting profit. **(4 marks)**

   d When the optimal solution is adopted determine which, if any, of the stages has available time which is unused. State the amount of unused time. **(2 marks)**

Challenge

In three dimensions, each of the following equations represents a plane.
$$x + y + 2z = 160$$
$$x - z = 25$$
$$y + 2z = 100$$

a By setting up and solving a suitable matrix equation, or otherwise, find the point of intersection of these three planes.

A linear programming problem in x, y and z is described as follows:

maximise $P = 2x + 3y - z$

subject to $x + y + 2z \leqslant 160$

               $x - z \leqslant 25$

               $y + 2z \geqslant 100$

                   $z \geqslant 15$

              $x, y \geqslant 0$

The feasible region for this problem is a tetrahedron.

b Find the coordinates of the vertices of this tetrahedron.

c By considering these vertices, solve the linear programming problem, giving the optimal values of P, x, y and z.

Hint A system of three equations in three unknowns can represent three planes. You can use matrices to the systems of equations and interpret the result geometrically.

← **Core Pure 1, Chapter 6**

Summary of key points

1. To formulate a problem as a **linear programming** problem:
   - Define the **decision variable** (x, y, z, etc.).
   - State the **objective** (maximise or minimise, together with an algebraic expression called the **objective function**).
   - Write the **constraints** as inequalities.

2. The region of a graph that satisfies all the constraints of a linear programming problem is called the **feasible region**.

3. To solve a linear programming problem, you need to find the point in the feasible region which maximises or minimises the objective function.

4. - For a **maximum point**, look for the last point covered by an objective line as it leaves the feasible region.
   - For a **minimum point**, look for the first point covered by an objective line as it enters the feasible region.

5. To find an optimal point using the **vertex method**:
   - First find the coordinates of each vertex of the feasible region.
   - Evaluate the objective function at each of these points.
   - Select the vertex that gives the optimal value of the objective function.

6. If a linear programming problem requires integer solutions, you need to consider points with integer coordinates near the optimal vertex.

The simplex algorithm

7

Prior knowledge check

1 Find the coordinates of the point where
   the line $2x + 3y = 11$ crosses the line
   $x + 2y = 4$.   ← **Pure Year 1, Chapter 5**

2 Find the maximum value of $P = 5x - 2y$
   given the constraints $x \leqslant 6$ and $3 \leqslant y \leqslant 8$.
   ← **Section 6.3**

3 Here is a system of three simultaneous
   equations:
   $$2x + 3y - z = 5$$
   $$3y + 2z = 0$$
   $$x - 4y = 5$$

   a Use the first two equations to show
      that $4x + 9y = 10$.

   b Hence find the values of x, y and z.
      ← **Pure Year 1, Chapter 3**

The simplex method allows you to solve linear
programming problems with more than two
variables. Large online retailers use methods
like this to optimise their shipping and
distribution networks.

7.1 Formulating linear programming problems

You need to be able to formulate real-life problems as linear programming problems. In the previous chapter you did this with two-variable situations.

- **To formulate a linear programming problem:**
  - **define your decision variables**
  - **write down the objective**
  - **write down the constraints.**

In this chapter you will consider more complicated linear programming problems.

Example 1

A grower specialising in cut flowers is considering cultivating four varieties of the new 'Sunlip' flowers A, B, C and D in one of his fields. He estimates the time, in hours per hectare, taken for each of four stages – sowing, thinning, picking and packing, for each variety.

For variety A sowing, thinning, picking and packing will take 3, 18, 20 and 24 hours per hectare respectively.

For variety B sowing, thinning, picking and packing will take 4, 17, 25 and 27 hours per hectare respectively.

For variety C sowing, thinning, picking and packing will take 3, 19, 26 and 28 hours per hectare respectively.

For variety D sowing, thinning, picking and packing will take 5, 16, 22 and 23 hours per hectare respectively.

For this crop he can devote up to 70 hours to sowing, 360 to thinning, 500 to picking and 550 to packing.

He estimates the total profit, in pounds per hectare, as 67, 63, 71 and 75 for varieties A, B, C and D respectively.

He has up to 20 hectares to use for this crop and wishes to maximise his profit.

Formulate this as a linear programming problem. Define your variables, state your objective and write your constraints as inequalities.

Time per hectare	A	B	C	D	Total time available
Sowing	3	4	3	5	70
Thinning	18	17	19	16	360
Picking	20	25	26	22	500
Packing	24	27	28	23	550
Profit	67	63	71	75	

It is clearer if you summarise the information in a table.

First define your decision variables.

Let x_A, x_B, x_C and x_D, be the number of hectares planted of varieties A, B, C and D respectively.

With more than two variables you can use x_A, x_B, x_C, ... or x_1, x_2, x_3, ... instead of letters of the alphabet. x_1, x_2, x_3, ... is particularly useful when there are a large number of variables.

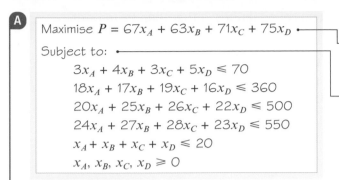

A

Maximise $P = 67x_A + 63x_B + 71x_C + 75x_D$

Subject to:

$3x_A + 4x_B + 3x_C + 5x_D \leqslant 70$
$18x_A + 17x_B + 19x_C + 16x_D \leqslant 360$
$20x_A + 25x_B + 26x_C + 22x_D \leqslant 500$
$24x_A + 27x_B + 28x_C + 23x_D \leqslant 550$
$x_A + x_B + x_C + x_D \leqslant 20$
$x_A, x_B, x_C, x_D \geqslant 0$

Next write down the objective. You need to state 'maximise' or 'minimise' and give the objective function.

Finally write down the constraints.

Watch out Remember to include non-negativity constraints. It is impossible to plant a negative number of hectares of any variety.

Example 2

In order to supplement his diet Andy wishes to take some Vitatab, Weldo, Xtramin and Yestivit tablets. Amongst other ingredients, the contents of vitamins A, B, C and iron, in milligrams per tablet, are shown in the table.

	A	B	C	Iron
Vitatab	10	10	20	4
Weldo	15	20	10	5
Xtramin	25	15	15	3
Yestivit	20	15	20	2

Andy wishes to take tablets to provide him with at least 80, 30, 60 and 14 milligrams of vitamins A, B, C and iron per day.

Because of other factors Andy wants at least 25% of the tablets he takes to be Vitatab and wants to take at least twice as many Weldo as Yestivit.

The costs of the tablets are 4, 6, 12 and 7 pence per tablet. Andy wishes to minimise the cost.

Formulate this as a linear programming problem, defining your variables, stating your objective and writing your constraints as inequalities.

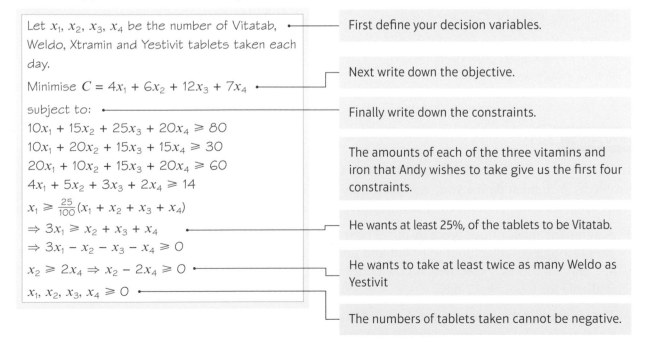

Let x_1, x_2, x_3, x_4 be the number of Vitatab, Weldo, Xtramin and Yestivit tablets taken each day.

First define your decision variables.

Minimise $C = 4x_1 + 6x_2 + 12x_3 + 7x_4$

Next write down the objective.

subject to:

Finally write down the constraints.

$10x_1 + 15x_2 + 25x_3 + 20x_4 \geqslant 80$
$10x_1 + 20x_2 + 15x_3 + 15x_4 \geqslant 30$
$20x_1 + 10x_2 + 15x_3 + 20x_4 \geqslant 60$
$4x_1 + 5x_2 + 3x_3 + 2x_4 \geqslant 14$

The amounts of each of the three vitamins and iron that Andy wishes to take give us the first four constraints.

$x_1 \geqslant \frac{25}{100}(x_1 + x_2 + x_3 + x_4)$
$\Rightarrow 3x_1 \geqslant x_2 + x_3 + x_4$
$\Rightarrow 3x_1 - x_2 - x_3 - x_4 \geqslant 0$

He wants at least 25%, of the tablets to be Vitatab.

$x_2 \geqslant 2x_4 \Rightarrow x_2 - 2x_4 \geqslant 0$

He wants to take at least twice as many Weldo as Yestivit

$x_1, x_2, x_3, x_4 \geqslant 0$

The numbers of tablets taken cannot be negative.

A It is often easier to implement an algorithm using systems of equations, rather than systems of inequalities.

■ **Inequalities can be transformed into equations using slack variables (so called because they represent the amount of slack between an actual quantity and the maximum possible value of that quantity).**

> **Hint** A slack variable acts like a sponge, absorbing spare capacity.

Example 3

Rewrite the inequality

$$x_1 + 3x_2 + 5x_3 \leqslant 23$$

as an equation, using the slack variable, r.

$$x_1 + 3x_2 + 5x_3 + r = 23$$

> The value of slack variable, r, tells us by how much $x_1 + 3x_2 + 5x_3$ is less than 23.
> $r = 23 - x_1 - 3x_2 - 5x_3$

Example 4

Rewrite the constraints for Example 1 as equations using slack variables r, s, t and u.

$$3x_A + 4x_B + 3x_C + 5x_D \leqslant 70$$
$$18x_A + 17x_B + 19x_C + 16x_D \leqslant 360$$
$$20x_A + 25x_B + 26x_C + 22x_D \leqslant 500$$
$$24x_A + 27x_B + 28x_C + 23x_D \leqslant 550$$
$$x_A + x_B + x_C + x_D \leqslant 20$$
$$x_A, x_B, x_C, x_D \geqslant 0$$

> Add the four slack variables.

$$3x_A + 4x_B + 3x_C + 5x_D + r = 70$$
$$18x_A + 17x_B + 19x_C + 16x_D + s = 360$$
$$20x_A + 25x_B + 26x_C + 22x_D + t = 500$$
$$24x_A + 27x_B + 28x_C + 23x_D + u = 550$$
$$x_A + x_B + x_C + x_D + v = 20$$
$$x_A, x_B, x_C, x_D, r, s, t, u, v \geqslant 0$$

> **Watch out** Notice that you alter the non-negativity constraint too. In order to satisfy the original inequalities, r, s, t and u must also be non-negative. In general, slack variables cannot be negative.

Exercise 7A

In questions **1** to **4**, formulate the problems as linear programming problems. You must define your variables, state your objective and write your constraints other than non-negativity as **equations**.

> **Hint** You will need to use slack variables to write your constraints as equations.

1 A company makes three types of metal box, round, square and rectangular. Each box has to pass through two machines to be cut and formed. The round, square and rectangular boxes need 4, 2 and 3 minutes respectively on the cutter and 2, 3 and 3 on the former. Both machines are available for 6 hours per day.

   The profit, in pence, made on each round, square and rectangular box is 12, 10 and 11 respectively. The company wishes to maximise its profit.

A **2** A company makes four different types of backpacks, *A*, *B*, *C* and *D*. Each type *A* uses 2.5 units of material, needs 10 minutes of cutting time and 5 minutes of stitching time. These figures, together with those for types *B*, *C* and *D* are shown in the table

	A	*B*	*C*	*D*
Material in units	2.5	3	2	4
Cutting time in minutes	10	12	8	15
Stitching time in minutes	5	7	4	9

There are 1400 units of material available each week, 150 hours per week available on the cutting machine and 80 hours available on the stitching machine.

Market research says that they will sell at most 500 backpacks each week.

The profit, in pounds, is 8, 7, 6 and 9 for types *A*, *B*, *C* and *D* respectively. The company wishes to maximise its profit.

3 The annual subscription to a bowls club is £40 for adults, £10 for children and £20 for seniors.

The total number of members is restricted to 100.

At most half the club must be children and at least a third must be adults.

The club wishes to maximise its income from subscriptions.

4 Mrs Brown was rather alarmed to discover from her children at bedtime that (a week ago) they had promised she would make small cakes for a cake sale at school the next day. Not wishing to let her children down, she puts the oven on and checks her cupboards and finds she has 3 kg of flour, 2 kg of butter and 1.5 kg of sugar, as well as other ingredients.
Mrs Brown finds three cake recipes for rock cakes, fairy cakes and muffins. The recipe for rock cakes uses 220 g of flour, 100 g butter and 50 g sugar and makes 10 cakes. The recipe for fairy cakes uses 100 g each of flour, butter and sugar and makes 18 cakes. The recipe for muffins uses 250 g of flour, 50 g butter and 75 g sugar and makes 12 muffins.
Mrs Brown wishes to maximise the total number of cakes she makes.

E **5** Roma is moving house. She needs to pack all her extensive collection of china into special cardboard boxes which will be sold to her by the removal company. There are three sizes of box, small, medium and large. The small boxes have a capacity of $0.1\,m^3$ and will hold a maximum weight of 3 kg. The medium boxes have a capacity of $0.3\,m^3$ and will hold a maximum weight of 8 kg. The large boxes have a capacity of $0.7\,m^3$ and will hold a maximum weight of 18 kg. An expert from the removal company informs her that she should allow for at least $28\,m^3$ packing capacity and for at least 600 kg.

Roma decides that at least half of the boxes she uses should be small and that she should use at least twice as many medium as large.

She will be able to fill the boxes she buys and the cost of each small, medium and large box is 30p, 50p and 80p respectively.

Roma wishes to minimise the cost of the boxes she buys.

Formulate this situation as a linear programming problem, giving your constraints as inequalities.

(5 marks)

7.2 The simplex method

A

In the previous chapter you found optimal solutions to linear programming problems by considering the vertices of the feasible region. With a two-variable problem, the feasible region can be represented as an area on a graph. With three variables, you would need to represent the feasible region as a 3D polyhedron. As the number of variables increases, it becomes impossible to visualise the problem graphically.

The simplex algorithm starts at one vertex, and moves between vertices in sequence, increasing the objective function and **one** of the variables each time, until it reaches an optimal solution.

- **The simplex method allows you to:**

  - **determine if a particular vertex, on the edge of the feasible region, is optimal,**

  - **decide which adjacent vertex you should move to in order to increase the value of the objective function.**

- **Slack variables are essential when using the simplex algorithm.**

> **Hint** You can think of the algorithm as being a set of signposts on a treasure hunt. As you reach each vertex the algorithm tells you if you have reached the optimal solution and if not, which way you should go to get to the next signpost.

Example 5

Explain the significance of the slack variables in the graphical representation of the linear programming problem below.

Maximise $P = 3x + 2y$

subject to:

$$5x + 7y + r = 70$$
$$10x + 3y + s = 60$$
$$x, y, r, s \geqslant 0$$

The feasible region, R, is shown.

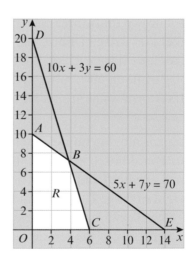

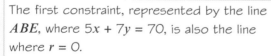

The first constraint, represented by the line ABE, where $5x + 7y = 70$, is also the line where $r = 0$.

The second constraint, represented by the line DBC, where $10x + 3y = 60$, is also the line where $s = 0$.

A

The four lines forming the boundaries of the feasible region can be seen as being formed by drawing the four lines $x = 0$, $y = 0$, $r = 0$ and $s = 0$. This is shown by the diagram.

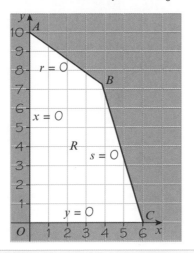

Problem-solving

At each vertex precisely two of the four variables and slack variables are zero.

Example 6

Explain, in detail, how the algebraic simplex algorithm is used to solve the following problem, relating each stage to the given graph.

Maximise $P = 3x + 2y$

subject to:

$$5x + 7y + r = 70$$
$$10x + 3y + s = 60$$
$$x, y, r, s \geqslant 0$$

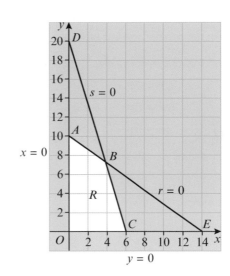

Our initial set of equations is

$$\begin{aligned} 5x + 7y + r &= 70 \quad (1) \\ 10x + 3y + s &= 60 \quad (2) \\ P - 3x - 2y &= 0 \quad (3) \end{aligned}$$

Start at $O(0, 0)$. This is the vertex where $x = y = 0$

Test the objective function, $P = 3x + 2y$, at this point and get $P = 0$. This is not optimal.

Note that the objective function has been rewritten so that the right-hand side is equal to 0. The reason for this will become apparent as you move on to the simplex tableau.

Look at the objective function. You can see that increasing x or y will increase the objective function.

Only increase one variable at a time so, keeping $y = 0$, increase x. (In effect this means that you move right along the x-axis.)

Keep going until you hit the line $r = 0$ or the line $s = 0$. (Here you can see which one you hit first, but in more than 2 dimensions it is harder to visualise the feasible region.)

From equation (1):   If $y = 0$ and $r = 0$, then $x = 14$

From equation (2):   If $y = 0$ and $s = 0$, then $x = 6$

So you get to the vertex formed by $y = s = 0$ first, given by equation (2).

Now look to see if you have reached the optimum point.

You know the values of y and s at this vertex so eliminate x from the equations.

Here are the current equations:

$$5x + 7y + r \qquad = 70 \qquad (1)$$
$$10x + 3y \quad + s = 60 \qquad (2)$$
$$P - 3x - 2y \qquad = 0 \qquad (3)$$

First divide equation (2) by 10 to make the coefficient of x one, getting equation (5).

This gives:

$$5x + 7y + r \qquad = 70 \quad (1)$$
$$x + \tfrac{3}{10}y \quad + \tfrac{1}{10}s = 6 \quad (5) = (2) \div 10$$
$$P - 3x - 2y \qquad = 0 \quad (3)$$

Eliminate the x terms in equations (1) and (3). Use equation (5) to do this.

To eliminate the $+5x$ in equation (1), **subtract** 5 copies of equation (5).

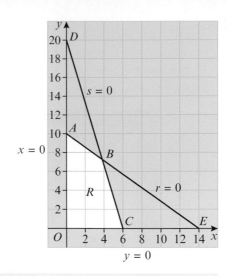

Increasing x will give a 'faster' increase. Increasing x by 1 adds 3 to the value of the function, whereas increasing y by 1 only adds 2 to the value of the function.

Once again the reason for this will become apparent once you move into the simplex tableau.

Use equation (2) because this was the equation that gave us the current vertex where $y = s = 0$

Problem-solving

The simplex method can be seen as an exercise in 'advanced' simultaneous linear equations.

To eliminate the $-3x$ in equation (3), **add** 3 copies of equation (5).

$$\tfrac{11}{2}y + r - \tfrac{1}{2}s = 40 \quad (4) = (1) - 5(5)$$
$$x + \tfrac{3}{10}y + \tfrac{1}{10}s = 6 \quad (5)$$
$$P - \tfrac{11}{10}y + \tfrac{3}{10}s = 18 \quad (6) = (3) + 3(5)$$

Look at equation (6).

When $y = s = 0$, $P = 18$. So the current profit is 18.

Make P the subject of equation (6):

$$P = 18 + \tfrac{11}{10}y - \tfrac{3}{10}s$$

You can see that increasing y will increase the profit, but increasing s will decrease the profit. So you need to increase y but keep $s = 0$. This means you will travel along the line $s = 0$ in the direction that makes y increase. In this case travel along $s = 0$ up and to the left.

Continue along the line $s = 0$ until you hit the next vertex.

You have just left the vertex $y = s = 0$, so you will hit either $s = r = 0$ or $s = x = 0$ next.

Put these values into the current set of equations:

$$\tfrac{11}{2}y + r - \tfrac{1}{2}s = 40 \quad (4)$$
$$x + \tfrac{3}{10}y + \tfrac{1}{10}s = 6 \quad (5)$$
$$P - \tfrac{11}{10}y + \tfrac{3}{10}s = 18 \quad (6)$$

When $s = r = 0$ equation (4) gives $y = \tfrac{80}{11} = 7\tfrac{3}{11}$

When $s = x = 0$ equation (5) gives $y = 20$

You get to $y = 7\tfrac{3}{11}$ before $y = 20$, so you get to the vertex $s = r = 0$ and the value given by equation (4). You know s and r at this vertex, so eliminate y from your equations.

Divide equation (4) by $\tfrac{11}{2}$ to reduce the coefficient of y to one, getting equation (7).

$$y + \tfrac{2}{11}r - \tfrac{1}{11}s = \tfrac{80}{11} \quad (7) = (4) \div \tfrac{11}{2}$$
$$x + \tfrac{3}{10}y + \tfrac{1}{10}s = 6 \quad (5)$$
$$P - \tfrac{11}{10}y + \tfrac{3}{10}s = 18 \quad (6)$$

Eliminate the terms in y from the other two equations.

A

To eliminate $+\frac{3}{10}y$ from equation (5), **subtract** $\frac{3}{10}$ copies of equation (7).

To eliminate $-\frac{11}{10}y$ from equation (6), **add** $\frac{11}{10}$ copies of equation (7).

$$y + \tfrac{2}{11}r - \tfrac{1}{11}s = \tfrac{80}{11} \quad (7)$$
$$x \quad - \tfrac{3}{55}r + \tfrac{7}{55}s = \tfrac{42}{11} \quad (8) = (5) - \tfrac{3}{10}(7)$$
$$P \quad + \tfrac{1}{5}r + \tfrac{1}{5}s = 26 \quad (9) = (6) + \tfrac{11}{10}(7)$$

Rearranging equation (9) to make P the subject

$$P = 26 - \tfrac{1}{5}r - \tfrac{1}{5}s$$

At this vertex $r = s = 0$, so $P = 26$

You can see that if you increase r or s you will decrease the profit, so you have reached the optimal solution.

Profit $= 26$ when $x = \frac{42}{11}$, $y = \frac{80}{11}$, $r = 0$ and $s = 0$.

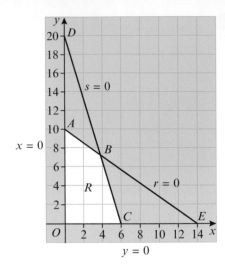

You can apply this method to a problem with 3 (non-slack) variables.

Example 7

Show, in detail, the application of the algebraic simplex algorithm to the linear programming problem below.

Maximise $P = 10x + 12y + 8z$

subject to:

$$2x + 2y \leqslant 5$$
$$5x + 3y + 4z \leqslant 15$$
$$x, y, z \geqslant 0$$

Introducing slack variables r and s, and forming equations you get:

$$2x + 2y \quad + r \quad = 5 \quad (1)$$
$$5x + 3y + 4z \quad + s = 15 \quad (2)$$
$$P - 10x - 12y - 8z \quad = 0 \quad (3)$$

Start at $(0, 0, 0)$. At this point $P = 0$, which is clearly not the maximum.

Rewrite equation (3) to make P the subject:

$$P = 10x + 12y + 8z$$

You can see that you can increase the profit by increasing x or y or z. Increase one variable at a time.

A

Choose to increase y, since this will give the greatest increase in the profit, and leave x and z still at zero.

As you increase y you must not leave the feasible region, so you must stop the first time you hit a vertex.

The vertices will occur

either when $x = z = r = 0$ or when $x = z = s = 0$

Work out the y value at each of these. The vertex with the lowest y value will be reached first.

From equation (1)   when $x = z = r = 0$, $y = \frac{5}{2}$

From equation (2)   when $x = z = s = 0$, $y = 5$

So you reach $\left(0, \frac{5}{2}, 0\right)$ first.

You now need to find out what the profit is at this point, and whether it is possible to increase it still further.

Since at this point the values of x, z and r are all zero, and you have already increased y to its maximum, you can use equation (1) to eliminate y from both of the other equations.

Divide equation (1) by 2 to reduce the coefficient of y to one. This creates equation (4).

$$\begin{aligned}
x + \quad y \quad\quad + \tfrac{1}{2}r \quad\quad &= \tfrac{5}{2} \quad (4) \quad\quad = (1) \div 2 \\
5x + \quad 3y + 4z \quad\quad + s &= 15 \quad (2) \\
P - 10x - 12y - 8z \quad\quad\quad &= 0 \quad (3)
\end{aligned}$$

> Because the coefficient of y is one, you can use multiples of this equation to eliminate y from the other equations. This row will correspond to the **pivot row** in the simplex tableau. You will encounter this in the next example.

Eliminate the y terms in the other equations, by adding or subtracting multiples of equation (4).

This gives the following set of equations.

$$\begin{aligned}
x + \quad y \quad\quad + \tfrac{1}{2}r \quad\quad &= \tfrac{5}{2} \quad (4) \\
2x \quad\quad + 4z - \tfrac{3}{2}r + s &= \tfrac{15}{2} \quad (5) \quad\quad = (2) - 3(4) \\
P + 2x \quad\quad - 8z + 6r \quad\quad &= 30 \quad (6) \quad\quad = (3) + 12(4)
\end{aligned}$$

Rewriting equation (6) to make P the subject:

$$P = 30 - 2x + 8z - 6r$$

you see that you can still increase the profit by increasing z.

> You must not increase x or r since these will decrease the profit, so you need to keep x and r zero and increase z.

Once again, keep the other variables in that equation zero, and determine which vertex you reach first.

(In this case (equation 4) will not give a value for z because there is no z term! This speeds things up.)

From equation (5) when $x = r = s = 0$ $z = \frac{15}{8}$

Divide equation (5) by 4 to reduce the coefficient of z to one giving equation (7).

A

$$x + y \quad + \tfrac{1}{2}r \quad = \tfrac{5}{2} \quad (4)$$

$$\tfrac{1}{2}x \quad + \quad z - \tfrac{3}{8}r + \tfrac{1}{4}s = \tfrac{15}{8} \quad (7) \quad = (5) \div 4$$

$$P + 2x \quad - 8z + 6r \quad = 30 \quad (6)$$

Eliminate z from the other equations using equation (7).

There is no z term in equation (4) so this remains unchanged.

To eliminate the $-8z$ in equation (6), **add** 8 copies of equation (7).

$$x + y \quad + \tfrac{1}{2}r \quad = \tfrac{5}{2} \quad (4) \text{ (unchanged)}$$

$$\tfrac{1}{2}x \quad + \quad z - \tfrac{3}{8}r + \tfrac{1}{4}s = \tfrac{15}{8} \quad (7)$$

$$P + 6x \quad + 3r + 2s = 45 \quad (8) = (6) + 8(7)$$

Rewrite equation (8) to make P the subject:

$$P = 45 - 6x - 3r - 2s$$

This tells you that at the current vertex, where $x = r = s = 0$, $P = 45$.

Increasing x or r or s will decrease the profit, so you cannot increase the profit further, so you have found the optimal solution.

$$\text{Profit} = 45 \text{ when } x = 0, y = \tfrac{5}{2}, z = \tfrac{15}{8}, r = 0 \text{ and } s = 0.$$

> The fact that $r = 0$ means the first constraint is at capacity, and the fact that $s = 0$ means that the second constraint is at capacity.

You can greatly simplify the method outlined in the examples above using a table called a **simplex tableau**.

In your exam you might need to use a simplex tableau to solve linear programming problems with a maximum of **four variables** and **four constraints** (in addition to any non-negativity constraints). Problems could require you either **maximise** or **minimise** an objective function.

This example uses a simplex tableau to solve the problem from Example 6.

Example ⑧

Solve the linear programming problem in Example 6 using simplex tableaux.

Maximise $P = 3x + 2y$

subject to:

> **Notation** The word **tableau** is French. The plural of tableau is **tableaux**.

$$5x + 7y + r = 70$$

$$10x + 3y + s = 60$$

$$x, y, r, s \geqslant 0$$

A Our initial tableau is:

Basic variable	x	y	r	s	Value
r	5	7	1	0	70
s	10	3	0	1	60
P	−3	−2	0	0	0

The first row shows the first constraint, the second row shows the second constraint and the final row shows the objective function.

The column marked 'basic variable' indicates the variables that are not currently at zero. Initially you start at the vertex (0, 0), so $x = y = 0$.

Any variables in a simplex tableau, that are not basic variables, have the value 0.

If $x = y = 0$ then $r = 70$ [from equation (1)] and $s = 60$ [from equation (2)]

You read across the tableau, so this initial tableau tells you that $r = 70$, $s = 60$ and $P = 0$. All other variables are zero, by definition, since they are not listed as basic variables.

Currently therefore

$$P = 0 \quad x = 0 \quad y = 0 \quad r = 70 \quad s = 60$$

This is called the **basic feasible solution** – all the constraints are satisfied, but applying the simplex method will lead to an improved solution. Notice that each column corresponding to a basic variable only contains 0s and a single 1 in the row corresponding to that variable.

Compare this with our first set of equations from the algebraic solution

$$5x + 7y + r \quad\quad = 70 \quad (1)$$
$$10x + 3y \quad\; + s = 60 \quad (2)$$
$$P - 3x - 2y \quad\quad = 0 \quad (3)$$

If you compare these equations with the initial tableau you should see that the columns marked x, y, r, s and value give the coefficients of those letters in the equations. You do not need to write down any of the letters at all in the tableau.

Now scan the objective (bottom) row of the tableau for the most negative number, in this case, −3.

This gives the **pivot column** as the x column.

For each of the other rows, calculate the θ values where

$\theta = $ (the term in the value column) ÷ (the term in the pivot column)

Basic variable	x	y	r	s	Value	θ values
r	5	7	1	0	70	70 ÷ 5 = 14
s	10	3	0	1	60	60 ÷ 10 = 6
P	−3	−2	0	0	0	

In the algebraic example you found that increasing x initially was the most effective way of increasing the profit.

Select the row containing the smallest (positive) θ value. (This will become the pivot row.)

In this case this is the second row.

Basic variable	x	y	r	s	Value	θ values
r	5	7	1	0	70	$70 \div 5 = 14$
s	10	3	0	1	60	$60 \div 10 = 6$
P	−3	−2	0	0	0	

We divide this row by the **pivot**, (which is the value in the pivot row and pivot column), to create the **pivot row**.

In this case we divide all the elements in row 2 by 10.

Basic variable	x	y	r	s	Value	Row operations
r	5	7	1	0	70	
x	1	$\frac{3}{10}$	0	$\frac{1}{10}$	6	R2 ÷ 10
P	−3	−2	0	0	0	

Use this pivot row to eliminate x from each of the other rows.

Basic variable	x	y	r	s	Value	Row operations
r	0	$\frac{11}{2}$	1	$-\frac{1}{2}$	40	R1 − 5R2
x	1	$\frac{3}{10}$	0	$\frac{1}{10}$	6	
P	0	$-\frac{11}{10}$	0	$\frac{3}{10}$	18	R3 + 3R2

Note that the basic variable entry has also been changed. You are leaving the starting vertex where $x = y = 0$ and increasing x, so x is no longer equal to zero.

The basic variable changes to the variable in the pivot column.

We replace the s at the start of the pivot row by x, the variable at the top of the pivot column.

So there have been changes to every cell in the pivot row.

Comparing this second tableau with the second set of equations from Example 6, you see that you are still matching the algebraic solution, but without having to write down all the algebra.

$$5x + 7y + r \qquad = 70 \quad (1)$$
$$x + \tfrac{3}{10}y \qquad + \tfrac{1}{10}s = 6 \quad (5) = (2) \div 10$$
$$P - 3x - 2y \qquad = 0 \quad (3)$$

We are aiming to get just one number 1 and make all other terms zero in the x column.

New R1 = old R1 − 5 × new R2

New R3 = old R3 + 3 × new R2

R1 − 5R2 is standard notation stating that for each entry in the row, you took the row 1 number and subtracted 5 times the row 2 entry in the same column.

So the calculations you did for row 1 were:

In column x: $5 − 5 × 1 = 0$ In column y: $7 − 5 × \frac{3}{10} = \frac{11}{2}$

In column r: $1 − 5 × 0 = 1$ In column s: $0 − 5 × \frac{1}{10} = -\frac{1}{2}$

In the value column $70 − 5 × 6 = 40$

Similarly R3 + 3R2 states that we took the row 3 entry and added three times the corresponding row 2 entry.

So the calculations we did for row 3 were:

In column x: $−3 + 3 × 1 = 0$ In column y: $−2 + 3 × \frac{3}{10} = -\frac{11}{10}$

In column r: $0 + 3 × 0 = 0$ In column s: $0 + 3 × \frac{1}{10} = \frac{3}{10}$

In the value column $0 + 3 × 6 = 18$

Compare this third tableau with the third set of equations

$$\frac{11}{2}y + r - \frac{1}{2}s = 40 \quad (4) \qquad = (1) - 5(5)$$

$$x + \frac{3}{10}y \quad + \frac{1}{10}s = 6 \quad (5)$$

$$P \quad - \frac{11}{10}y \quad + \frac{3}{10}s = 18 \quad (6) \qquad = (3) + 3(5)$$

Now repeat the process again.

First we look for the most negative entry in the objective (bottom) row. In this case it is the $-\frac{11}{10}$ in the y column. This gives the new pivot column.

Basic variable	x	y	r	s	Value
r	0	$\frac{11}{2}$	1	$-\frac{1}{2}$	40
x	1	$\frac{3}{10}$	0	$\frac{1}{10}$	6
P	0	$-\frac{11}{10}$	0	$\frac{3}{10}$	18

Second calculate the new θ values.

Basic variable	x	y	r	s	Value	θ values
r	0	$\frac{11}{2}$	1	$-\frac{1}{2}$	40	$40 \div \frac{11}{2} = \frac{80}{11} = 7\frac{3}{11}$
x	1	$\frac{3}{10}$	0	$\frac{1}{10}$	6	$6 \div \frac{3}{10} = 20$
P	0	$-\frac{11}{10}$	0	$\frac{3}{10}$	18	

Third we select the smallest, positive θ value. This lies in the first row, so this will become the next pivot row.

Basic variable	x	y	r	s	Value	θ values
r	0	$\frac{11}{2}$	1	$-\frac{1}{2}$	40	$40 \div \frac{11}{2} = \frac{80}{11} = 7\frac{3}{11}$
x	1	$\frac{3}{10}$	0	$\frac{1}{10}$	6	$6 \div \frac{3}{10} = 20$
P	0	$-\frac{11}{10}$	0	$\frac{3}{10}$	18	

Fourth divide the row by the pivot to create the pivot row. In this case we divide row 1 by $\frac{11}{2}$, not forgetting to change the basic variable.

Basic variable	x	y	r	s	Value	Row operations
y	0	1	$\frac{2}{11}$	$-\frac{1}{11}$	$\frac{80}{11}$	R1 $\div \frac{11}{2}$
x	1	$\frac{3}{10}$	0	$\frac{1}{10}$	6	
P	0	$-\frac{11}{10}$	0	$\frac{3}{10}$	18	

Here are the fourth set of equations determined in Example 6

$$y + \frac{2}{11}r - \frac{1}{11}s = \frac{80}{11} \quad (7) \qquad = (4) \div \frac{11}{2}$$

$$x + \frac{3}{10}y \quad + \frac{1}{10}s = 6 \quad (5)$$

$$P \quad - \frac{11}{10}y \quad + \frac{3}{10}s = 18 \quad (6)$$

Finally, eliminate the pivot term from the other two rows, using the pivot row to do so. In this case eliminate y from the x and P rows.

Basic variable	x	y	r	s	Value	Row operations
y	0	1	$\frac{2}{11}$	$-\frac{1}{11}$	$\frac{80}{11}$	
x	1	0	$-\frac{3}{55}$	$\frac{7}{55}$	$\frac{42}{11}$	R2 $-\frac{3}{10}$R1
P	0	0	$\frac{1}{5}$	$\frac{1}{5}$	26	R3 $+\frac{11}{10}$R1

Here is the corresponding set of equations from Example 6

$$y + \tfrac{2}{11}r - \tfrac{1}{11}s = \tfrac{80}{11} \quad (7)$$
$$x - \tfrac{3}{55}r + \tfrac{7}{55}s = \tfrac{42}{11} \quad (8) \quad = (5) - \tfrac{3}{10}(7)$$
$$P + \tfrac{1}{5}r + \tfrac{1}{5}s = 26 \quad (9) \quad = (6) + \tfrac{11}{10}(7)$$

Watch out In your exam you might be asked to write down the equations from a given tableau.

Look along the objective row for the most negative. All numbers in this row are non-negative so you know that you have reached the optimal solution.

The standard simplex method is designed to **maximise** the value of the objective function. However, it is also possible to **minimise** an objective function by maximising its **negative**.

Basic variable	x	y	r	s	Value
y	0	1	$\frac{2}{11}$	$-\frac{1}{11}$	$\frac{80}{11}$
x	1	0	$-\frac{3}{55}$	$\frac{7}{55}$	$\frac{42}{11}$
P	0	0	$\frac{1}{5}$	$\frac{1}{5}$	26

Looking at the basic variable column and the value column we see that

$P = 26$, $y = \frac{80}{11}$ and $x = \frac{42}{11}$ and all other variables, and slack variables, are zero.

So our full solution is

$$P = 26, x = \tfrac{42}{11}, y = \tfrac{80}{11}, r = 0, \text{ and } s = 0$$

You should state the values of P, each variable and each slack variable, as your final answer.

- **The simplex method always starts from a basic feasible solution, at the origin, and then progresses with each iteration to an adjacent point within the feasible region until the optimal solution is found.**

Example 9

Minimise $P = 3x - y$

subject to:

$$2x + y \leqslant 12$$
$$x + 4y \leqslant 8$$
$$x \geqslant 0, y \geqslant 0$$

Problem-solving

Minimising P is equivalent to maximising Q, so you can use the standard simplex tableau method.

$P = 3x - y$

Define $Q = -P = -3x + y$

A

Introducing slack variables, r and s, the problem may be expressed as:

Maximise $Q = -3x + y$

subject to:

$$2x + y + r = 12$$
$$x + 4y + s = 8$$
$$x, y, r, s \geqslant 0$$

> To put the objective function into the tableau, it must first be rearranged as $Q + 3x - y = 0$.

The initial tableau is:

Basic variable	x	y	r	s	Value
r	2	1	1	0	12
s	1	4	0	1	8
Q	3	−1	0	0	0

> The basic feasible solution shown in this tableau is:
> $x = 0, y = 0, r = 12, s = 8, Q = 0$.

> The only negative value in the objective row is −1 which is in column y. This becomes the pivot column.

Basic variable	x	y	r	s	Value	θ values
r	2	1	1	0	12	12 ÷ 1 = 12
s	1	4	0	1	8	8 ÷ 4 = 2
Q	3	−1	0	0	0	

The smallest positive θ value is 2 in row 2, so this becomes the pivot row.

Basic variable	x	y	r	s	Value	θ values
r	2	1	1	0	12	12 ÷ 1 = 12
s	1	4	0	1	8	8 ÷ 4 = 2
Q	3	−1	0	0	0	

> Divide the values in the pivot row by the pivot and replace the basic variable in the pivot row with the variable in the pivot column. In this case, s will be replaced by y.

The value in the pivot column and pivot row is 4 which becomes the pivot.

Basic variable	x	y	r	s	Value	Row operations
r	2	1	1	0	12	
y	$\frac{1}{4}$	1	0	$\frac{1}{4}$	2	R2 ÷ 4
Q	3	−1	0	0	0	

Basic variable	x	y	r	s	Value	Row operations
r	$\frac{7}{4}$	0	1	$-\frac{1}{4}$	10	R1 − R2
y	$\frac{1}{4}$	1	0	$\frac{1}{4}$	2	
Q	$\frac{13}{4}$	0	0	$\frac{1}{4}$	2	R3 + R2

> Use row operations with the pivot row to give zeros in the pivot column (apart from the 1 in the pivot position).

A

There are no negative values in the objective row, so the solution is optimal.

Maximum value of $Q = 2$, when $x = 0$, $y = 2$, $r = 10$, $s = 0$

So, the minimum value of $P = -2$, when $x = 0$, $y = 2$, $r = 10$, $s = 0$

Your solution should give the **minimum** value of P, where $P = -Q$. The values of x, y, r and s will be the same for the minimum value of P as they are for the maximum value of Q.

The simplex algorithm can **only** be used to solve linear programming problems in which all the constraints, other than the non-negativity conditions, are of the form $a_1x_1 + a_2x_2 + \ldots + a_nx_n \leqslant K$.

These constraints can all be converted into equations by adding a non-negative slack variable.

Problems involving constraints of the form $a_1x_1 + a_2x_2 + \ldots + a_nx_n \geqslant K$ need to be treated slightly differently, and methods for solving problems such as these are given later in this chapter.

- **Using a simplex tableau to solve a maximising linear programming problem, where the constraints are given as equalities.**

1 Draw the tableaux.
  You need a basic variable column on the left, one column for each variable (including the slack variables) and a value column. You need one row for each constraint and the bottom row for the objective function.

2 Create the initial tableau.
  Enter the coefficients of the variables in the appropriate column and row.

3 Look along the objective row for the most negative entry: this indicates the pivot column.

4 Calculate the θ values, for each of the constraint rows, where
  $\theta =$ (the term in the value column) $\div$ (the term in the pivot column)

5 Select the row with the smallest, positive θ value to become the pivot row.

6 The element in the pivot row and pivot column is the pivot.

7 Divide the row found in step **5** by the pivot, and change the basic variable at the start of the row to the variable at the top of the pivot column. This is now the pivot row.

8 Use the pivot row to eliminate the pivot's variable from the other rows.
  This means that the pivot column now contains one 1 and zeros.

9 Repeat steps **3** to **8** until there are no more negative numbers in the objective row.

10 The tableau is now optimal and the non-zero values can be read off using the basic variable column and value column.

- **The steps for solving a minimising linear programming problem are identical to those given above apart from:**
  - **First, define a new objective function that is the negative of the original objective function.**
  - **After you have maximised this new objective function, write your solution as the negative of this value, which will minimise the original objective function.**

In the following example the simplex tableau method is used to solve the three-variable problem from Example 7.

Example 10

A

a Use simplex tableaux to solve the linear programming problem below (from Example 7).

Maximise $P = 10x + 12y + 8z$

subject to:

$$2x + 2y \leqslant 5$$
$$5x + 3y + 4z \leqslant 15$$
$$x, y, z \geqslant 0$$

b Verify your solution using the original problem.

---

a Introducing slack variables r and s, and forming equations:

$$2x + 2y \quad\quad + r \quad\quad = 5 \quad (1)$$
$$5x + 3y + 4z \quad\quad + s = 15 \quad (2)$$
$$P - 10x - 12y - 8z \quad\quad = 0 \quad (3)$$

Step 1, Step 2 •——————————————————

Basic variable	x	y	z	r	s	Value
r	2	2	0	1	0	5
s	5	3	4	0	1	15
P	−10	−12	−8	0	0	0

The steps refer to the algorithm on page 188.

Step 3

Basic variable	x	y	z	r	s	Value
r	2	2	0	1	0	5
s	5	3	4	0	1	15
P	−10	−12	−8	0	0	0

The most negative entry in the objective row is −12, this becomes the pivot column.

Step 4

Basic variable	x	y	z	r	s	Value	θ values
r	2	2	0	1	0	5	$5 \div 2 = 2\frac{1}{2}$
s	5	3	4	0	1	15	$15 \div 3 = 5$
P	−10	−12	−8	0	0	0	

Now calculate the θ values.

Step 5, Step 6

Basic variable	x	y	z	r	s	Value	θ values
r	2	2	0	1	0	5	$5 \div 2 = 2\frac{1}{2}$
s	5	3	4	0	1	15	$15 \div 3 = 5$
P	−10	−12	−8	0	0	0	

The smallest positive θ value lies in the r row, so this will become the pivot row. The pivot is 2.

A

Step 7

Basic variable	x	y	z	r	s	Value	Row operations
y	1	1	0	$\frac{1}{2}$	0	$2\frac{1}{2}$	R1 ÷ 2
s	5	3	4	0	1	15	R2
P	−10	−12	−8	0	0	0	R3

Divide the first row by the pivot (2) and change the basic variable. This is now the pivot row.

Step 8

Basic variable	x	y	z	r	s	Value	Row operations
y	1	1	0	$\frac{1}{2}$	0	$2\frac{1}{2}$	
s	2	0	4	$-1\frac{1}{2}$	1	$7\frac{1}{2}$	R2 − 3R1
P	2	0	−8	6	0	30	R3 + 12R1

Use this pivot row to eliminate y from the other two rows.

Basic variable	x	y	z	r	s	Value
y	1	1	0	$\frac{1}{2}$	0	$2\frac{1}{2}$
s	2	0	4	$-1\frac{1}{2}$	1	$7\frac{1}{2}$
P	2	0	−8	6	0	30

The y column now contains just one 1 (where the pivot was) and zeros.

Repeat Steps 3 to 8

Basic variable	x	y	z	r	s	Value	θ values
y	1	1	0	$\frac{1}{2}$	0	$2\frac{1}{2}$	$2\frac{1}{2} \div 0 \rightarrow \infty$
s	2	0	4	$-1\frac{1}{2}$	1	$7\frac{1}{2}$	$7\frac{1}{2} \div 4 = 1\frac{7}{8}$
P	2	0	−8	6	0	30	

Identify the pivot column and calculate the θ values.

Basic variable	x	y	z	r	s	Value	θ values
y	1	1	0	$\frac{1}{2}$	0	$2\frac{1}{2}$	$2\frac{1}{2} \div 0 \rightarrow \infty$
s	2	0	4	$-1\frac{1}{2}$	1	$7\frac{1}{2}$	$7\frac{1}{2} \div 4 = 1\frac{7}{8}$
P	2	0	−8	6	0	30	

This means that the next pivot row will be the second row.

Basic variable	x	y	z	r	s	Value	Row operations
y	1	1	0	$\frac{1}{2}$	0	$2\frac{1}{2}$	R1
z	$\frac{1}{2}$	0	1	$-\frac{3}{8}$	$\frac{1}{4}$	$1\frac{7}{8}$	R2 ÷ 4
P	2	0	−8	6	0	30	R3

Now divide all elements in the second row by 4 and change the basic variable to z.

Basic variable	x	y	z	r	s	Value	Row operations
y	1	1	0	$\frac{1}{2}$	0	$2\frac{1}{2}$	R1 no change
z	$\frac{1}{2}$	0	1	$-\frac{3}{8}$	$\frac{1}{4}$	$1\frac{7}{8}$	
P	6	0	0	3	2	45	R3 + 8R2

Eliminate z from the other rows, using pivot row 2.

A

Basic variable	Value
y	$2\frac{1}{2}$
z	$1\frac{7}{8}$
P	45

There are no negatives in the objective (bottom) row, so our tableau is optimal. Read off the values of the basic variables y, z and P from the value column.

The optimal solution is

$$P = 45,\ x = 0,\ y = 2\frac{1}{2},\ z = 1\frac{7}{8},\ r = 0,\ s = 0.$$

b Objective function:

$$10x + 12y + 8z = 0 + 30 + 15 = 45 = P\ \checkmark$$

Constraints:

$$2x + 2y = 0 + 5 = 5 \leqslant 5\ \checkmark$$

$$5x + 3y + 4z = 0 + 7\frac{1}{2} + 7\frac{1}{2} = 15 \leqslant 15\ \checkmark$$

$$x,\ y,\ z \geqslant 0\ \checkmark$$

The simplex tableau algorithm is much quicker than the algebraic simplex method. Although the algebra is there, it is hidden.

Check that your values of P, x, y and z satisfy the original objective function, and check that each constraint is satisfied.

Reducing the number of tableaux

In the above example you wrote each line of each tableau twice. Once you are more comfortable with the algorithm you may be able to reduce the number of tableaux by combining all the row operations into one tableau. The important thing is to make sure that the pivot row is written down first. A complete solution to the problem would look like this.

Watch out This is optional. In your exam sufficient tableaux will be provided for the full, non-reduced solution.

$$
\begin{array}{llllll}
2x + & 2y & & + r & & = 5 & (1) \\
5x + & 3y + & 4z & & + s & = 15 & (2) \\
P - 10x - & 12y - & 8z & & & = 0 & (3)
\end{array}
$$

Basic variable	x	y	z	r	s	Value	θ values
r	2	2	0	1	0	5	$5 \div 2 = 2\frac{1}{2}$*
s	5	3	4	0	1	15	$15 \div 3 = 5$
P	−10	−12	−8	0	0	0	

We sometimes indicate the smallest θ value by putting *.

Basic variable	x	y	z	r	s	Value	Row operations
y	1	1	0	$\frac{1}{2}$	0	$2\frac{1}{2}$	R1 ÷ 2
s	2	0	4	$-1\frac{1}{2}$	1	$7\frac{1}{2}$	R2 − 3R1
P	2	0	−8	6	0	30	R3 + 12R1

The first row to be written in this tableau is the pivot row, row 2. The pivot row from the above tableau is then used to work out the other rows.

New R1 = old R1 ÷ 2

A

Basic variable	x	y	z	r	s	Value	θ values
y	1	1	0	$\frac{1}{2}$	0	$2\frac{1}{2}$	$2\frac{1}{2} \div 0 \rightarrow \infty$
s	2	0	4	$-1\frac{1}{2}$	1	$7\frac{1}{2}$	$7\frac{1}{2} \div 4 = 1\frac{7}{8}$
P	2	0	-8	6	0	30	

Basic variable	x	y	z	r	s	Value	Row operations
y	1	1	0	$\frac{1}{2}$	0	$2\frac{1}{2}$	R1 no change
z	$\frac{1}{2}$	0	1	$-\frac{3}{8}$	$\frac{1}{4}$	$1\frac{7}{8}$	R2 ÷ 4
P	6	0	0	3	2	45	R3 + 8R2

The optimal solution is

$$P = 45,\ x = 0,\ y = 2\tfrac{1}{2},\ z = 1\tfrac{7}{8},\ r = 0,\ s = 0$$

Example 11

a Use the simplex tableau method to solve the following linear programming problem.

Maximise $P = 3x + 4y - 5z$

subject to:

$$2x - 3y + 2z + r = 4$$
$$x + 2y + 4z + s = 8$$
$$y - z + t = 6$$
$$x, y, z, r, s, t \geqslant 0$$

b State the values of the objective function and every variable at the optimal point.

c Write down the equations given by your optimal tableau.

d Use the profit equation you wrote down in part **c** to explain why your final tableau is optimal.

a

Basic variable	x	y	z	r	s	t	Value	θ values
r	2	-3	2	1	0	0	4	$-\frac{4}{3}$
s	1	2	4	0	1	0	8	4
t	0	1	-1	0	0	1	6	6
P	-3	-4	5	0	0	0	0	

Note The most negative entry in the objective row lies in the y column, so this is the pivot column.

Use the smallest, **positive** θ value, so although $-\frac{4}{3}$ is the smallest, you can not use it as a pivot.

The smallest, positive θ value is 4, so row 2 will become the pivot row.

The pivot is the 2 in row 2 column y, and the basic variable will change to y.

A

Basic variable	x	y	z	r	s	t	Value	Row operations
r	2	−3	2	1	0	0	4	R1
y	$\frac{1}{2}$	1	2	0	$\frac{1}{2}$	0	4	R2 ÷ 2
t	0	1	−1	0	0	1	6	R3
P	−3	−4	5	0	0	0	0	R4

Now divide row 2 by the pivot, 2.

Basic variable	x	y	z	r	s	t	Value	Row operations
r	$\frac{7}{2}$	0	8	1	$\frac{3}{2}$	0	16	R1 + 3R2
y	$\frac{1}{2}$	1	2	0	$\frac{1}{2}$	0	4	
t	$-\frac{1}{2}$	0	−3	0	$-\frac{1}{2}$	1	2	R3 − R2
P	−1	0	13	0	2	0	16	R4 + 4R2

Now eliminate y from the other rows. Use the pivot row, R2 in the tableau above.
To eliminate the −3 from row 1 you need to add 3 copies of row 2, so R1 + 3R2.
To eliminate the 1 from row 3 you need to subtract row 2, so R3 − R2.
To eliminate the −4 from row 4 you need to add 4 copies of row 2, so R4 + 4R2.

Basic variable	x	y	z	r	s	t	Value	Row operations
r	$\frac{7}{2}$	0	8	1	$\frac{3}{2}$	0	16	$\frac{32}{7} = 4\frac{4}{7}$
y	$\frac{1}{2}$	1	2	0	$\frac{1}{2}$	0	4	8
t	$-\frac{1}{2}$	0	−3	0	$-\frac{1}{2}$	1	2	−4
P	−1	0	13	0	2	0	16	

The only negative entry in the objective row lies in the x column, so this is the pivot column.
Use the smallest, **positive** θ value, so although −4 is the smallest, you cannot use it as a pivot.
The smallest, positive θ value is $4\frac{4}{7}$, so row 1 will become the pivot row.
The pivot is the $\frac{7}{2}$ in row 1 column x.

Basic variable	x	y	z	r	s	t	Value	θ values
x	1	0	$\frac{16}{7}$	$\frac{2}{7}$	$\frac{3}{7}$	0	$\frac{32}{7}$	R1 ÷ $\frac{7}{2}$
y	$\frac{1}{2}$	1	2	0	$\frac{1}{2}$	0	4	
t	$-\frac{1}{2}$	0	−3	0	$-\frac{1}{2}$	1	2	
P	−1	0	13	0	2	0	16	

Now divide row 1 by the pivot, $\frac{7}{2}$.

Basic variable	x	y	z	r	s	t	Value	Row operations
x	1	0	$\frac{16}{7}$	$\frac{2}{7}$	$\frac{3}{7}$	0	$\frac{32}{7}$	
y	0	1	$\frac{6}{7}$	$-\frac{1}{7}$	$\frac{2}{7}$	0	$\frac{12}{7}$	R2 − $\frac{1}{2}$R1
t	0	0	$-\frac{13}{7}$	$\frac{1}{7}$	$-\frac{2}{7}$	1	$\frac{30}{7}$	R3 + $\frac{1}{2}$R1
P	0	0	$\frac{107}{7}$	$\frac{2}{7}$	$\frac{17}{7}$	0	$\frac{144}{7}$	R4 + R1

Eliminate x from the other rows. Use the pivot row, R1.
To eliminate the $\frac{1}{2}$ from row 2 you need to subtract $\frac{1}{2}$ of row 1, so R2 − $\frac{1}{2}$R1.
To eliminate the $-\frac{1}{2}$ from row 3 you need to add $\frac{1}{2}$ row 1, so R3 + $\frac{1}{2}$R1.
To eliminate the −1 from row 4 you need to add row 1, so R4 + R1.

There are no negatives in the objective row, so you have an optimal solution.

A

b $P = \frac{144}{7}$, $x = \frac{32}{7}$, $y = \frac{12}{7}$, $z = 0$, $r = 0$, $s = 0$, $t = \frac{30}{7}$

Basic variable	x	y	z	r	s	t	Value	Row operations
x	1	0	$\frac{16}{7}$	$\frac{2}{7}$	$\frac{3}{7}$	0	$\frac{32}{7}$	R1 ÷ $\frac{7}{2}$
y	0	1	$\frac{6}{7}$	$-\frac{1}{7}$	$\frac{2}{7}$	0	$\frac{12}{7}$	R2 $- \frac{1}{2}$R1
t	0	0	$-\frac{13}{7}$	$\frac{1}{7}$	$-\frac{2}{7}$	1	$\frac{30}{7}$	R3 $+ \frac{1}{2}$R1
P	0	0	$\frac{107}{7}$	$\frac{2}{7}$	$\frac{17}{7}$	0	$\frac{144}{7}$	R4 + R1

Using the first and last columns, we read off the values of P, x, y and t. All other variables are zero.

c $\quad x + \frac{16}{7}z + \frac{2}{7}r + \frac{3}{7}s \quad = \frac{32}{7}$

$\quad y + \frac{6}{7}z - \frac{1}{7}r + \frac{2}{7}s \quad = \frac{12}{7}$

$\quad -\frac{13}{7}z + \frac{1}{7}r - \frac{2}{7}s + t = \frac{30}{7}$

$P + \frac{107}{7}z + \frac{2}{7}r + \frac{17}{7}s \quad = \frac{144}{7}$

d If we rearrange the profit equation, making P the subject we get:

$\quad P = \frac{144}{7} - \frac{107}{7}z - \frac{2}{7}r - \frac{17}{7}s$

Increasing z or r or s would decrease the profit, so the solution is optimal.

Watch out If you have to write a profit equation you need to write '$P +$' and then the rest is read off the other coefficients from the tableau. Remember: the P 'pushes in at the front' of the profit equation.

Problem-solving

You need to rearrange the profit equation to make P the subject to show why your solution is optimal.

Exercise 7B

Solve the linear programming problems in questions **1** to **6** using the simplex tableau algorithm.

1 Maximise $\quad P = 5x + 6y + 4z$
subject to

$$x + 2y + r = 6$$
$$5x + 3y + 3z + s = 24$$
$$x, y, z, r, s \geq 0$$

2 Maximise $\quad P = 3x + 4y + 10z$
subject to

$$x + 2y + 2z + r = 100$$
$$x + 4z + s = 40$$
$$x, y, z, r, s \geq 0$$

3 Maximise $\quad P = 3x + 5y + 2z$
subject to

$$3x + 4y + 5z + r = 10$$
$$x + 3y + 10z + s = 5$$
$$x - 2y + t = 1$$
$$x, y, z, r, s, t \geq 0$$

A **4** Maximise $P = 4x_1 - 3x_2 + 2x_3 + 3x_4$
P subject to

$$x_1 + 4x_2 + 3x_3 + x_4 + r = 95$$
$$2x_1 + x_2 + 2x_3 + 3x_4 + s = 67$$
$$x_1 + 3x_2 + 2x_3 + 2x_4 + t = 75$$
$$3x_1 + 2x_2 + x_3 + 2x_4 + u = 72$$
$$x_1, x_2, x_3, x_4, r, s, t, u \geqslant 0$$

> **Hint** There are 4 decision variables and 4 slack variables. Your initial tableau will need rows for r, s, t and u, and columns for $x_1, x_2, x_3, x_4, r, s, t$ and u.

P **5** Minimise $P = 3x + 6y - 32z$
subject to

$$x + 6y + 24z + r = 672$$
$$3x + y + 24z + s = 336$$
$$x + 3y + 16z + t = 168$$
$$2x + 3y + 32z + u = 352$$
$$x, y, z, r, s, t, u \geqslant 0$$

P **6** Minimise $P = 2x - 2y + 3z$
subject to

$$4x + 2y + z + r = 2$$
$$2x + 4y + 2z + s = 8$$
$$3x + 3y + 4z + t = 4$$
$$x, y, z, r, s, t \geqslant 0$$

P **7** For each question **1** to **6**:

a verify, using the original equations and constraints, that your solution is feasible,

b write down the final set of equations given by your optimal tableau,

c use the profit equation, written in part **b**, to explain why your solution is optimal.

E **8** The following tableau was obtained as part of the solution of a maximising linear programming problem.

Basic variable	x	y	z	r	s	t	Value
z	2	$\frac{1}{2}$	1	$\frac{1}{2}$	0	0	$\frac{3}{2}$
s	-1	$\frac{3}{2}$	0	$-\frac{1}{2}$	1	0	$\frac{15}{2}$
t	-4	$\frac{5}{2}$	0	$-\frac{3}{2}$	0	1	$\frac{3}{2}$
P	3	4	0	1	0	0	3

a Write down the values of x, y and z as indicated by this tableau. **(2 marks)**

b Write down the profit equation from the tableau. **(2 marks)**

A 9 For a linear programming problem, in x, y and z, to maximise profit P, the tableau below is the
E   initial tableau.

Basic variable	x	y	z	r	s	t	Value
r	4	1	2	1	0	0	2
s	1	2	1	0	1	0	8
t	2	4	3	0	0	1	4
P	−1	2	−2	0	0	0	0

a Write down the profit equation represented by the initial tableau. **(1 mark)**

b Continue to solve the linear programming problem, indicating the pivot rows that you use and
making your row operations clear. **(9 marks)**

c State the final values of the objective function, and of each variable. **(3 marks)**

7.3 Problems requiring integer solutions

You can use the simplex tableau method to solve linear programming problems requiring integer
solutions. You need to consider integer-valued points near your optimal solution.

Example 12

Solve the linear programming problem from Example 8, given that integer solutions are required.

Maximise $P = 3x + 2y$

subject to:

$$5x + 7y \leqslant 70$$
$$10x + 3y \leqslant 60$$
$$x, y \geqslant 0$$

In Example 8 we found the following solution

$$P = 26, \qquad x = \frac{42}{11} = 3\frac{9}{11}, \qquad y = \frac{80}{11} = 7\frac{3}{11}$$

Test points around this optimal solution.

Point	$5x + 7y \leqslant 70$	$10x + 3y \leqslant 60$	In feasible region?	$P = 3x + 2y$
(3, 7)	15 + 49 ⩽ 70	30 + 21 ⩽ 60	Yes	9 + 14 = 23
(3, 8)	15 + 56 > 70		No	
(4, 7)	20 + 49 ⩽ 70	40 + 21 > 60	No	
(4, 8)	20 + 56 > 70		No	

So our best integer solution is

$$P = 23, x = 3, y = 7$$

Problem-solving

For each point, you need to test whether the point
lies in the feasible region, and evaluate P.

← Section 6.4

Example (13)

A

Solve the linear programming problem from Example 10, given that integer solutions are required.

Maximise $P = 10x + 12y + 8z$

subject to:

$$2x + 2y \leqslant 5$$
$$5x + 3y + 4z \leqslant 15$$
$$x, y, z \geqslant 0$$

In Example 10 we found the following solution

$$P = 45, \qquad x = 0, \qquad y = 2\tfrac{1}{2}, \qquad z = 1\tfrac{7}{8}$$

Test points around this optimal solution.

Point	$2x + 2y \leqslant 5$	$5x + 3y + 4z \leqslant 15$	In feasible region?	$P = 10x + 12y + 8z$
(0, 2, 1)	$0 + 4 < 5$	$0 + 6 + 4 < 15$	Yes	$0 + 24 + 8 = 32$
(0, 2, 2)	$0 + 4 < 5$	$0 + 6 + 8 < 15$	Yes	$0 + 24 + 16 = 40$
(0, 3, 1)	$0 + 6 > 5$		No	
(0, 3, 2)	$0 + 6 > 5$		No	

So our best integer solution is

$$P = 40, x = 0, y = 2, z = 2$$

- **When integer solutions are needed, test points around the optimal solution to find a set of points which fits the constraints and gives a maximum for the objective function.**

Exercise (7C)

P **1** Solve the following linear programming problems using the simplex tableau, given that integer solutions are required for x, y and z.

a Maximise $P = 3x + 2y$

subject to:

$$2x + 4y \leqslant 60$$
$$3x + 5y \leqslant 40$$
$$x, y \geqslant 0$$

b Maximise $P = 10x + 12y + 8z$

subject to:

$$x + 2z \leqslant 10$$
$$4x + 3y - 4z \leqslant 8$$
$$x, y, z \geqslant 0$$

2 A cake company produces cake boxes to sell to its customers. Three types of cake box are produced, Supreme, Dreamtime and Perfection. The cakes used are categorised as iced doughnuts, French whirls or treacle tarts.

Each Supreme box contains 2 iced doughnuts, 4 French whirls and 3 treacle tarts.

Each Dreamtime box contains 3 iced doughnuts, 2 French whirls and 4 treacle tarts.

Each Perfection box contains 1 iced doughnut, 3 French whirls and 2 treacle tarts.

The cake company can prepare at most 80 iced doughnuts, at most 140 French whirls and at most 96 treacle tarts each day.

The profits on Supreme, Dreamtime and Perfection boxes are £12, £20 and £16 respectively. The cake company wishes to maximise its profit.

Let x, y and z be the numbers of Supreme, Dreamtime and Perfection boxes respectively, produced each day.

a Formulate this situation as a linear programming problem, giving your constraints as inequalities. **(5 marks)**

b State the further restriction that applies to the values of x, y and z in this context. **(1 mark)**

The simplex algorithm is used to solve this problem. After one iteration, the tableau is:

Basic variable	x	y	z	r	s	t	Value
r	$-\frac{1}{4}$	0	$-\frac{1}{2}$	1	0	$-\frac{3}{4}$	8
s	$\frac{5}{2}$	0	2	0	1	$-\frac{1}{2}$	92
y	$\frac{3}{4}$	1	$\frac{1}{2}$	0	0	$\frac{1}{4}$	24
P	3	0	-6	0	0	5	480

c State, with a reason, which value will become the pivot for the second iteration. **(2 marks)**

d Complete one further iteration of the simplex algorithm, and explain why the resulting tableau gives an optimal solution. **(5 marks)**

e State the number of each type of box the company should prepare in order to maximise its profit. **(1 mark)**

Challenge

$P = 5x + 2y + 4z$

subject to
$$2x + y + kz \leqslant 10$$
$$x + 4y + z \leqslant 12$$
$$4x - 2y + 3z \leqslant 28$$

where k is a constant.

Given that an optimal solution is obtained after one iteration of the simplex algorithm, find the range of possible values of k.

7.4 Two-stage simplex method

A The simplex method described above only works when all the constraints, other than the non-negativity conditions, are ⩽ constraints. These can be converted into equations using slack variables. The method relies on starting at the origin and using the slack variables as **basic variables** to give a **basic feasible solution**. This solution is then improved upon by moving systematically to other vertices of the feasible region until the optimum solution is found.

Problems that include ⩾ constraints have no obvious basic feasible solution since the origin is not in the feasible region. These constraints also require different treatment to convert them into equations.

You could convert the inequality $x + 3y \geqslant 10$ into an equation by **subtracting** a variable, s. This would give you:

$$x + 3y - s = 10, \text{ where } s \geqslant 0$$

Here, s is referred to as a **surplus variable**. Notice that surplus variables are also subject to non-negativity conditions and must be subtracted to produce the equation. Surplus variables may be used to convert any ⩾ inequality into an equation. However, this doesn't solve the problem of using the simplex method with x and y as **non-basic variables**. This would require $x = y = 0$, giving $-s = 10$ which makes s negative. To avoid this problem, a new type of variable called an **artificial variable** is introduced.

$$x + 3y - s + a = 10, \text{ where } s, a \geqslant 0$$

This constraint can now be written in a simplex tableau using x, y and s as non-basic variables and a as a basic variable with value 10.

Example 14

Use surplus and artificial variables to write the following inequalities as equations.

$$3x - 2y \geqslant 7$$
$$4x + 5y - 3z \geqslant 9$$

$$3x - 2y - s_1 + a_1 = 7$$
$$4x + 5y - 3z - s_2 + a_2 = 9$$

Introduce an artificial variable for each surplus variable.

Regardless of the number of variables, each ⩾ inequality requires just one surplus variable and one artificial variable to convert it into an equation. Make sure you use different labels for the different surplus and artificial variables.

A ■ **The two-stage simplex method for problems that include ⩾ constraints:**

1 Use slack, surplus and artificial variables, as necessary, to write all the constraints as equations.

2 Define a new objective function to minimise the sum of all the artificial variables.

3 Use the simplex method to solve this problem.

4 If the minimum sum of the artificial values is 0 then the solution found is a basic feasible solution of the original problem, which is then the starting point for the second stage. Use the simplex method again to solve this problem.

5 If the minimum sum of the artificial variables is not 0 then the original problem has no feasible solution.

Notation This method is called the **two-stage simplex method** because you have two separate applications of the simplex algorithm. Steps **1** to **3** given here constitute the first stage.

If the original problem has a feasible solution then solving the first stage (steps **1** to **3** above), provides a basic feasible solution that you can use in the second stage. In this case, you will find that the new objective function has minimum value zero. This implies that all the artificial variables are zero and can be ignored for the second stage. If the new objective function is non-zero then this indicates that the original problem has no feasible solution and there is no need to proceed to the second stage.

Example 15

Maximise $P = 3x - 2y + z$

subject to:

$$x + y + 2z \leqslant 10$$
$$2x - 3y + z \geqslant 5$$
$$x + y \geqslant 8$$
$$x, y, z \geqslant 0$$

Converting the inequalities into equations:

$$x + y + 2z + s_1 = 10$$
$$2x - 3y + z - s_2 + a_1 = 5$$
$$x + y - s_3 + a_2 = 8$$

Let $I = -(a_1 + a_2)$

Substituting for a_1 and a_2 from the equations above gives the new objective as

maximise $I = -(13 - 3x + 2y - z + s_2 + s_3)$

rearranging

$$I - 3x + 2y - z + s_2 + s_3 = -13$$

s_1 is a slack variable, s_2 and s_3 are surplus variables and a_1 and a_2 are artificial variables.

In stage 1 you need to minimise $a_1 + a_2$

This is equivalent to maximising $-(a_1 + a_2)$

The artificial variables will always be basic variables in the initial tableau. I should be expressed in terms of non-basic variables so that the tableau is in the right form. The columns for a_1 and a_2 should each only contain zeros and a single 1.

A

The initial tableau is now given as:

Basic variable	x	y	z	s_1	s_2	s_3	a_1	a_2	Value
s_1	1	1	2	1	0	0	0	0	10
a_1	2	−3	1	0	−1	0	1	0	5
a_2	1	1	0	0	0	−1	0	1	8
P	−3	2	−1	0	0	0	0	0	0
I	−3	2	−1	0	1	1	0	0	−13

Basic variable	x	y	z	s_1	s_2	s_3	a_1	a_2	Value
s_1	1	1	2	1	0	0	0	0	10
x	1	$-\frac{3}{2}$	$\frac{1}{2}$	0	$-\frac{1}{2}$	0	$\frac{1}{2}$	0	$\frac{5}{2}$
a_2	1	1	0	0	0	−1	0	1	8
P	−3	2	−1	0	0	0	0	0	0
I	−3	2	−1	0	1	1	0	0	−13

Basic variable	x	y	z	s_1	s_2	s_3	a_1	a_2	Value	Row operations
s_1	0	$\frac{5}{2}$	$\frac{3}{2}$	1	$\frac{1}{2}$	0	$-\frac{1}{2}$	0	$\frac{15}{2}$	R1 − R2
x	1	$-\frac{3}{2}$	$\frac{1}{2}$	0	$-\frac{1}{2}$	0	$\frac{1}{2}$	0	$\frac{5}{2}$	R2 ÷ 2
a_2	0	$\frac{5}{2}$	$-\frac{1}{2}$	0	$\frac{1}{2}$	−1	$-\frac{1}{2}$	1	$\frac{11}{2}$	R3 − R2
P	0	$-\frac{5}{2}$	$\frac{1}{2}$	0	$-\frac{3}{2}$	0	$\frac{3}{2}$	0	$\frac{15}{2}$	R4 + 3R2
I	0	$-\frac{5}{2}$	$\frac{1}{2}$	0	$-\frac{1}{2}$	1	$\frac{3}{2}$	0	$-\frac{11}{2}$	R5 + 3R2

Basic variable	x	y	z	s_1	s_2	s_3	a_1	a_2	Value	Row operations
s_1	0	0	2	1	0	1	0	−1	2	R1 − $\frac{5}{2}$R3
x	1	0	$\frac{1}{5}$	0	$-\frac{1}{5}$	$-\frac{3}{5}$	$\frac{1}{5}$	$\frac{3}{5}$	$\frac{29}{5}$	R2 + $\frac{3}{2}$R3
y	0	1	$-\frac{1}{5}$	0	$\frac{1}{5}$	$-\frac{2}{5}$	$-\frac{1}{5}$	$\frac{2}{5}$	$\frac{11}{5}$	R3 × $\frac{2}{5}$
P	0	0	0	0	−1	−1	1	1	13	R4 + $\frac{5}{2}$R3
I	0	0	0	0	0	0	1	1	0	R5 + $\frac{5}{2}$R3

There are no negative values in the bottom row, so this tableau is optimal.

$I = 0$ so a basic feasible solution has been found for the original problem.

Problem-solving

Notice that the last two rows of the table contain the original objective function and the temporary objective function needed for stage 1. It is useful to put P in the tableau at this point so that it will automatically be in an appropriate form for stage 2 if a basic feasible solution is found.

The most negative number in the I row is −3, so the x column is the pivot column.

Calculating the θ values, the smallest is given by $5 \div 2$ so the pivot is 2.

x now replaces a_1 as a basic variable. Each value in the pivot row is divided by the pivot.

Row operations are now carried out so that the remaining cells in the x column contain only 0s.

The new pivot is in R3 of the y column. y replaces a_2 as a basic variable.

Watch out Make sure that the value of I is 0. If this is not the case, then the original problem has no feasible solution.

A

Basic variable	x	y	z	s_1	s_2	s_3	Value
s_1	0	0	2	1	0	1	2
x	1	0	$\frac{1}{5}$	0	$-\frac{1}{5}$	$-\frac{3}{5}$	$\frac{29}{5}$
y	0	1	$-\frac{1}{5}$	0	$\frac{1}{5}$	$-\frac{2}{5}$	$\frac{11}{5}$
P	0	0	0	0	-1	-1	13

The I row and the artificial variable columns can now be removed from the tableau.

The basic feasible solution for the second stage is:

$$x = \frac{29}{5},\ y = \frac{11}{5},\ s_1 = 2,\ z = s_2 = s_3 = 0$$

Continuing in the same way produces the following tableaux.

Basic variable	x	y	z	s_1	s_2	s_3	Value
s_1	0	0	2	1	0	1	2
x	1	1	0	0	0	-1	8
s_2	0	5	-1	0	1	-2	11
P	0	5	-1	0	0	-3	24

With each iteration, the value of P should increase until the optimal solution is found.

Basic variable	x	y	z	s_1	s_2	s_3	Value
s_3	0	0	2	1	0	1	2
x	1	1	2	1	0	0	10
s_2	0	5	3	2	1	0	15
P	0	5	5	3	0	0	30

There are no negative values in the bottom row, so the optimal solution has been found.

The maximum value of P is 30

This occurs when $x = 10$, $s_2 = 15$, $s_3 = 2$ and $y = z = s_1 = 0$

Example 16

Verify that the following linear programming problem has no feasible solution.

Maximise $P = 3x - 2y + z$

subject to:

$$x + y + 2z \leqslant 8$$
$$2x - 3y + z \geqslant 5$$
$$x + y \geqslant 10$$
$$x, y, z \geqslant 0$$

Converting the inequalities to equations:

$$x + y + 2z + s_1 = 8$$
$$2x - 3y + z - s_2 + a_1 = 5$$
$$x + y - s_3 + a_2 = 10$$

A

Changing the objective function to minimise $a_1 + a_2$.

$$a_1 + a_2 = 5 - 2x + 3y - z + s_2 + 10 - x - y + s_3$$
$$= 15 - 3x + 2y - z + s_2 + s_3$$

So minimising $a_1 + a_2$ is equivalent to maximising

$$I = -(15 - 3x + 2y - z + s_2 + s_3)$$

Rearranging gives $I - 3x + 2y - z + s_2 + s_3 = -15$

The first tableau becomes:

Basic variable	x	y	z	s_1	s_2	s_3	a_1	a_2	Value
s_1	1	1	2	1	0	0	0	0	8
a_1	2	-3	1	0	-1	0	1	0	5
a_2	1	1	0	0	0	-1	0	1	10
I	-3	2	-1	0	1	1	0	0	-15

The most negative value in the bottom row corresponds to the x column.

x now replaces a_1 as a basic variable.

The smallest θ value is given by $5 \div 2$ so the pivot is 2.

Carrying out row operations gives the second tableau.

Basic variable	x	y	z	s_1	s_2	s_3	a_1	a_2	Value	Row operations
s_1	0	$\frac{5}{2}$	$\frac{3}{2}$	1	$\frac{1}{2}$	0	$-\frac{1}{2}$	0	$\frac{11}{2}$	R1 − R2
x	1	$-\frac{3}{2}$	$\frac{1}{2}$	0	$-\frac{1}{2}$	0	$\frac{1}{2}$	0	$\frac{5}{2}$	R2 ÷ 2
a_2	0	$\frac{5}{2}$	$-\frac{1}{2}$	0	$\frac{1}{2}$	-1	$-\frac{1}{2}$	1	$\frac{15}{2}$	R3 − R2
I	0	$-\frac{5}{2}$	$\frac{1}{2}$	0	$-\frac{1}{2}$	1	$\frac{3}{2}$	0	$-\frac{15}{2}$	R4 + 3R2

The most negative value in the bottom row corresponds to the y column.

y now replaces s_1 as a basic variable.

The smallest θ value is given by $\frac{11}{2} \div \frac{5}{2}$ and the pivot is $\frac{5}{2}$

Carrying out row operations gives the third tableau.

Basic variable	x	y	z	s_1	s_2	s_3	a_1	a_2	Value	Row operations
y	0	1	$\frac{3}{5}$	$\frac{2}{5}$	$\frac{1}{5}$	0	$-\frac{1}{5}$	0	$\frac{11}{5}$	R1 × $\frac{2}{5}$
x	1	0	$\frac{7}{5}$	$\frac{3}{5}$	$-\frac{1}{5}$	0	$\frac{1}{5}$	0	$\frac{29}{5}$	R2 + $\frac{3}{2}$R1
a_2	0	0	-2	-1	0	-1	0	1	2	R3 − $\frac{5}{2}$R1
I	0	0	2	1	0	1	1	0	-2	R4 + $\frac{5}{2}$R1

There are no negative values in the bottom row of the main part of the tableau, so the maximum value of I is $-2 \neq 0$

This tells us that the original problem has no feasible solution.

The slack and artificial variables are basic variables. All other variables are zero. You are told in the question that there is no feasible solution, so there won't be a second stage. This means that you don't need to include a row for P.

This means there are no values of x, y and z that satisfy all of the initial constraints.

Exercise 7D

A
P

1 Use the two-stage simplex method to solve the following problems.

  a Maximise $P = 2x + y$

   subject to: $x - y \geqslant 11$
   $x + 3y \leqslant 15$
   $x, y \geqslant 0$

  b Minimise $C = x - 3y + z$

   subject to: $x + 2y \leqslant 12$
   $2x - y - z \geqslant 10$
   $x + y + z \geqslant 6$
   $x, y, z \geqslant 0$

  c Maximise $P = 3x - y + 2z$

   subject to: $5x + z \leqslant 16$
   $3x + y + z \leqslant 12$
   $x - y + 4z \geqslant 9$
   $x, y, z \geqslant 0$

E/P 2 Show that the following linear programming problem has no feasible solution.

Maximise $P = 2x_1 + x_2$

subject to: $5x_1 + 6x_2 \leqslant 75$
$3x_1 + 4x_2 \geqslant 52$
$x_1, x_2 \geqslant 0$ **(6 marks)**

E/P 3 **a** Define what is meant by:

   i a surplus variable

   ii an artificial variable.   **(2 marks)**

   The following tableau is obtained during the first stage of a two-stage simplex process to maximise P:

Basic variable	x	y	z	s_1	s_2	s_3	a_1	a_2	Value
s_1	0	0	2	1	0	1	0	-1	2
x	1	0	$\frac{1}{2}$	0	$-\frac{1}{2}$	$-\frac{3}{2}$	$\frac{1}{2}$	$\frac{3}{2}$	$\frac{29}{2}$
y	0	1	$-\frac{1}{2}$	0	$\frac{1}{2}$	-1	$-\frac{1}{2}$	1	$\frac{11}{2}$
P	0	0	0	0	-1	-2	1	1	13
I	0	0	0	0	0	0	1	1	0

  b Explain why this linear programming problem has a feasible solution.   **(1 mark)**

  c Complete the second stage of a two-stage simplex process to obtain an optimal solution to this problem.   **(8 marks)**

7.5 The Big-M method

A

The first stage of the two-stage simplex method involves minimising the sum of the artificial variables. If there is a feasible solution then this sum will be zero. The Big-M method involves a different but equivalent approach which, again, leads to the sum of the artificial variables being zero.

Hint Note that when the sum of the artificial variables is zero, each artificial variable must be zero since none of them can be negative.

In the Big-M method, M is used to represent an **arbitrarily large number**. This guarantees that a term such as $M - 100$ will be positive and a term such as $75 - 0.5M$ will be negative. You can compare terms involving M in order to select the pivot column.

Example 17

Solve the following linear programming problem using the Big-M method. Maximise $P = x - y + z$ subject to:

$$2x + y + z \leqslant 20$$
$$x - 2y - z \leqslant 7$$
$$x \geqslant 4$$
$$x, y, z \geqslant 0$$

$2x + y + z + s_1 = 20$
$x - 2y - z + s_2 = 7$
$x - s_3 + a_1 = 4$
$x, y, z, s_1, s_2, s_3, a_1 \geqslant 0$

$P = x - y + z - Ma_1$
$a_1 = 4 - x + s_3$

Start by writing the inequalities as equations using slack, surplus and artificial variables.

Now modify the objective function by subtracting the term Ma_1.

This gives
$P = x - y + z - M(4 - x + s_3)$
$\quad = (1 + M)x - y + z - Ms_3 - 4M$

Rearranging gives
$P - (1 + M)x + y - z + Ms_3 = -4M$

You can now write the initial tableau as

Problem-solving

Because M is large and positive, maximizing this modified objective function will now automatically push a_1 towards zero. This means you can essentially maximise P and minimise a_1 with one application of the algorithm.

Basic variable	x	y	z	s_1	s_2	s_3	a_1	Value
s_1	2	1	1	1	0	0	0	20
s_2	1	-2	-1	0	1	0	0	7
a_1	1	0	0	0	0	-1	1	4
P	$-(1 + M)$	1	-1	0	0	M	0	$-4M$

Notice that x, y and z are non-basic variables which means that the starting point is the origin. This is not a feasible solution of the original problem because the artificial variable, a_1, is non-zero. The Big-M method always starts from an infeasible solution. It may take several iterations for the solution to become feasible and may take a few more for it to become optimal.

A

The most negative value in the objective row is $-(1 + M)$ so the x column is the pivot column.

> Remember M is large and positive.

The smallest positive θ value is $4 \div 1$ so the pivot is the 1 in the a_1 row.

x now becomes a basic variable in place of a_1.

Using the pivot to make the other values in the pivot column zero gives the next tableau.

Basic variable	x	y	z	s_1	s_2	s_3	a_1	Value	Row operations
s_1	0	1	1	1	0	2	-2	12	R1 − 2R3
s_2	0	-2	-1	0	1	1	-1	3	R2 − R3
x	1	0	0	0	0	-1	1	4	R3
P	0	1	-1	0	0	-1	$1 + M$	4	R4 + $(1 + M)$R3

There is a choice of pivot column, since -1 appears in the z column and the s_3 column.

Choosing the s_3 column, the minimum **positive** θ value is given by $3 \div 1$ so the pivot is the 1 in the s_2 row.

s_3 now becomes a basic variable in place of s_2.

Using the pivot to make the other values in the pivot column zero gives the next tableau.

Basic variable	x	y	z	s_1	s_2	s_3	a_1	Value	Row operations
s_1	0	5	3	1	-2	0	0	6	R1 − 2R2
s_3	0	-2	-1	0	1	1	-1	3	R2
x	1	-2	-1	0	1	0	0	7	R3 + R2
P	0	-1	-2	0	1	0	M	7	R4 + R2

There are still some negative values in the objective row. The most negative is -2 so the z column becomes the pivot column.

The only positive θ value is $6 \div 3 = 2$ so 3 is the pivot and z becomes a basic variable in place of s_1.

Dividing the pivot row by 3 gives:

Basic variable	x	y	z	s_1	s_2	s_3	a_1	Value	Row operations
z	0	$\frac{5}{3}$	1	$\frac{1}{3}$	$-\frac{2}{3}$	0	0	2	$\frac{1}{3}$R1
s_3	0	-2	-1	0	1	1	-1	3	
x	1	-2	-1	0	1	0	0	7	
P	0	-1	-2	0	1	0	M	7	

A

Using the pivot to make the other values in the pivot column zero gives the next tableau.

Basic variable	x	y	z	s_1	s_2	s_3	a_1	Value	Row operations
z	0	$\frac{5}{3}$	1	$\frac{1}{3}$	$-\frac{2}{3}$	0	0	2	
s_3	0	$-\frac{1}{3}$	0	$\frac{1}{3}$	$\frac{1}{3}$	1	-1	5	R2 + R1
x	1	$-\frac{1}{3}$	0	$\frac{1}{3}$	$\frac{1}{3}$	0	0	9	R3 + R1
P	0	$\frac{7}{3}$	0	$\frac{2}{3}$	$-\frac{1}{3}$	0	M	11	R4 + 2R1

There is only one negative value in the objective row now, so s_2 becomes the pivot column and the $\frac{1}{3}$ in the s_3 row becomes the pivot.

Dividing the pivot row by $\frac{1}{3}$ gives the new pivot row.

Using row operations to put the pivot column in the right form gives the next tableau.

Basic variable	x	y	z	s_1	s_2	s_3	a_1	Value	Row operations
z	0	1	1	1	0	2	-2	12	R1 + $\frac{2}{3}$R2
s_2	0	-1	0	1	1	3	-3	15	
x	1	0	0	0	0	-1	1	4	R3 $-\frac{1}{3}$R2
P	0	2	0	1	0	1	$M-1$	16	R4 + $\frac{1}{3}$R2

All values in the objective row are now positive so the solution is optimal.

The solution is:

$$x = 4, \ y = 0, \ z = 12, \ s_1 = 0, \ s_2 = 15, \ s_3 = 0, \ a_1 = 0, \ P = 16$$

Checking against the original equations:

$$2x + y + z + s_1 = 8 + 0 + 12 + 0 = 20 \ \checkmark$$
$$x - 2y - z + s_2 = 4 - 0 - 12 + 15 = 7 \ \checkmark$$
$$x - s_3 + a_1 = 4 - 0 + 0 = 4 \ \checkmark$$

Notice that 16 is the largest value that has been found for P.

- **Linear programming problems that include $\geqslant$ constraints may also be solved using the Big-M method instead of the two-stage simplex method. The Big-M method uses an arbitrarily large number, M, in the objective function. Its purpose is to drive the artificial variables towards 0.**

A ■ **The Big-M method uses the following steps:**

1 Introduce a slack variable for each constraint of the form $\leqslant$.
2 Introduce a surplus variable and an artificial variable for each constraint of the form $\geqslant$.
3 For each artificial variable $a_1, a_2, a_3 \dots$ subtract $M(a_1 + a_2 + a_3 \dots)$ from the objective function, where M is an arbitrarily large number.
4 Eliminate the artificial variables from your objective function so that the variables remaining in your objective function are non-basic variables.
5 Formulate an initial tableau, and apply the simplex method in the normal way.

> **Hint** In the Big-M method, any artificial variables will always start as basic variables.

The Big-M method can also be used to minimise the value of the objective function.

Example 18

Use the Big-M method to solve this linear programming problem.
Minimise $P = x - y + z$
subject to:

$$2x + y + z \leqslant 20$$
$$x - 2y - z \leqslant 7$$
$$x \geqslant 4$$
$$x, y, z \geqslant 0$$

This example uses the same constraints as the previous example, but this time the objective is to **minimise** $P = x - y + z$.

$$2x + y + z + s_1 = 20$$
$$x - 2y - z + s_2 = 7$$
$$x - s_3 + a_1 = 4$$
$$x, y, z, s_1, s_2, s_3, a_1 \geqslant 0$$
$$-P = -x + y - z$$

The inequalities are converted to equations exactly as in the previous example.

The objective is to minimise P, but this is equivalent to maximising $-P$.

Using Big-M, you can minimise P by maximising Q where

$$Q = -x + y - z - Ma_1$$

Writing a_1 in terms of non-basic variables gives

$$Q = -x + y - z - M(4 - x + s_3)$$
$$Q = -x + y - z - 4M + Mx - Ms_3$$

Rearranging gives

$$Q + x - y + z - Mx + Ms_3 = -4M$$
$$Q + (1 - M)x - y + z + Ms_3 = -4M$$

This means your non-basic variables will be x, y, z and s_3.

You can now write the initial tableau as

Basic variable	x	y	z	s_1	s_2	s_3	a_1	Value
s_1	2	1	1	1	0	0	0	20
s_2	1	-2	-1	0	1	0	0	7
a_1	1	0	0	0	0	-1	1	4
Q	$1 - M$	-1	1	0	0	M	0	$-4M$

A

The most negative value in the objective row is $1 - M$ so the x column becomes the pivot column.

The smallest value of θ is given by $4 \div 1$ so the 1 in the a_1 row becomes the pivot.

x now replaces a_1 as a basic variable.

Using row operations to put the pivot column in the standard form gives the next tableau.

Basic variable	x	y	z	s_1	s_2	s_3	a_1	Value	Row operations
s_1	0	1	1	1	0	2	−2	12	R1 − 2R3
s_2	0	−2	−1	0	1	1	−1	3	R2 − R3
x	1	0	0	0	0	−1	1	4	
Q	0	−1	1	0	0	1	$M - 1$	−4	R4 − $(1 - M)$R3

The only negative value in the objective row is in the y column. The least positive θ value is given by $12 \div 1$ so the 1 in the s_1 row becomes the pivot.

y replaces s_1 as a basic variable.

Using row operations to put the pivot column in the standard form gives the next tableau.

Basic variable	x	y	z	s_1	s_2	s_3	a_1	Value	Row operations
y	0	1	1	1	0	2	−2	12	
s_2	0	0	1	2	1	5	−5	27	R2 + 2R1
x	1	0	0	0	0	−1	1	4	R3
Q	0	0	2	1	0	3	$M - 3$	8	R4 + R1

All of the values in the objective row are now positive so the solution is optimal.

> Max $Q = 8$
> Min $P = -8$
> $x = 4$, $y = 12$, $z = 0$, $s_1 = 0$, $s_2 = 27$, $s_3 = 0$, $a_1 = 0$

Checking against the original equations:

> $2x + y + z + s_1 = 8 + 12 + 0 + 0 = 20$ ✓
> $x - 2y - z + s_2 = 4 - 24 - 0 + 27 = 7$ ✓
> $x - s_3 + a_1 = 4 - 0 + 0 = 4$ ✓

A The next example demonstrates how the Big-M method can be used to solve the same problem that was tackled in Example 15 using the two-stage simplex method.

Example 19

a Set up the initial tableau for solving the following linear programming problem using the Big-M method:

Maximise $P = 3x - 2y + z$

subject to:

$$x + y + 2z \leqslant 10$$
$$2x - 3y + z \geqslant 5$$
$$x + y \geqslant 8$$
$$x, y, z \geqslant 0$$

b Find the value of the first pivot.

c Carry out one iteration and state the value of each variable at that point.

d Explain why the solution given by the first iteration is not feasible.

a Converting the inequalities into equations:

$$x + y + 2z + s_1 = 10$$
$$2x - 3y + z - s_2 + a_1 = 5$$
$$x + y - s_3 + a_2 = 8$$

The objective function is now modified to

$$P = 3x - 2y + z - M(a_1 + a_2)$$
$$a_1 = 5 - 2x + 3y - z + s_2$$

and

$$a_2 = 8 - x - y + s_3$$

adding gives

$$a_1 + a_2 = 13 - 3x + 2y - z + s_2 + s_3$$

using this expression

$$P = 3x - 2y + z - M(13 - 3x + 2y - z + s_2 + s_3)$$
$$P = (3 + 3M)x - (2 + 2M)y + (1 + M)z - Ms_2 - Ms_3 - 13M$$
$$P = 3(1 + M)x - 2(1 + M)y + (1 + M)z - Ms_2 - Ms_3 - 13M$$

Rearranging:

$$P - 3(1 + M)x + 2(1 + M)y - (1 + M)z + Ms_2 + Ms_3 = -13M$$

The first tableau may now be written as:

Basic variable	x	y	z	s_1	s_2	s_3	a_1	a_2	Value
s_1	1	1	2	1	0	0	0	0	10
a_1	2	-3	1	0	-1	0	1	0	5
a_2	1	1	0	0	0	-1	0	1	8
P	$-3(1 + M)$	$2(1 + M)$	$-(1 + M)$	0	M	M	0	0	$-13M$

Problem-solving

When there is more than one artificial variable, you need to multiply M by the sum of all the artificial variables. Since you are trying to maximise P, and M is very large, this has the effect of minimising $a_1 + a_2$ just as in the first stage of the simplex method. You need to rewrite $a_1 + a_2$ in terms of the non-basic variables so that the tableau is in the standard form.

Rearrange to give each coefficient in terms of M.

A

b The most negative value in the objective row is $-3(1 + M)$ in the x column.

Comparing the θ ratios, the smallest is $5 \div 2$, so the pivot is 2.

Basic variable	x	y	z	s_1	s_2	s_3	a_1	a_2	Value
s_1	1	1	2	1	0	0	0	0	10
a_1	2	-3	1	0	-1	0	1	0	5
a_2	1	1	0	0	0	-1	0	1	8
P	$-3(1+M)$	$2(1+M)$	$-(1+M)$	0	M	M	0	0	$-13M$

c Dividing the pivot row by 2 and replacing the basic variable a_1 with x gives:

Basic variable	x	y	z	s_1	s_2	s_3	a_1	a_2	Value	Row operations
s_1	1	1	2	1	0	0	0	0	10	
x	1	$-\frac{3}{2}$	$\frac{1}{2}$	0	$-\frac{1}{2}$	0	$\frac{1}{2}$	0	$\frac{5}{2}$	$\frac{1}{2}$R2
a_2	1	1	0	0	0	-1	0	1	8	
P	$-3(1+M)$	$2(1+M)$	$-(1+M)$	0	M	M	0	0	$-13M$	

Using row operations to put column x in standard form gives:

Basic variable	x	y	z	s_1	s_2	s_3	a_1	a_2	Value	Row operations
s_1	0	$\frac{5}{2}$	$\frac{3}{2}$	1	$\frac{1}{2}$	0	$-\frac{1}{2}$	0	$\frac{15}{2}$	R1 − R2
x	1	$-\frac{3}{2}$	$\frac{1}{2}$	0	$-\frac{1}{2}$	0	$\frac{1}{2}$	0	$\frac{5}{2}$	
a_2	0	$\frac{5}{2}$	$-\frac{1}{2}$	0	$\frac{1}{2}$	-1	$-\frac{1}{2}$	1	$\frac{11}{2}$	R3 − R2
P	0	$-\frac{5}{2}(1+M)$	$\frac{1}{2}(1+M)$	0	$-\frac{1}{2}(3+M)$	M	$\frac{3}{2}(1+M)$	0	$\frac{1}{2}(15-11M)$	R4 + $3(1+M)$R2

The pivot column is now in standard form and this completes the first iteration.

From the table:

$$x = \tfrac{5}{2},\ y = 0,\ z = 0,\ s_1 = \tfrac{15}{2},\ s_2 = 0,\ s_3 = 0,\ a_1 = 0,\ a_2 = \tfrac{11}{2}$$

The values of the basic variables are shown in the right-hand column. All other variables have value zero.

d The solution given by this tableau is not feasible because $a_2 = \frac{11}{2}$ is an artificial variable which must be zero in a feasible solution.

Exercise 7E

E/P 1 Here is the initial tableau for solving a linear programming problem using the Big-M method.

Basic variable	x	y	z	s_1	s_2	a_1	Value	Row operations
s_1	1	1	2	1	0	0	6	
a_1	2	-3	1	0	-1	1	5	
P	$-(1+M)$	$2(1+M)$	$-2(1+M)$	0	$-M$	0	$-15M$	

A

  a Explain why the tableau does not represent a feasible solution. **(1 mark)**

  b Locate the pivot and explain how you made your choice. **(2 marks)**

  c Explain without doing any calculations how you know that the solution given by the next iteration will not be feasible. **(2 marks)**

  d Carry out one iteration, showing your row operations. **(4 marks)**

E/P **2** The following linear programming problem is to be solved using the Big-M method.

Maximise $P = 4x + 2y - z$

subject to: $x + 3y + z \leqslant 100$

              $3x - y \leqslant 52$

        $x \geqslant 20, \ y, \ z \geqslant 0$

  a Write the constraints as equations using slack, surplus and artificial variables as appropriate. **(3 marks)**

  b Write the initial tableau in standard form. **(3 marks)**

  c Explain what M represents. **(1 mark)**

  d Identify the first pivot. **(2 marks)**

E/P **3** Here is a linear programming problem to be solved using the Big-M method.

Minimise $C = 4x + 3y$

subject to: $3x - y \leqslant 110$

            $x + 2y \geqslant 45$

            $x, \ y \geqslant 0$

  a Explain why the problem cannot be solved by the standard simplex method. **(1 mark)**

  b Express the inequalities as equations making use of slack, surplus and artificial variables. **(3 marks)**

  c Set up the initial tableau in standard form. **(5 marks)**

E/P **4** The Big-M method is to be used to solve the following linear programming problem.

Maximise $P = 3x + 5y - z$

subject to: $x + y + z \leqslant 20$

           $3x + y + 2z \geqslant 24$

           $x, \ y, \ z \geqslant 0$

  a Rewrite the constraints as equations making use of slack, surplus and artificial variables. **(2 marks)**

  b Modify the objective in preparation for using the Big-M method, using non-basic variables. **(3 marks)**

  c Represent the problem in an initial tableau in standard form. **(2 marks)**

  d Carry out iterations to solve the problem. State the value of each variable in your solution. **(8 marks)**

Mixed exercise 7

A
E/P

1 In a particular factory 3 types of product, A, B and C, are made. The number of each of the products made is x, y and z respectively and P is the profit in pounds. There are two machines involved in making the products which have only a limited time available. These time limitations produce two constraints.

In the process of using the simplex algorithm the following tableau is obtained, where r and s are slack variables.

Basic variable	x	y	z	r	s	Value
z	$\frac{1}{3}$	0	1	-8	1	75
y	$\frac{2}{11}$	1	0	$\frac{17}{11}$	0	56
P	$\frac{3}{2}$	0	0	$\frac{3}{4}$	0	840

a Give one reason why this tableau can be seen to be optimal (final). **(1 mark)**

b By writing out the profit equation, or otherwise, explain why a further increase in profit is not possible under these constraints. **(2 marks)**

c From this tableau deduce
  i the maximum profit,
  ii the optimum number of type A, B and C that should be produced to maximise the profit.

E/P 2 A sweet manufacturer produces packets of orange and lemon flavoured sweets.
The manufacturer can produce up to 25 000 orange sweets and up to 36 000 lemon sweets per day.

  Small packets contain 5 orange and 5 lemon sweets.
  Medium packets contain 8 orange and 6 lemon sweets.
  Large packets contain 10 orange and 15 lemon sweets.

The manufacturer makes a profit of 14p, 20p and 30p on each of the small, medium and large packets respectively. He wishes to maximise his total daily profit.

Use x, y and z to represent the number of small, medium and large packets respectively, produced each day.

a Formulate this information as a linear programming problem, making your objective function and constraints clear. Change any inequalities to equations using r and s as slack variables.
  (5 marks)

The tableau below is obtained after one complete iteration of the simplex algorithm.

Basic variable	x	y	z	r	s	Value
r	$1\frac{2}{3}$	4	0	1	$-\frac{2}{3}$	1000
z	$\frac{1}{3}$	$\frac{2}{5}$	1	0	$\frac{1}{15}$	2400
P	-4	-8	0	0	2	72 000

b Start from this tableau and continue the simplex algorithm by increasing y, until you have either completed two complete iterations or found an optimal solution. **(4 marks)**

A From your final tableau

   c **i** write down the numbers of small, medium and large packets indicated.

     ii write down the profit.

     iii state whether this is an optimal solution, giving your reason. **(3 marks)**

E/P **3** Tables are to be bought for a new restaurant. The owners may buy small, medium and large tables that seat 2, 4 and 6 people respectively.

The owners require at most 20% of the total number of tables to be medium sized.
The tables cost £60, £100 and £160 respectively for small, medium and large. The owners have a budget of £2000 for buying tables.

Let the number of small, medium and large tables be x, y and z respectively.

a Write down 5 inequalities implied by the constraints. Simplify these where appropriate.

                                                              (5 marks)

The owners wish to maximise the total seating capacity, S, of the restaurant.

b Write down the objective function for S in terms of x, y and z. **(2 marks)**

c Explain why it is not appropriate to use a graphical method to solve this problem. **(1 mark)**

It is decided to use the simplex algorithm to solve this problem.

d Show that a possible initial tableau is

Basic variable	x	y	z	r	t	Value
r	−1	4	−1	1	0	0
t	3	5	8	0	1	100
S	−2	−4	−6	0	0	0

                                                             (3 marks)

It is decided to increase z first.

e Show that, after one complete iteration, the tableau becomes

Basic variable	x	y	z	r	t	Value
r	$-\frac{5}{8}$	$\frac{37}{8}$	0	1	$\frac{1}{8}$	$\frac{25}{2}$
z	$\frac{3}{8}$	$\frac{5}{8}$	1	0	$\frac{1}{8}$	$\frac{25}{2}$
S	$\frac{1}{4}$	$-\frac{1}{4}$	0	0	$\frac{3}{4}$	75

                                                             (3 marks)

f Perform one further complete iteration. **(3 marks)**

g Explain how you can decide if your tableau is now final. **(1 mark)**

h Find the number of each type of table the restaurant should buy and their total cost. **(2 marks)**

E/P **4** Kuddly Pals Co. Ltd. make two types of soft toy: bears and cats. The quantity of material needed and the time taken to make each type of toy is given in the table.

Toy	Material (m²)	Time (minutes)
Bear	0.05	12
Cat	0.08	8

Each day the company can process up to 20 m² of material and there are 48 worker hours available to assemble the toys.

A Let x be the number of bears made and y the number of cats made each day.

a Show that this situation can be modelled by the inequalities

$$5x + 8y \leqslant 2000$$
$$3x + 2y \leqslant 720$$

in addition to $x \geqslant 0$, $y \geqslant 0$. **(2 marks)**

The profit made on each bear is £1.50 and on each cat £1.75. Kuddly Pals Co. Ltd. wishes to maximise its daily profit.

b Set up an initial simplex tableau for this problem. **(3 marks)**

c Solve the problem using the simplex algorithm. **(3 marks)**

The diagram shows a graphical representation of the feasible region.

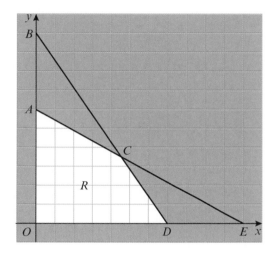

d Relate each stage of the simplex tableau to the corresponding point in the diagram. **(2 marks)**

E/P **5** A clocksmith makes three types of luxury wristwatch. The mechanism for each watch is assembled by hand by a skilled watchmaker and then the complete watch is formed, weatherproofed and packaged for sale by a fitter.

The table shows the times, in minutes, for each stage of the process.

Watch type	Watchmaker	Fitter
A	54	60
B	72	36
C	36	48

The watchmaker works for a maximum of 30 hours per week and the fitter for a maximum of 25 hours per week.

Let the number of type A, B and C watches made per week be x, y and z.

a Show that the above information leads to the two inequalities

$$3x + 4y + 2z \leqslant 100$$
$$5x + 3y + 4z \leqslant 125$$

(2 marks)

A The profit made on type A, B and C watches is £12, £24 and £20 respectively.

  b Write down an expression for the profit, P, in pounds, in terms of x, y and z. **(2 marks)**

The clocksmith wishes to maximise his weekly profit. It is decided to use the simplex algorithm to solve this problem.

  c Write down the initial tableau using r and s as the slack variables. **(3 marks)**

  d Increasing y first, show that after two complete iterations of the simplex algorithm the tableau becomes

Basic variable	x	y	z	r	s	Value
y	$\frac{1}{5}$	1	0	$\frac{2}{5}$	$-\frac{1}{5}$	15
z	$\frac{11}{10}$	0	1	$-\frac{3}{10}$	$\frac{2}{5}$	20
P	$\frac{74}{5}$	0	0	$\frac{18}{5}$	$\frac{16}{5}$	760

(5 marks)

  e Give a reason why this tableau is optimal (final). **(1 mark)**

  f Write down the numbers of each type of watch that should be made to maximise the profit. State the maximum profit. **(2 marks)**

(E/P) 6 A craftworker makes three types of wooden animals for sale in wildlife parks. Each animal has to be carved and then sanded.

  Each Lion takes   2 hours to carve and 25 minutes to sand.

  Each Giraffe takes $2\frac{1}{2}$ hours to carve and 20 minutes to sand.

  Each Elephant takes $1\frac{1}{2}$ hours to carve and 30 minutes to sand.

Each day the craftworker wishes to spend at most 8 hours carving and at most 2 hours sanding.

Let x be the number of Lions, y the number of Giraffes and z the number of Elephants he produces each day.

The craftworker makes a profit of £14 on each Lion, £12 on each Giraffe and £13 on each Elephant. He wishes to maximise his profit, P.

  a Model this as a linear programming problem, simplifying your expressions so that they have integer coefficients. **(4 marks)**

It is decided to use the simplex algorithm to solve this problem.

  b Explaining the purpose of r and s, show that the initial tableau can be written as:

Basic variable	x	y	z	r	s	Value
r	4	5	3	1	0	16
t	5	4	6	0	1	24
P	−14	−12	−13	0	0	0

(3 marks)

  c Increasing x first, work out the next complete tableau, where the x column includes two zeros. **(2 marks)**

  d Explain what this first iteration means in practical terms. **(2 marks)**

A 7 Here is a linear programming problem that is to be solved using the two-stage simplex method.

E

Maximise $P = x + 3y$

subject to: $3x + 2y \leqslant 15$

$2x + 5y \leqslant 20$

$y \geqslant 2$

$x \geqslant 0$

a Express the constraints as equations using slack, surplus and artificial variables. **(2 marks)**

b Write a new objective function for the first stage in terms of non-basic variables. **(3 marks)**

c Write the initial tableau for the first stage in standard form. **(3 marks)**

d Complete the first stage and explain what the solution represents. **(4 marks)**

e Set up the tableau for the second stage. **(2 marks)**

P 8 The following tableau was obtained after a number of iterations of the Big-M method.

Basic variable	x	y	z	s_1	s_2	s_3	a_1	Value
s_1	2	0	1	1	2	0	0	150
y	1	1	−1	0	1	0	0	180
a_1	1	0	0	0	0	1	1	70
P	−1	0	1	0	M	$M + 5$	$2M$	230

a State the value of the pivot, making your reasons clear.

b Complete the solution, justifying that it is optimal. You should clearly state your row operations.

E/P 9 This is the first tableau for a linear programming problem to be solved using the Big-M method.

Basic variable	x	y	z	s_1	s_2	s_3	a_1	Value	Row operations
s_1	1	1	4	1	0	0	0	4	
s_2	2	3	1	0	1	0	0	5	
a_1	2	−1	0	0	0	−1	1	8	
P	$-(1 + 2M)$	$(1 + M)$	−4	0	0	M	0	$-8M$	

a Explain why this tableau does not represent a feasible solution. **(2 marks)**

b State the value of the next pivot, making your reasons clear. **(2 marks)**

c Carry out one further iteration of the simplex algorithm, showing your row operations. **(4 marks)**

E/P 10 A bakery has the capacity to produce 800 pies a day. The bakery makes three types of pies, A, B, and C.

Pie A will return a profit of £1.00 per pie. Pie B will return a profit of £0.80 per pie. Pie C will return a profit of £0.60 per pie.

In order to satisfy demand, the factory must produce at least 200 of pie C.

The factory manager formulates this problem as:

Let x, y, z represent the numbers of pies A, B, C made respectively.

Maximise $P = 100x + 80y + 60z$

subject to: $\quad x + y + z \leqslant 800$

$$z \geqslant 200$$

$$x, y, z \geqslant 0$$

a Explain why the basic simplex algorithm cannot be used to solve the linear programming problem. **(1 mark)**

b Without working, state the optimal solution to this linear programming problem. **(1 mark)**

Based on available ingredients and on customer demand, the factory manager also implements the following constraints

$$2x + 2y + z \leqslant 1200$$

$$4y + 5z \geqslant 1000$$

The Big-M method is to be used to solve this linear programming problem.

c Define what M represents in this context. You must use correct mathematical language in your answer. **(1 mark)**

d Set up the initial tableau for this problem. Show your modified objective function and clearly state the basic variables. **(4 marks)**

e State, with a reason, which value you would use as your first pivot, and explain how you would carry out one iteration of the algorithm. You do not need to apply your method. **(3 marks)**

Challenge

A linear programming problem is given as follows:

Maximise $P = 3x + 4y + 2z$

subject to: $\quad 2x + 4y \leqslant 10$

$$x + 2y + z = 12$$

$$4y + 2z \geqslant 15$$

$$x, y, z \geqslant 0$$

Solve this linear programming problem using an appropriate method.

Hint You will need to think carefully about how you deal with the second constraint, which contains an equality.

Summary of key points

A

1 To formulate a **linear programming** problem:
   - define your decision variables
   - write down the objective function
   - write down the constraints.

2 Inequalities can be transformed into **equations** using **slack variables** (so called because they represent the amount of slack between an actual quantity and the maximum possible value of that quantity).

3 The **simplex method** allows you to:
   - determine if a particular vertex, on the edge of the feasible region, is optimal
   - decide which adjacent vertex you should move to in order to increase the value of the objective function.

4 **Slack variables** are essential when using the simplex algorithm.

5 The simplex method always starts from a basic feasible solution, at the origin, and then progresses with each iteration to an adjacent point within the feasible region until the optimal solution is found.

6 To use a **simplex tableau** to solve a maximising linear programming problem, where the constraints are given as equalities:
   - Draw the tableaux: you need a basic variable column on the left, one column for each variable (including the slack variables) and a value column. You need one row for each constraint and the bottom row for the objective function.
   - Create the initial tableau: enter the coefficients of the variables in the appropriate column and row.
   - Look along the objective row for the most negative entry: this indicates the pivot column.
   - Calculate θ, for each of the constraint rows, where

$$\theta = \frac{\text{the term in the value column}}{\text{the term in the pivot column}}$$

   - Select the row with the smallest, positive θ value to become the pivot row.
   - The element in the pivot row and pivot column is the pivot.
   - Divide all of the elements in the pivot row by the pivot, and change the basic variable at the start of the row to the variable at the top of the pivot column.
   - Use the pivot row to eliminate the pivot's variable from the other rows: this means that the pivot column now contains one 1 and zeros.
   - Repeat bullets 3 to 8 until there are no negative numbers in the objective row.
   - The tableau is now optimal and the non-zero values can be read off using the basic variable column and value column.

A

7   The steps for solving a minimising linear programming problem are identical to those given above apart from:

   - First, define a new objective function that is the negative of the original objective function.
   - After you have maximised this new objective function, write your solution as the negative of this value, which will minimise the original objective function.

8   When integer solutions are needed, test points around the optimal solution to find a set of points which fit the constraints and give a maximum for the objective function.

9   The **two-stage simplex method** for problems that include $\geqslant$ constraints:

   - Use **slack**, **surplus** and **artificial** variables, as necessary, to write all the constraints as equations.
   - Define a new objective function to minimise the sum of all the artificial variables.
   - Use the simplex method to solve this problem.
   - If the minimum sum of the artificial values is 0 then the solution found is a basic feasible solution of the original problem, which is then the starting point for the second stage. Use the simplex method again to solve this problem.
   - If the minimum sum of the artificial variables is not 0 then the original problem has no feasible solution.

10 Linear programming problems that include $\geqslant$ may also be solved using the **Big-M method** instead of the two-stage simplex method. The Big-M method uses an arbitrarily large number, M, in the objective function. Its purpose is to drive the artificial variables towards 0.

11 The **Big-M method** uses the following steps:

   - Introduce a slack variable for each constraint of the form $\leqslant$.
   - Introduce a surplus variable and an artificial variable for each constraint of the form $\geqslant$.
   - For each artificial variable $a_1, a_2, a_3 \ldots$ subtract $M(a_1 + a_2 + a_3 \ldots)$ from the objective function, where M is an arbitrarily large number.
   - Eliminate the artificial variables from your objective function so that the variables remaining in your objective function are non-basic variables.
   - Formulate an initial tableau, and apply the simplex method in the normal way.

Critical path analysis

Objectives

After completing this chapter you should be able to:

* Model a project by an activity network using a precedence table
  → **pages 222–226**

* Use dummy activities
  → **pages 226–229**

* Identify and calculate early and late event times in activity networks
  → **pages 230–232**

* Identify critical activities
  → **pages 232–235**

* Calculate the total float of an activity
  → **pages 236–237**

* Calculate and use Gantt (cascade) charts
  → **pages 238–241**

* Construct resource histograms
  → **pages 242–249**

* Construct scheduling diagrams
  → **pages 249–253**

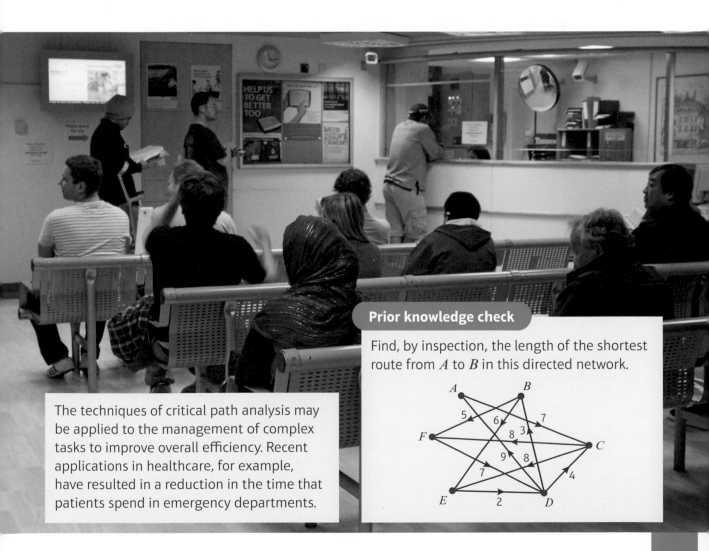

The techniques of critical path analysis may be applied to the management of complex tasks to improve overall efficiency. Recent applications in healthcare, for example, have resulted in a reduction in the time that patients spend in emergency departments.

Prior knowledge check

Find, by inspection, the length of the shortest route from A to B in this directed network.

8.1 Modelling a project

Imagine that you have responsibility for the completion of a complex project. The total amount of work to be done is divided into separate activities and some of these cannot be started until others have been completed. You are expected to organise the activities efficiently in order to avoid any unnecessary delay.

In order to plan the project effectively you would need to represent the activities in some way that makes any dependencies clear. It would be helpful to make use of some suitable notation and to apply a systematic approach.

> **Notation** When one activity cannot proceed until another activity is completed, it is referred to as a **dependency**.

- **A precedence table, or dependence table, is a table which shows which activities must be completed before others are started.**

Example 1

The manufacture of a sofa involves the construction of a wooden frame which is then fitted with springs. The frame is then covered with padding and fabric. Cushions are cut out of the same fabric which must then be stitched and filled. The assembly of the sofa is completed by attaching the cushions. It is then inspected before being wrapped in a protective covering ready for shipping.

Represent the information above in a systematic way that makes any dependencies clear.

This process may be broken down into separate activities.

Typically, these are labelled A, B, C, D, ... for ease of reference later.

A Build wooden frame
B Cut out fabric for cushions
C Stitch and fill cushions
D Attach springs to wooden frame
E Cover frame
F Complete assembly
G Inspect
H Wrap

The activities may now be written in a precedence table.

Activity	Depends on
A	–
B	–
C	B
D	A
E	D
F	C, E
G	F
H	G

E depends on D. The fact that E depends on A (because D depends on A) is clear from the entry above and is not shown again, i.e. only immediate dependence is shown in the table.

Both activities C and E must be completed before work on activity F may be started.

Example 2

The production of a new textbook may be broken down into activities A to G.

Activities A and B do not depend on any other activities.

Both activities C and D can only be started once A has been completed.

Activity E cannot be started until activity B has been completed and activity F cannot be started until activities C and E have been completed. Activity G can only begin once all other activities have been completed.

Draw a precedence table to represent this information.

Activity	Depends on
A	–
B	–
C	A
D	A
E	B
F	C, E
G	D, F

This appears to be the tricky part to complete. It's a lot easier when you realise that you must include just those activities that are not already written in this column.

The production of a precedence table goes some way towards representing a project in a form that can help you to coordinate the activities effectively. However, a diagram may be a lot easier to understand, especially if the project is more complex.

The information provided in a precedence table may be transferred to an **activity network** to give a visual representation of the project.

Links You have already seen how network diagrams may be used to represent and help analyse a variety of problem types. ← **Chapter 3**

There are two types of activity network but only the **activity on arc** type will be used here.

- In an activity on arc network, the activities are represented by arcs and the completion of those activities, known as events, are shown as nodes.
  - Each arc is labelled with an activity letter. The beginning and end of an activity are shown at the ends of the arc and an arrow is used to define the direction. The convention is to use straight lines for arcs.
  - The nodes are numbered starting from 0 for the first node which is called the **source node**.

    Watch out Sometimes the source node is labelled 1 instead of 0.

  - Number each node as it is added to the network.
  - The final node is called the **sink node**.

Example 3

Draw an activity network for the precedence table given in Example 2.

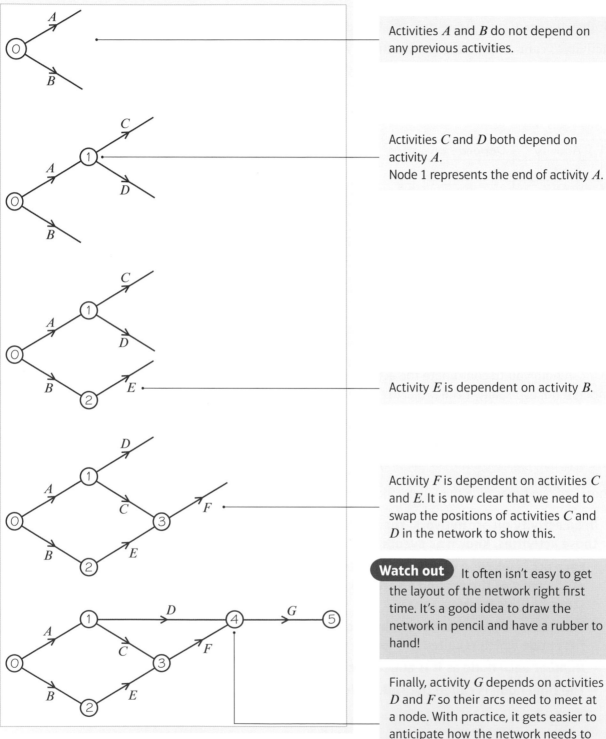

Activities A and B do not depend on any previous activities.

Activities C and D both depend on activity A.
Node 1 represents the end of activity A.

Activity E is dependent on activity B.

Activity F is dependent on activities C and E. It is now clear that we need to swap the positions of activities C and D in the network to show this.

Watch out It often isn't easy to get the layout of the network right first time. It's a good idea to draw the network in pencil and have a rubber to hand!

Finally, activity G depends on activities D and F so their arcs need to meet at a node. With practice, it gets easier to anticipate how the network needs to be set out and so there is less need for adjustment.

Exercise 8A

1 The steps involved in starting a car and moving forwards in a straight line are given below.

  A Check that car is in neutral.

  B Start engine.

  C Depress clutch.

  D Select first gear.

  E Check that it is safe to move off.

  F Release the handbrake.

  G Raise the clutch and depress the accelerator.

  Draw a precedence table for this process.

Hint There is more than one possible solution.

2 The development of a commercial computer program is divided into activities *A* to *J*.

  Activity *A* does not depend on any other activity.

  Activities *B*, *C* and *D* all require that Activity *A* is completed before they can start.

  Activities *E* and *F* depend on activity *B*.

  Activity *G* cannot be started until activities *C* and *E* have been completed.

  Activity *H* requires the completion of activity *D*, while activity *I* requires that both activities *F* and *G* are completed first.

  Activity *J* requires the completion of all activities before it may be started.

  a Draw a precedence table to represent the development of the computer program.

  b Use the precedence table to draw the corresponding activity network.

Ⓔ 3 The precedence table for a project is shown below.

  Draw the corresponding activity network.

Activity	Depends on
A	–
B	–
C	*A*
D	*A*
E	*B*
F	*B*
G	*D*
H	*D*
I	*C, E*
J	*F*
K	*G, I, J*
L	*H, K*

(3 marks)

E **4** Here is an activity network for a project.

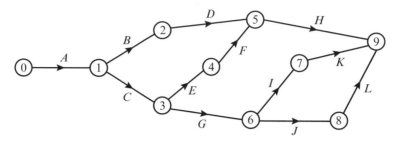

Draw a precedence table to represent the project. **(3 marks)**

8.2 Dummy activities

You need to understand the use of dummy activities.

The precedence table given below appears to be very simple and yet the corresponding activity network cannot be completed using the methods described so far.

Activity	Depends on
A	–
B	–
C	A, B
D	A

Activities A and B do not depend on any other activities and so they are linked to the source node.

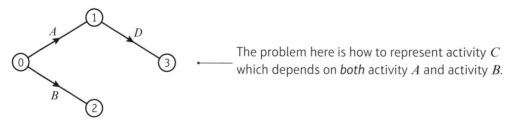

The problem here is how to represent activity C which depends on *both* activity A and activity B.

To resolve this problem you introduce a **dummy activity** between events 1 and 2 to show that activity C depends on activity A as well as activity B.

- **A dummy activity has no time or cost. Its sole purpose is to show dependencies between activities.**

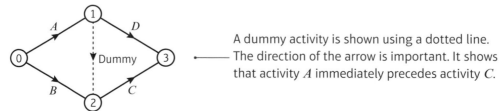

A dummy activity is shown using a dotted line. The direction of the arrow is important. It shows that activity A immediately precedes activity C.

Example 4

Draw an activity network for this precedence table. Use exactly two dummies.

Activity	Immediately preceding activities
A	–
B	A
C	A
D	A
E	B
F	B, C
G	D, F
H	D
I	G, H

The label for this column may be written in different ways.

Activity *E* depends on activity *B* only, but activity *F* depends on **both** activity *B* and activity *C*. This indicates the need for a dummy.

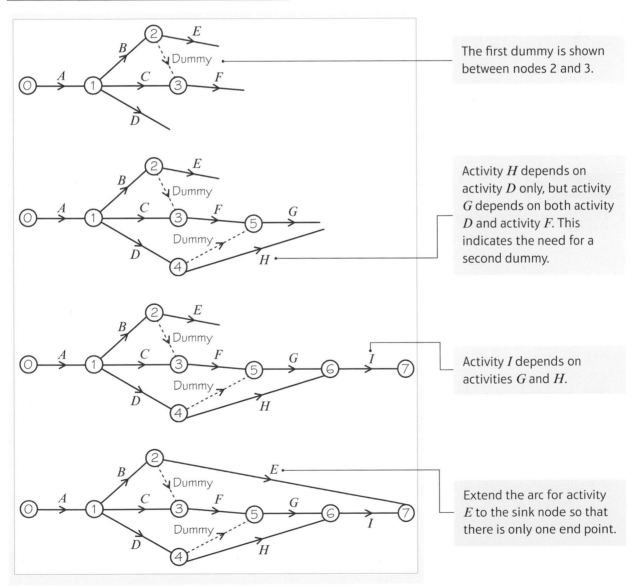

The first dummy is shown between nodes 2 and 3.

Activity *H* depends on activity *D* only, but activity *G* depends on both activity *D* and activity *F*. This indicates the need for a second dummy.

Activity *I* depends on activities *G* and *H*.

Extend the arc for activity *E* to the sink node so that there is only one end point.

227

- **Every activity must be uniquely represented in terms of its events. This requires that there can be at most one activity between any two events.**

Once again, a dummy may be required to satisfy this condition. Here is an example to show how this works.

This diagram shows part of an activity network. Suppose that event S depends on activities P, Q and R.

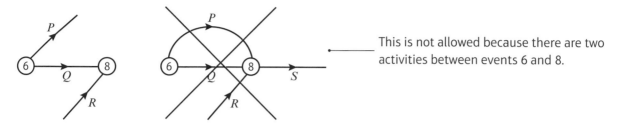

This is not allowed because there are two activities between events 6 and 8.

Using a dummy allows the dependence to be shown while ensuring that all activities are uniquely determined by their events.

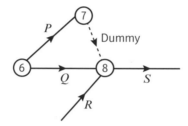

Exercise 8B

1 This activity network contains a dummy.

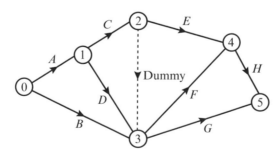

Draw a precedence table for the network.

(E) 2 A project is modelled by the activity network shown opposite.

   a Explain the significance of the dotted line from events 2 to 4. **(2 marks)**

   b Draw a precedence table for this network. **(3 marks)**

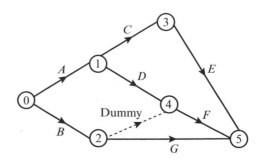

E **3** Draw an activity network to represent the precedence table below.

Your network should contain exactly one dummy.

Activity	Must be preceded by
A	–
B	–
C	–
D	A
E	C
F	A, B, E
G	C
H	D

(3 marks)

4 a Draw an activity network for this precedence table using exactly two dummies.

Activity	Depends on
P	–
Q	–
R	P
S	P
T	P, Q

(3 marks)

b Explain the purpose of each dummy. **(2 marks)**

E/P **5** The precedence table for a project is shown below. Draw the activity network described in the table, using the minimum number of dummies.

Activity	Depends on
A	–
B	–
C	A, B
D	B
E	B
F	C
G	C, D
H	E

(3 marks)

8.3 Early and late event times

Activities within a project typically take different lengths of time to complete.

■ **The length of time an activity takes to complete is known as its duration. You can add weights to the arcs in an activity network to represent these times.**

The activity network opposite was used in Exercise 8B question **2** but now each activity has a figure in brackets representing its duration in hours.

Activity A takes 2 hours to complete.———▶ $A(2)$

The dummy takes 0 hours to complete. ——

Each node (vertex), of the network represents an event. It is useful to consider two separate times associated with each event.

■ **The early event time is the earliest time of arrival at the event allowing for the completion of all preceding activities.**

■ **The late event time is the latest time that the event can be left without extending the time needed for the project.**

The activity network is now adapted to show this information by using ⊟ at each vertex.

┌── Early event time
└── Late event time

■ **The early event times are calculated starting from 0 at the source node and working towards the sink node. This is called a forward pass, or forward scan.**

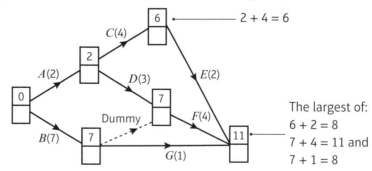

The largest of:
$6 + 2 = 8$
$7 + 4 = 11$ and
$7 + 1 = 8$

■ **The late event times are calculated starting from the sink node and working backwards towards the source node. This is called a backward pass or backward scan.**

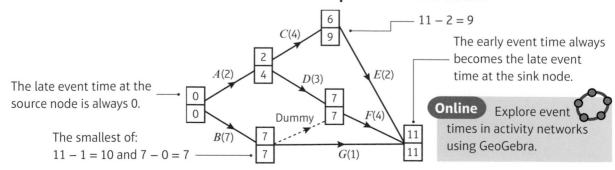

The late event time at the source node is always 0.

The smallest of:
$11 - 1 = 10$ and $7 - 0 = 7$

The early event time always becomes the late event time at the sink node.

Online Explore event times in activity networks using GeoGebra.

Example 5

The diagram shows part of an activity network.

Calculate the value of x.

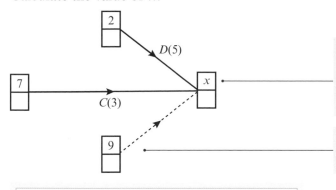

The largest of:
2 + 5 = 7
7 + 3 = 10
9 + 0 = 9

The dotted line indicates a dummy. This has a weight of 0.

$x = 10$

Example 6

The diagram shows part of an activity network.

Calculate the value of y.

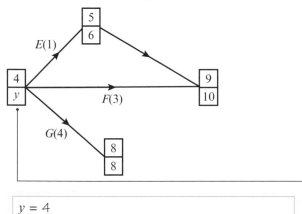

For each event, the late time is greater than or equal to the early time. You can use this fact to check your answers.

The smallest of:
6 − 1 = 5
10 − 3 = 7
8 − 4 = 4

$y = 4$

Exercise 8C

Answer templates for questions marked * are available at www.pearsonschools.co.uk/d1maths

1 The diagram shows part of an activity network.
Calculate the value of x.

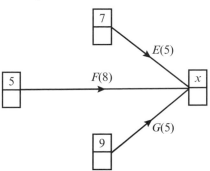

(E) **2** The activity network for a project is given below.

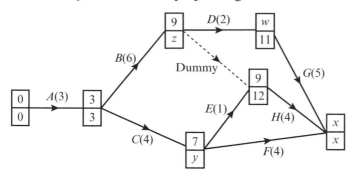

Watch out Activities A, C and E can be completed in 8 hours, but there is a dummy showing that you cannot start activity H until activity B has been completed.

The time in hours needed to complete each activity is shown in brackets.

Early and late times are shown at each vertex.

Calculate the values of w, x, y and z. **(8 marks)**

(E) **3*** The activity network for a project is given below.

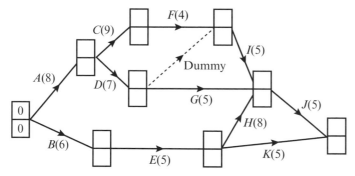

The time in days needed to complete each activity is shown in brackets.

Calculate the early and late times at each vertex. **(5 marks)**

8.4 Critical activities

You need to be able to identify critical activities.

- **An activity is described as a critical activity if any increase in its duration results in a corresponding increase in the duration of the whole project.**

- **A path from the source node to the sink node which entirely follows critical activities is called a critical path. A critical path is the longest path contained in the network.**

- **At each node (vertex) on a critical path the early event time is equal to the late event time.**

Watch out It is possible for a project to have more than one critical path, in which case the total project time is the same on each one.

Example 7

The diagram shows an activity network with early and late event times shown at the nodes. Identify the critical activities.

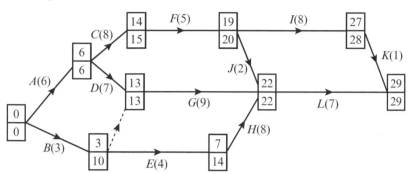

Online Explore critical paths using GeoGebra.

The critical activities are A, D, G, and L. Adding the durations of the critical activities gives the total duration for the project, as shown by the early and late event times at the sink node.

Check: $6 + 7 + 9 + 7 = 29$.

Problem-solving

Notice that at each vertex on the critical path, the early and late event times are equal. These indicate the critical events. Elsewhere, there is a difference between the two values.

■ **An activity connecting two critical events isn't necessarily a critical activity.**

Example 8

Part of an activity network is shown below including the early and late event times given in hours. Which are the critical activities?

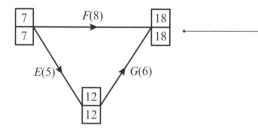

$7 + 5 = 12$ (critical)
$12 + 6 = 18$ (critical)
$7 + 8 \neq 18$ (not critical)

The critical activities are E and G. F is not a critical activity even though it connects two critical events. Any increase in the duration of activity E or activity G will increase the total time for the project, whereas the duration of activity F may be increased by up to 3 hours and have no effect on the total time.

Watch out In your exam, you may need to find more than one critical path and to identify all of the critical activities.

Example 9

Find the critical paths in this activity network and identify the critical activities.

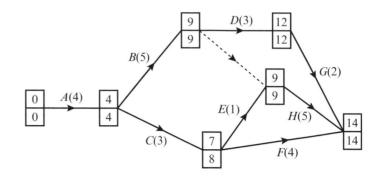

There are two critical paths: *A B D G* and *A B H*.
The critical activities are *A*, *B*, *D*, *G* and *H*.

The critical activities are those that lie on either critical path.

Exercise 8D

Answer templates for questions marked * are available at www.pearsonschools.co.uk/d1maths

P **1** Part of an activity network is shown opposite including the early and late event times given in hours.
Activities *J* and *K* are critical.

Find the values of *x*, *y* and *z*.

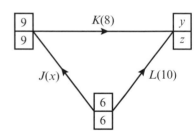

E/P **2** The diagram shows an activity network with early and late event times, in hours, shown at the vertices.

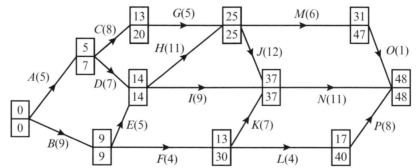

a Identify the critical activities. **(2 marks)**

b Name an activity that links two critical events but is not critical.
Explain your reasoning. **(3 marks)**

E/P **3*** The activity network for a project is shown below. Activity times are given in days and are shown in brackets.

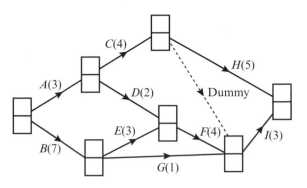

a Copy and complete the activity network to show the early and late event times. **(4 marks)**

b Explain why *G* is not a critical activity. **(2 marks)**

c Determine the critical activities and the length of the critical path. **(3 marks)**

E **4*** The diagram below is the activity network relating to an engineering project. The number in brackets on each arc gives the time taken, in days, to complete the activity.

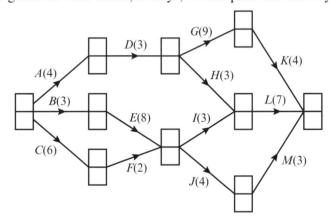

a Complete the precedence table below.

Event	Immediately preceding activity
A	–
B	–
C	–
D	
E	
F	
G	
H	
I	
J	
K	
L	
M	

(3 marks)

b Complete the diagram by calculating the early time and the late time for each event. **(4 marks)**

c Determine the critical activities and the length of the critical path. **(2 marks)**

8.5 The float of an activity

You can determine the total float of activities.

- **The total float of an activity is the amount of time that its start may be delayed without affecting the duration of the project.**

    total float = latest finish time – duration – earliest start time

- **The total float of any critical activity is 0.**

Example **10**

Determine the total float of each activity in this activity network.

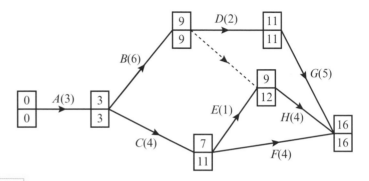

Activity A

Total float = 0

This is a critical activity. Any delay in the start time will affect the duration of the project.

Activities B, D and G

Total float = 0

These are all critical activities so the total float of each one is 0.

Activity C

| 3 | C(4) | 7 |
| 3 | | 11 |

Total float = 11 – 4 – 3 = 4

3 is the earliest time that the activity can start.
11 is the latest time that the activity can be finished by.
4 is the duration of the activity.

Activity E

| 7 | E(1) | 9 |
| 11 | | 12 |

Total float = 12 – 1 – 7 = 4

7 is the earliest time that the activity can start.
12 is the latest time that the activity can be finished by.
1 is the duration of the activity.

Activity F

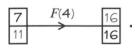

Total float = 16 – 4 – 7 = 5

7 is the earliest time that the activity can start.
16 is the latest time that the activity can be finished by.
4 is the duration of the activity.

Activity H

| 9 | H(4) | 16 |
| 12 | | 16 |

Total float = 16 – 4 – 9 = 3

9 is the earliest time that the activity can start.
16 is the latest time that the activity can be finished by.
4 is the duration of the activity.

Exercise 8E

1 Determine the total float of each activity in this activity network.

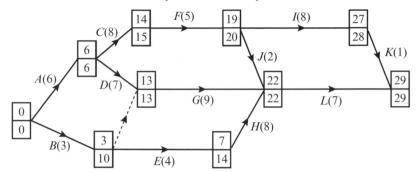

E/P **2** The diagram shows part of an activity network with activity times measured in hours.

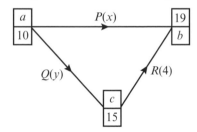

P is a critical activity.

Q has a total float of 3 hours.

a Work out the values of *a*, *b*, *x* and *y*. **(4 marks)**

b What is the minimum possible value of *c*? **(1 mark)**

c What is the maximum possible value of the total float of *R*? **(2 marks)**

E **3** A project's activity network is shown below.

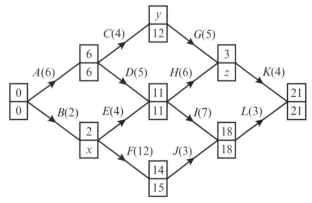

a Find the values of *x*, *y* and *z*. **(2 marks)**

b List the critical activities. **(1 mark)**

c Calculate the total float of activity *E*. **(1 marks)**

8.6 Gantt charts

You need to be able to construct and use Gantt (cascade) charts.

- **A Gantt (cascade) chart provides a graphical way to represent the range of possible start and finish times for all activities on a single diagram.**

The number scale shows *elapsed* time. So, the first hour is shown between 0 and 1 on the scale, the second hour is shown between 1 and 2 and so on.

The critical activities are shown as rectangles in a line at the top.

Example 11

Here is an activity network for a project.
Early and late event times are shown in hours at the nodes.
Draw a Gantt chart to represent the project.

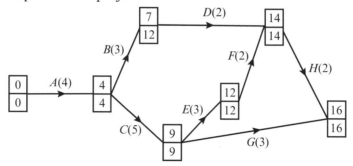

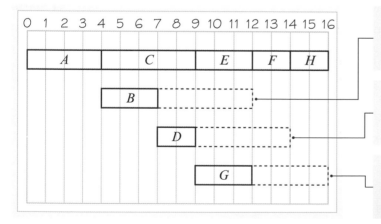

Activity *B* has duration 3 hours, an earliest start time of 4 hours and a latest finish time of 12 hours.

Activity *D* has duration 2 hours, an earliest start time of 7 hours and a latest finish time of 14 hours.

Activity *G* has duration 3 hours, an earliest start time of 9 hours and a latest finish time of 16 hours.

The Gantt chart illustrates clearly that there is no flexibility in the timing of the critical activities. It also illustrates the degree of flexibility in the timing of each of the remaining activities.

The total float of each activity is represented by the range of movement of its rectangle on the chart.

Watch out You must clearly distinguish each activity from its total float in your Gantt (cascade) chart. You can use dotted lines as above, or you can shade in the total float.

Exercise (8F)

(E) **1** The diagram shows an activity network for a project. Early and late event times are shown in days at the nodes. Draw a Gantt chart to represent the project.

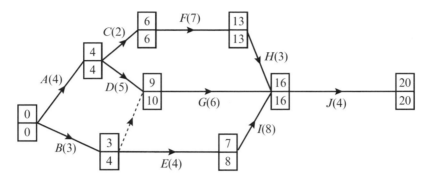

(4 marks)

(E) **2** An activity network for a project is shown below.

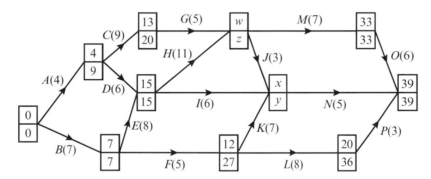

   a Calculate the values of w, x, y and z.   **(4 marks)**

   b List the critical activities.   **(2 marks)**

   c Calculate the total float for activities G and N.   **(2 marks)**

   d Draw a Gantt chart to represent the project.   **(4 marks)**

The overview of a project provided by a Gantt chart allows you to determine which activities must be happening at any given time and those that may be happening at a given time. In practice, once a project is underway, this provides a useful means of checking that non-critical activities have not been delayed to the point that they have become critical.

Example (12)

The Gantt chart overleaf represents a project that must be completed within 25 days. Given that the project is on time,

a determine 3 activities that **must** be happening at midday on day 10.

b determine 1 additional activity that **may** be happening at midday on day 10.

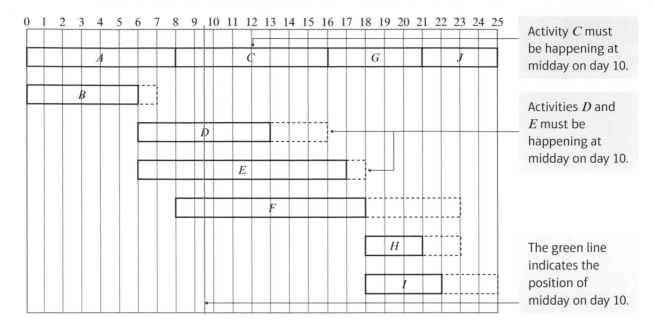

Activity *C* must be happening at midday on day 10.

Activities *D* and *E* must be happening at midday on day 10.

The green line indicates the position of midday on day 10.

a The activities that **must** be happening at midday on day 10 are *C*, *D* and *E*.

b The activity that **may** be happening at midday on day 10 is *F*, because its duration is 10 days and it has to finish at the end of day 23.

Problem-solving

Activity *F* **may** be happening at midday on day 10, but the total float of 5 days shows that the start could be delayed until day 14.

Exercise 8G

(P) 1 Refer to the Gantt chart shown in Example 12 for this question.

  a Which activities **must** be happening at midday on day 8?

  b Which activities **must** be happening at midday on day 21?

  c Which activities **may** be happening at midday on day 22?

(E/P) 2 The Gantt chart below represents an engineering project. An engineer decides to carry out some spot checks on the progress of the project.

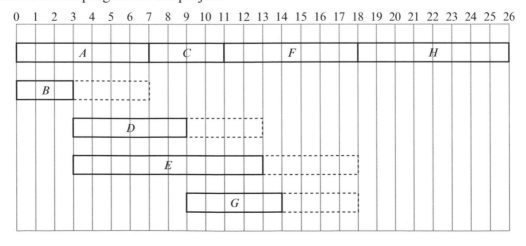

The first spot check takes place at 12 noon on day 8.

a Determine which activities **must** be happening at this time. **(2 marks)**

The second spot check takes place at 12 noon on day 15.

b List all the activities which **may** be happening at this time. **(2 marks)**

(E/P) **3 a** Draw a Gantt chart to represent the activity network below. **(4 marks)**

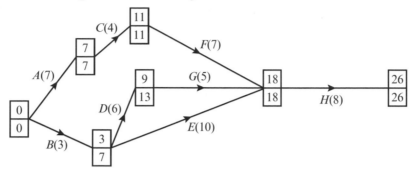

Use your chart to determine:

b which activities **may** be happening at midday on day 5, **(2 marks)**

c which activities **must** be happening at midday on day 7. **(2 marks)**

(E/P) **4** The activity network below shows the activities that are needed in order to complete a railway line laying project. The number in brackets on each arc represents the time, in days, to complete the activity.

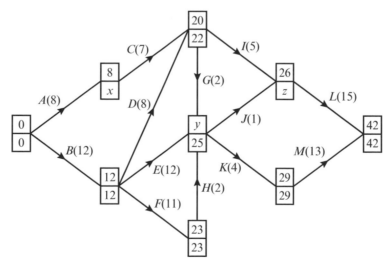

a Find the values of x, y and z. **(2 marks)**

b Identify the critical path. **(1 mark)**

c Calculate the total float on activities A and C. **(1 mark)**

d Draw a Gantt chart for the project. **(4 marks)**

The foreman checks on how well the project is progressing. He makes a check on Day 21 and notices that Activity I has not yet started.

e Can the project still be completed on time? Explain your answer. **(1 mark)**

241

8.7 Resource histograms

You need to be able to consider the number of **workers** needed to complete a project. You will be told the number of workers that are needed for each activity. Workers follows these rules:

- **No worker can carry out more than one activity simultaneously.**
- **Once a worker, or workers, have started an activity, they must complete it.**

> **Watch out** In many cases each activity will only require one worker, but you may also need to consider situations where more than worker is needed for certain activities.

- **Once a worker, or workers, have finished an activity, they immediately become available to start another activity.**

A **resource histogram** shows the number of workers that are active at any given time. The convention, when constructing the diagram, is to assume that each activity starts at the earliest time possible. However, once drawn, it may be possible to use the diagram to identify which activities may be delayed in order to minimise the number of workers required.

In some cases, the number of workers available is less than the number required to complete the project in the minimum possible time. The start and finish times of some activities may have to be delayed in order to meet this constraint, which then extends the time needed for project completion.

- **The process of adjusting the start and finish times of the activities in order to take into account constraints on resources is called resource levelling.**

One way to draw a resource histogram is to start from a Gannt (cascade) chart.

Example 13

Here is the Gannt chart from question **2** of exercise 8G.

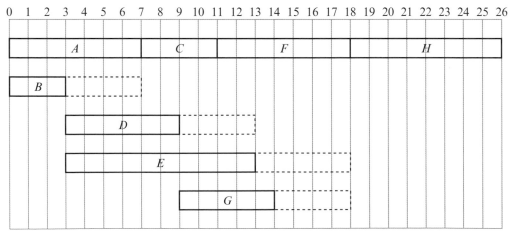

A The number of workers required for each activity is shown in this table.

Activity	Number of workers
A	2
B	1
C	1
D	2
E	1
F	1
G	2
H	1

Draw a resource histogram to show the number of workers required each day when each activity begins at its earliest time.

From end of day	To end of day	Activities	Number of workers
0	3	A, B	3
3	7	A, D, E	5
7	9	C, D, E	4
9	11	C, E, G	4
11	13	F, E, G	4
13	14	F, G	3
14	18	F	1
18	26	H	1

Combine the information given in the Gantt (cascade) chart with the information about the number of workers required for each activity. Work out how many workers are needed on each day of the project. This is the information that will be shown on the resource histogram.

This number of workers assumes that each activity starts at the earliest possible time. This might not be the most efficient way to complete the project.

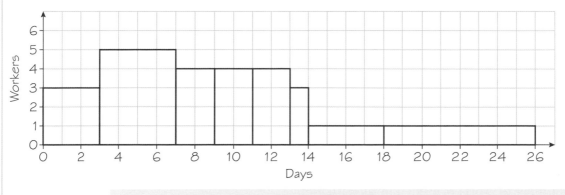

With practice, there is no need to draw the table; you can put the information directly onto the resource histogram.

Watch out Make sure there are no gaps in your resource histogram.

A You can show all of the information from the table in the previous example on your resource histogram by writing letters to show which activities are being completed on each day:

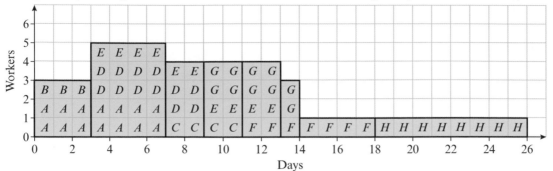

Notice that whenever activities A, D or G are being carried out they require **two rows** because they each need two workers.

In the example above **five workers** are needed between days 3 and 7. However, by delaying the start of activity E by four days, it is possible to complete the project in the same time with only four workers. Activity E has a total float of 5 so it will still be finished by its late time.

Notation The process of adjusting start times to optimise the allocation of workers is called **resource levelling**.

Example 14

The diagram shows a project modelled by an activity network. The number on each arc gives the time, in days, required to complete each activity, and the number of workers needed for each activity is shown in the table.

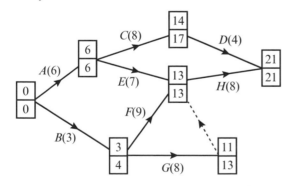

Activity	Workers
A	1
B	2
C	2
D	1
E	1
F	1
G	2
H	1

A project manager determines that each activity should start at the earliest possible time.

a Draw a resource histogram to show the number of workers required on each day, and state the total number of workers needed in this situation.

Due to budget cuts, only four workers are available for the project.

b Use resource levelling to show how the project can be completed with just 4 workers, and state the time required to complete the project in this case.

A

a

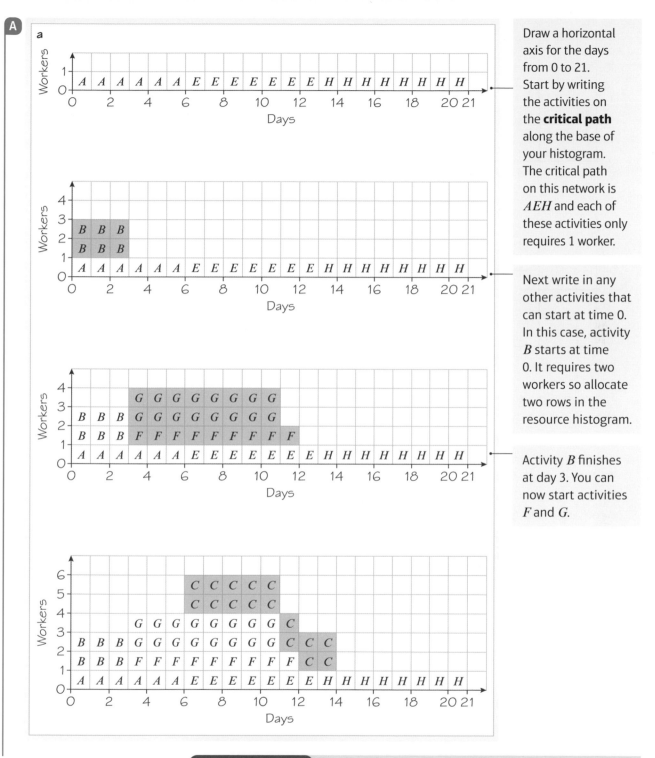

Draw a horizontal axis for the days from 0 to 21. Start by writing the activities on the **critical path** along the base of your histogram. The critical path on this network is *AEH* and each of these activities only requires 1 worker.

Next write in any other activities that can start at time 0. In this case, activity *B* starts at time 0. It requires two workers so allocate two rows in the resource histogram.

Activity *B* finishes at day 3. You can now start activities *F* and *G*.

Problem-solving

As you complete your resource histogram, look at any points where an activity ends and consider what other activities could start at that point. Activity *A* ends at day 6. At this point you can start activities *C* and *E*. *E* has already been included as it is on the critical path, so write in activity *C*. In each column, use as few rows as possible.

245

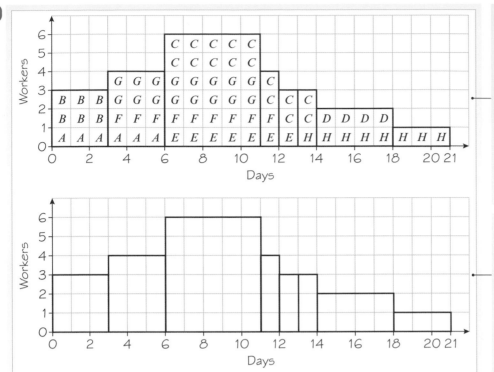

The resource histogram shows that 6 workers are required to complete the project in a time of 21 days if each activity starts at its early time.

b To reduce the number of workers to 4 we must consider delaying activity C, E, F or G which are currently active in the third bar of the resource histogram. Activities F and G become active in the second bar of the resource histogram.

Problem-solving

The only point where more than 4 workers are needed is between days 6 and 11, so consider the activities that are being carried out in this period.

Activity	Total float
C	$17 - 8 - 6 = 3$
E	$13 - 7 - 6 = 0$
F	$13 - 9 - 3 = 1$
G	$13 - 8 - 3 = 2$

Activity C has the largest total float and requires 2 workers. Delaying activity C by 5 days brings the number of workers required down to 4 and requires the total project time to be extended by just 2 days to 23 days.

The only activity left to include is D. The earliest it can start is immediately after activity C finishes. Write it in, then draw bars around your finished resource histogram, and add a vertical scale.

You can show your finished resource histogram with or without the activity letters. In this diagram a vertical line has been added after 13 days to show that activity H starts here.

Calculate the total float of each of activities C, E, F and G to determine how much they could be delayed without delaying the project.

You need to delay activity C two days beyond its given late time, which means the project will be delayed by 2 days.

Exercise 8H

1 Here is a Gantt (cascade) chart for a project that takes 22 days to complete. Given that each activity requires one worker, and that each activity begins at its earliest time,

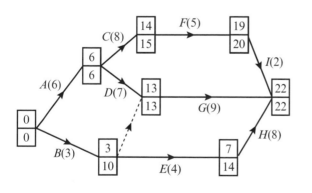

a draw a resource histogram (4 marks)

b state the number of workers required to complete the project. (1 mark)

2 The diagram shows an activity network. The numbers in brackets show the duration of each activity in days, and the table shows the number of workers required for each activity. The early and late times for each event are shown.

Activity	Workers
A	1
B	1
C	2
D	1
E	1
F	2
G	2
H	1
I	1

a Given that each activity starts at the earliest possible time, draw a resource histogram for this activity network. (4 marks)

b Using your resource histogram, state the number of workers required to complete the project on time. (1 mark)

3 The diagram shows an activity network. The numbers in brackets show the duration of each activity in weeks, and the table shows the number of workers required for each activity. The early and late times for each event are shown.

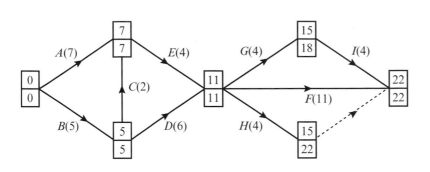

Activity	Workers
A	1
B	1
C	2
D	1
E	1
F	2
G	2
H	1
I	1

A manager decides that each activity should start at the earliest possible time.

a Draw a resource histogram to show the number of workers required on each day, and state the total number of workers needed in this situation. **(4 marks)**

b Use resource levelling to show how the project can be completed using 4 workers and state the amount of time by which the project must be extended. **(4 marks)**

4 The activity network opposite shows the activities involved in assembling a particular vehicle on a factory production line.

The times shown are in hours and the latest start times and earliest finish times are shown at the vertices.

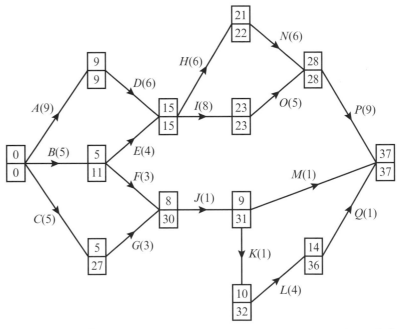

a Draw a Gantt (cascade) chart for this network. **(4 marks)**

The number of workers required for each activity is shown in the table.

Number of workers required	1	2	3
Activity	B, C, E, F, G, H, L, Q	A, D, K, J, M, N, O	I, P

b Draw a resource histogram for this project, assuming that each activity starts at the earliest possible time. State the number of workers needed in this instance. **(4 marks)**

A   A manager says that she can complete the project on time with only 4 workers by delaying the start of activity N by 2 hours.

    c Explain why she is not correct.   **(1 mark)**

    d Given that there are only 3 workers available, work out the earliest time that the project can be completed.   **(3 marks)**

8.8   Scheduling diagrams

The process of assigning workers to activities is known as scheduling. You need to be able to construct a scheduling diagram to show which workers should be assigned to which activity.

> **Note**   When constructing scheduling diagrams, you can assume that each activity requires only one worker.

- **When you are scheduling a project:**
  - **you should always use the first available worker**
  - **if there is a choice of activities for a worker, assign the one with the lowest value for its late time.**

Example   15

The diagram shows a Gantt chart for a project.

Schedule the activities to be completed in the critical time by the minimum number of workers.

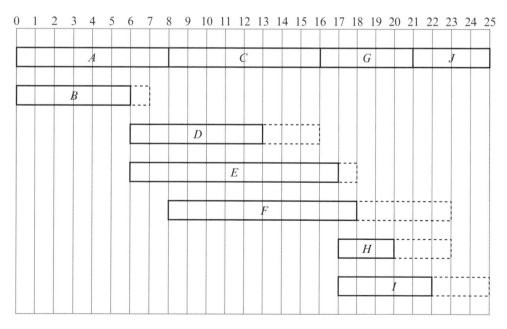

A

This Gantt chart information may be transferred to a scheduling diagram in stages.

One worker is needed to complete the critical activities.

Worker 2 starts with activity B and, on completion, starts activity D. A third worker is needed to start activity E on time.

Delay activity F as late as possible, so it may be started by worker 2 on completion of activity D.

All of the activities have now been assigned and so the scheduling diagram is complete.

Shading is used to indicate any periods of inactivity for each worker.

By starting activity H as soon as possible and activity I on completion of activity H, both activities may be completed by worker 3.

Watch out Check your final diagram to make sure that all the dependency conditions are satisfied. If one activity has been delayed, this might affect the earliest possible start time for other activities.

A In Example 15, a minimum of three workers were required to complete the project within the critical time of 25 days. In general, you can calculate a **lower bound** for the number of workers by considering the sum of the activity times (or, in other words, the total weight of the network).

- **The lower bound for the number of workers needed to complete a project within its critical time is the smallest integer greater than or equal to:**

$$\frac{\text{sum of all of the activity times}}{\text{critical time of the project}}$$

Consider the network in Example 15:

Activity	A	B	C	D	E	F	G	H	I	J	**Total**
Duration	8	6	8	7	11	10	5	3	5	4	**67**

The sum of all of the activity times is 67 days.

This gives a lower bound as the smallest integer greater than or equal to $\frac{67}{25}$

$\frac{67}{25} = 2.68$ and so a lower bound is 3.

This simply means that it is impossible to complete the project in the critical time using fewer than 3 workers. However, since the calculation takes no account of the degree of overlap of the activities, it doesn't guarantee that 3 workers is sufficient.

When the number of available workers is fewer than the minimum required to complete a project within its critical time, the information shown on a Gantt chart cannot be relied upon as some activities will be delayed further than shown. In this situation it is better to construct the scheduling diagram direct from the activity network because special care is needed to ensure that the dependency conditions are satisfied.

Example 16

The diagram shows an activity network representing a project with a minimum time of 31 days.
Use a scheduling diagram to find the new completion time for the project given that only two workers are available.

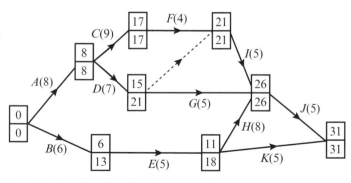

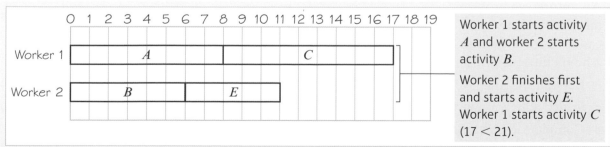

Worker 1 starts activity *A* and worker 2 starts activity *B*.

Worker 2 finishes first and starts activity *E*. Worker 1 starts activity *C* (17 < 21).

A

0 1 2 3 4 5 6 7 8 9 10 11 12 13 14 15 16 17 18 19 20 21 22

Worker 1 | A | C | F

Worker 2 | B | E | D

Worker 2 starts activity D and worker 1 starts activity F.

Continuing in the same way produces the following scheduling diagram.

0 1 2 3 4 5 6 7 8 9 10 11 12 13 14 15 16 17 18 19 20 21 22 23 24 25 26 27 28 29 30 31 32 33 34

Worker 1 | A | C | F | H | J

Worker 2 | B | E | D | G | I | K

The minimum time needed to complete the project using two workers is 34 days.

Watch out When Worker 2 finishes activity G, the next available activity is I. Check to make sure that activities D and F have been completed before allocating this activity.

Exercise 8I

(E/P) **1** The Gantt chart represents a project which must be completed in its critical time of 22 hours.

0 1 2 3 4 5 6 7 8 9 10 11 12 13 14 15 16 17 18 19 20 21 22

A | D | F | J

B

C

E

G

H

I

a Given that the total duration of all of the activities is 64 hours, calculate a lower bound for the number of workers needed to complete the project in the minimum time.
(2 marks)

An unforeseen problem means that Activity B cannot be started until 2 hours into the project.

b Explain why this will not result in a delay for the whole project. **(1 mark)**

c Which activities **must** be happening 17.5 hours into the project? **(1 mark)**

d Complete a scheduling diagram to complete the project in 22 hours.
State the number of workers required. **(4 marks)**

A **2** The diagram shows an activity network representing a project. The numbers on each arc
E represent the duration of each activity in days, and the early and late times are shown.

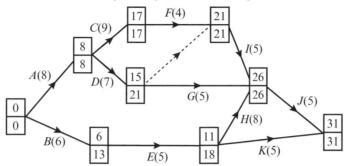

a Draw a Gantt chart to represent the project. **(4 marks)**

b Schedule the project to be completed by the minimum number of workers in the
critical time. State the number of workers required. **(4 marks)**

E/P **3** The diagram shows an activity network representing a project. The numbers on each arc
represent the duration of each activity in hours, and the early and late times are shown.

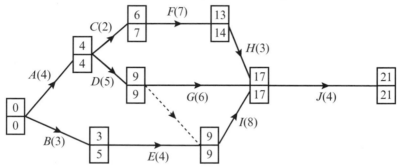

Given that only two workers are available for this project, construct a scheduling
diagram and determine the minimum time needed to complete the project. **(5 marks)**

Mixed exercise **8**

**Answer templates for questions marked * are available
at www.pearsonschools.co.uk/d1maths**

E **1** The precedence table for activities involved in
producing a computer game is shown opposite.

An activity on arc network is to be drawn to
model this production process.

  a Explain why it is necessary to use at least two
  dummies when drawing the activity network.
  (2 marks)

  b Draw the activity network using exactly two
  dummies. **(4 marks)**

Activity	Must be preceded by
A	–
B	–
C	B
D	A, C
E	A
F	E
G	E
H	G
I	D, F
J	G, I
K	G, I
L	H, K

(E) **2 a** Draw the activity network described in this precedence table, using activity on arc and exactly two dummies. **(3 marks)**

  b Explain why each of the two dummies is necessary. **(2 marks)**

Activity	Immediately preceding activities
A	–
B	–
C	A
D	B
E	B, C
F	B, C

(E/P) **3** An engineering project is modelled by the activity network shown on the right. The activities are represented by the arcs. The number in brackets on each arc gives the time, in days, to complete the activity. Each activity requires one worker. The project is to be completed in the shortest time.

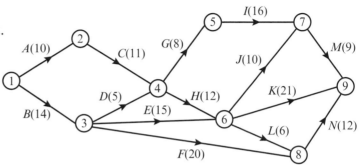

  a Calculate the early time and late time for each event. **(3 marks)**

  b State the critical activities. **(2 marks)**

  c Find the total float of activities *D* and *F*. You must show your working. **(2 marks)**

  d Draw a cascade (Gantt) chart for this project. **(4 marks)**

The chief engineer visits the project on day 15 and day 25 to check the progress of the work. Given that the project is on schedule,

  e which activities **must** be happening on each of these two days? **(1 mark)**

(A) (E/P) **4** A project is modelled by the activity network shown on the right. The activities are represented by the arcs. The number in brackets on each arc gives the time, in hours, to complete the activity. The numbers in circles are the event numbers. Each activity requires one worker.

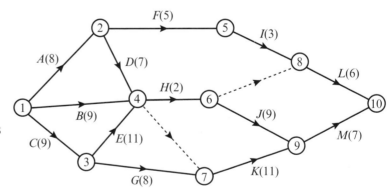

  a Explain the purpose of the dotted line from event 6 to event 8. **(1 mark)**

  b Calculate the early time and late time for each event. **(3 marks)**

  c Calculate the total float of activities *D*, *E* and *F*. **(2 marks)**

  d Determine the critical activities. **(1 mark)**

  e Given that the sum of all the times of the activities is 95 hours, calculate a lower bound for the number of workers needed to complete the project in the minimum time. You must show your working. **(2 marks)**

  f Given that workers may not share an activity, schedule the activities so that the process is completed in the shortest time using the minimum number of workers. **(3 marks)**

A **5*** The network shows the activities that need to be undertaken to complete a project. Each activity
E/P is represented by an arc. The number in brackets is the duration of the activity in days. The early
and late event times are to be shown at each vertex and some have been completed for you.

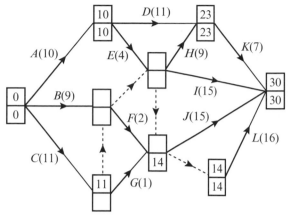

a Calculate the missing early and late times. **(2 marks)**

b Explain what is meant by a critical path. **(1 mark)**

c List the two critical paths for this network. **(2 marks)**

The sum of all the activity times is 110 days and each activity requires just one worker.
The project must be completed in the minimum time.

d Calculate a lower bound for the number of workers needed to complete the project
in the minimum time. You must show your working. **(2 marks)**

e List the activities that must be happening on day 20. **(2 marks)**

f Comment on your answer to part **e** with regard to the lower bound you found in part **d**.
(2 marks)

g Schedule the activities, using the minimum number of workers, so that the project is
completed in 30 days. **(3 marks)**

E/P **6*** A project is modelled by this activity network. The number in brackets on each arc represents the
time, in days, to complete the activity.

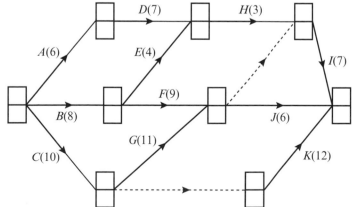

a Work out the early and late event times for each activity. State the minimum time
needed to complete the project. **(3 marks)**

b Calculate a lower bound for the number of workers required to complete the project
on time. **(1 mark)**

A   c Draw a Gantt (cascade) chart to represent the project.   **(4 marks)**

d Assuming that each activity only requires one worker, draw a resource histogram.   **(3 marks)**

e Use resource levelling to show that the project can be completed with 4 workers without extending the total time needed.   **(2 marks)**

E/P **7*** A group of workers are involved in a building project. The number in brackets on each arc represents the time, in days, to complete the activity.

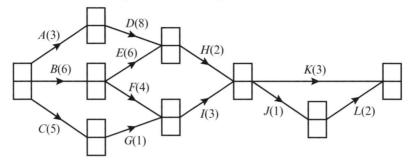

a Work out the early and late event times for each activity. State the minimum time needed to complete the project.   **(4 marks)**

b Find the critical path(s).   **(2 marks)**

The number of workers required for each activity is shown in the table.

Activity	A	B	C	D	E	F	G	H	I	J	K	L
Number of workers required	4	4	5	2	5	2	6	3	5	3	2	4

c Given that each activity starts as early as possible and assuming that there is no limit on the number of workers available, draw a resource histogram for the project, indicating clearly which activities are taking place at any given time.   **(4 marks)**

d The building manager decides that there will only be 8 workers available at any time. Use resource levelling to construct a new resource histogram, showing how the project can be completed with the minimum extra time. State the minimum extra time required.   **(3 marks)**

Challenge

The activity network below shows the tasks that need to be carried out in order to complete a bathroom renovation project. The numbers in brackets show the duration of each activity in days, and the early and late event times are shown at each vertex. The nodes are numbered from 0 to 5.

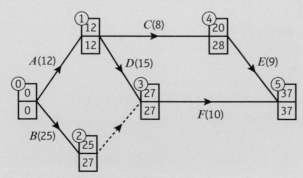

a Identify the activities which lie on the critical path, and calculate the total float of each of the remaining activities.

The homeowner wishes to reduce the total time to completion of the project. The builder identifies four activities which can be speeded up at additional cost, and creates a table showing the cost per day of reducing each activity's duration:

Activity	Additional cost of reduction (£ / day)	Maximum possible reduction (days)
B	£100	8
D	£200	10
E	£400	2
F	£150	3

Hint So for example, the duration of activity B could be reduced by 8 days at a cost of £800, but it could not be reduced further.

b Find the most cost-effective way of reducing the total project time to:

i 33 days   **ii** 30 days

In each case, you should state which activities have been reduced, and by how much, and give the total additional cost.

c Find the shortest possible time in which the project could be completed, and explain why the project time cannot be further reduced.

The builder decides to formulate a linear programming problem to optimise the cost of reducing the total project time to 28 days.

She defines x_1, x_2, x_3, x_4 and x_5 to be the early times at vertices 1 to 5 respectively, and y_B, y_D, y_E and y_F to be the number of days by which activities B, D, E and F are reduced.

d Write down a suitable objective function to minimise.

e Write down constraints which will ensure that the dependencies in the project are maintained and that the solution satisfies all of the conditions given above. You should not attempt to solve the linear programming problem.

Hint For part **e**, you will need constraints:
- For the duration of each activity, taking into account any reduction
- For the maximum possible reductions
- For the early time of the sink node, x_5

Summary of key points

1 A **precedence table**, or **dependence table**, is a table that shows which **activities** must be completed before others are started.

2 In an **activity on arc network**, the activities are represented by **arcs** and the completion of those activities, known as **events**, are shown as **nodes**.
   - Each arc is labelled with an activity letter.
   - The beginning and end of an activity are shown at the ends of the arc and an arrow is used to define the direction. The convention is to use straight lines for arcs.
   - The nodes are numbered starting from 0 for the first node, which is called the **source node**.
   - Sometimes the source node is labelled 1 instead of 0.
   - Number each node as it is added to the network.
   - The final node is called the **sink node**.

3 A **dummy activity** has no time or cost. Its sole purpose is to show dependencies between activities.

4 Every activity must be **uniquely represented** in terms of its events. This requires that there can be **at most one activity** between any two events.

5 The length of time an activity takes to complete is known as its duration. You can add weights to the arcs in an activity network to represent these times.

6 · The **early event time** is the earliest time of arrival at the event allowing for the completion of all preceding activities.

· The **late event time** is the latest time that the event can be left without extending the time needed for the project.

· The early event times are calculated starting from 0 at the source node and working towards the sink node. This is called a **forward pass** or **forward scan**.

· The late event times are calculated starting from the sink node and working backwards towards the source node. This is called a **backward pass** or **backward scan**.

7 · An activity is described as a **critical activity** if any increase in its duration results in a corresponding increase in the duration of the whole project.

· A path from the source node to the sink node which entirely follows critical activities is called a **critical path**. A critical path is the longest path contained in the network.

· At each node (vertex) on a critical path **the early event time is equal to the late event time**.

8 An activity connecting two critical events isn't necessarily a critical activity.

9 · The **total float** of an activity is the amount of time that its start may be delayed without affecting the duration of the project.

Total float = latest finish time – duration – earliest start time

· The total float of any critical activity is 0.

10 A **Gantt** (cascade) **chart** provides a graphical way to represent the range of possible start and finish times for all activities on a single diagram.

11 You need to be able to consider the number of **workers** needed to complete a project. You will be told the number of workers that are needed for each activity.

· No worker can carry out more than one activity simultaneously.

· Once a worker, or workers, have started an activity, they must complete it.

· Once a worker, or workers, have finished an activity, they immediately become available to start another activity.

12 The process of adjusting the start and finish times of the activities in order to take into account constraints on resources is called **resource levelling**.

13 When you are scheduling a project:

· you should always use the first available worker

· if there is a choice of activities for a worker, assign the one with the lowest value for its late time.

14 The lower bound for the number of workers needed to complete a project within its critical time is the smallest integer greater than or equal to:

$$\frac{\text{sum of all of the activity times}}{\text{critical time of the project}}$$

Review exercise

Answer templates for questions marked * are available on www.pearsonschools.co.uk/d1maths

(E/P) 1 Two fertilisers are available, a liquid X and a powder Y. A bottle of X contains 5 units of chemical A, 2 units of chemical B and $\frac{1}{2}$ unit of chemical C. A packet of Y contains 1 unit of A, 2 units of B and 2 units of C. A professional gardener makes her own fertiliser. She requires at least 10 units of A, at least 12 units of B and at least 6 units of C.

She buys x bottles of X and y packets of Y.

a Write down the inequalities which model this situation. **(4)**

b Draw a graph to illustrate the feasible region, R. **(2)**

A bottle of X costs £2 and a packet of Y costs £3.

c Write down an expression, in terms of x and y, for the total cost £T. **(1)**

d Using your graph, obtain the values of x and y that give the minimum value of T. Make your method clear and calculate the minimum value of T. **(2)**

← Sections 6.1, 6.2, 6.3, 6.4

(E) 2 A company produces two types of self-assembly wooden bedroom suites, the 'Oxford' and the 'York'. After the pieces of wood have been cut and finished, all the materials have to be packaged. The table below shows the time, in hours, needed to complete each stage of the process and the profit made, in pounds, on each type of suite.

	Oxford	York
Cutting	4	6
Finishing	3.5	4
Packaging	2	4
Profit (£)	300	500

The times available each week for cutting, finishing and packaging are 66, 56 and 40 hours respectively.

The company wishes to maximise its profit.

Let x be the number of Oxford, and y be the number of York suites made each week.

a Write down the objective function. **(1)**

b In addition to

$$2x + 3y \leqslant 33,$$
$$x \geqslant 0,$$
$$y \geqslant 0,$$

find two further inequalities to model the company's situation. **(2)**

c Illustrate all the inequalities, indicating clearly the feasible region. **(2)**

d Explain how you would locate the optimal point. **(1)**

e Determine the number of Oxford and York suites that should be made each week and the maximum profit gained. **(2)**

It is noticed that when the optimal solution is adopted, the time needed for one of the three stages of the process is less than that available.

f Identify this stage and state by how many hours the time may be reduced. **(2)**

← Sections 6.1, 6.2, 6.3, 6.4

(E/P) 3 The Young Enterprise Company 'Decide', is going to produce badges to sell to decision maths students. It will produce two types of badge.

Badge 1 reads 'I made the decision to do maths' and Badge 2 reads 'Maths is the right decision'.

'Decide' must produce at least 200 badges and has enough material for 500 badges.

Market research suggests that the number produced of Badge 1 should be between 20% and 40% of the total number of badges made.

The company makes a profit of 30p on each Badge 1 sold and 40p on each Badge 2. It will sell all that it produces, and wishes to maximise its profit.

Let x be the number produced of Badge 1 and y be the number of Badge 2.

a Formulate this situation as a linear programming problem, simplifying your inequalities so that all the coefficients are integers. **(4)**

b On suitable axes, construct and clearly label the feasible region. **(3)**

c Using your graph, advise the company on the numbers of each badge it should produce. State the maximum profit 'Decide' will make. **(3)**

← **Sections 6.1, 6.2, 6.3, 6.4**

(E) **4** Becky's bird food company makes two types of bird food. One type is for bird feeders and the other for bird tables. Let x kg represent the quantity of food for bird feeders made each day, and y kg represent the quantity of food for bird tables made each day. Due to restrictions in the production process, and known demand, the following constraints apply:

$$x + y \leqslant 12$$
$$y < 2x$$
$$2y \geqslant 7$$
$$y + 3x \geqslant 15$$

a Show these constraints on a diagram and label the feasible region R. **(3)**

The objective is to minimise $C = 2x + 5y$.

b Solve this problem, making your method clear. Give, as fractions, the value of C and the amount of each type of food that should be produced. **(4)**

Another objective (for the same constraints given above) is to maximise $P = 3x + 2y$.

c Solve this problem, making your method clear. State the value of P and the amount of each type of food that should be produced. Your answers must be whole numbers **(3)**

← **Sections 6.1, 6.2, 6.3, 6.4**

(E/P) **5** The company EXYCEL makes two types of battery, X and Y. Machinery, workforce and predicted sales determine the number of batteries EXYCEL make. The company decides to use a graphical method to find its optimal daily production of X and Y.

The constraints are modelled in the diagram below.

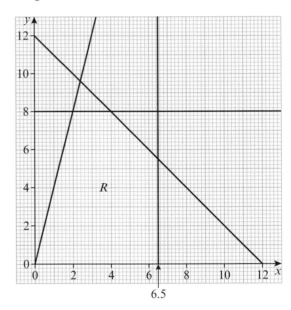

The feasible region, R, is indicated.

x = the number (in thousands) of type X batteries produced each day,

y = the number (in thousands) of type Y batteries produced each day.

The profit of each type X battery is 40p and on each type Y battery is 20p.

The company wishes to maximise its daily profit.

a Write this as a linear programming problem, in terms of x and y, stating the objective function and all the constraints. **(4)**

b Find the optimal number of batteries of each type to be made each day. Your answers must be a whole number of batteries and you must show your working clearly. **(4)**

c Find the daily profit, in £, made by EXYCEL. **(1)**

← **Sections 6.1, 6.2, 6.3, 6.4**

(A) 6 Flatland UK Ltd makes three types of carpet, the Lincoln, the Norfolk and the Suffolk. The carpets all require units of black, green and red wool.

For each roll of carpet, the Lincoln requires 1 unit of black, 1 of green and 3 of red, the Norfolk requires 1 unit of black, 2 of green and 2 of red, and the Suffolk requires 2 units of black, 1 of green and 1 of red.

There are up to 30 units of black, 40 units of green and 50 units of red available each day. Profits of £50, £80 and £60 are made on each roll of Lincoln, Norfolk and Suffolk respectively. Flatland UK Ltd wishes to maximise its profit.

Let the number of rolls of the Lincoln, Norfolk and Suffolk made daily be x, y and z respectively.

a Formulate the above situation as a linear programming problem, listing clearly the constraints as inequalities in their simplest form, and stating the objective function. **(4)**

This problem is to be solved using the simplex algorithm. The most negative number in the profit row is taken to indicate the pivot column at each stage.

b Stating your row operations, show that after one complete iteration the tableau becomes **(4)**

(A)

Basic variable	x	y	z	r	s	t	Value
r	$\frac{1}{2}$	0	$1\frac{1}{2}$	1	$-\frac{1}{2}$	0	10
y	$\frac{1}{2}$	1	$\frac{1}{2}$	0	$\frac{1}{2}$	0	20
t	2	0	0	0	-1	1	10
P	-10	0	-20	0	40	0	1600

c Explain the practical meaning of the value 10 in the top row. **(1)**

d i Perform one further complete iteration of the simplex algorithm. **(3)**

ii State whether your answer to part **d i** is optimal. Give a reason for your answer.

iii Interpret your current tableau, giving the value of each variable. **(1)**

← **Sections 7.1, 7.2**

(E/P) 7 T42 Co. Ltd produces three different blends of tea, Morning, Afternoon and Evening. The teas must be processed, blended and then packed for distribution. The table below shows the time taken, in hours, for each stage of the production of a tonne of tea. It also shows the profit, in hundreds of pounds, on each tonne.

	Processing	Blending	Packing	Profit (£100)
Morning blend	3	1	2	4
Afternoon blend	2	3	4	5
Evening blend	4	2	3	3

The total times available each week for processing, blending and packing are 35, 20 and 24 hours respectively. T42 Co. Ltd wishes to maximise the weekly profit.

Let x, y and z be the number of tonnes of Morning, Afternoon and Evening blend produced each week.

a Formulate the above situation as a linear programming problem, listing clearly the objective function, and the constraints as inequalities. **(4)**

A An initial simplex tableau for the above situation is

Basic variable	x	y	z	r	s	t	Value
r	3	2	4	1	0	0	35
s	1	3	2	0	1	0	20
t	2	4	3	0	0	1	24
P	−4	−5	−3	0	0	0	0

b Solve this linear programming problem using the simplex algorithm. Take the most negative number in the profit row to indicate the pivot column at each stage. **(5)**

T42 Co. Ltd wishes to increase its profit further and is prepared to increase the time available for processing or blending or packing or any two of these three.

c Use your answer to part **b** to advise the company as to which stage(s) should be allocated increased time. **(2)**

← Sections 7.1, 7.2

E/P **8** A company makes three sizes of lamps, small, medium and large. The company is trying to determine how many of each size to make in a day, in order to maximise its profit. As part of the process the lamps need to be sanded, painted, dried and polished. A single machine carries out these tasks and is available 24 hours per day. A small lamp requires one hour on this machine, a medium lamp 2 hours and a large lamp 4 hours.

Let x = number of small lamps made per day

y = number of medium lamps made per day

z = number of large lamps made per day

where $x \geqslant 0$, $y \geqslant 0$ and $z \geqslant 0$.

a Write the information about this machine as a constraint. **(3)**

b i Rewrite your constraint from part **a** using a slack variable s. **(1)**

A **ii** Explain what s means in practical terms. **(1)**

Another constraint and the objective function give the following simplex tableau. The profit P is stated in euros.

Basic variable	x	y	z	r	s	Value
r	3	5	6	1	0	50
s	1	2	4	0	1	24
P	−1	−3	−4	0	0	0

c Write down the profit on each small lamp. **(1)**

d Use the simplex algorithm to solve this linear programming problem. **(5)**

e Explain why the solution to part **d** is not practical. **(1)**

f Find a practical solution which gives a profit of 30 euros. Verify that it is feasible. **(3)**

← Sections 7.1, 7.2, 7.3

E **9** A carpenter makes small, medium and large chests of drawers. The small size requires $2\frac{1}{2}$ m of board, the medium size 10 m of board and the large size 15 m of board. The times required to produce a small chest, a medium chest and a large chest are 10 hours, 20 hours and 50 hours respectively. In a given year there are 300 m of board available and 1000 production hours available.

Let the number of small, medium and large chests made in the year be x, y and z respectively.

a Show that the above information leads to the inequalities

$$x + 4y + 6z \leqslant 120$$
$$x + 2y + 5z \leqslant 100$$

The profits made on small, medium and large chests are £10, £20 and £28 respectively. **(2)**

b Write down an expression for the profit £P in terms of x, y and z. **(2)**

A The carpenter wishes to maximise his profit. The simplex algorithm is to be used to solve this problem.

c Write down the initial tableau using r and s as slack variables. **(2)**

d Use two iterations of the simplex algorithm to obtain the following tableau. In the first iteration you should increase y. **(3)**

Basic variable	x	y	z	r	s	Value
y	0	1	$\frac{1}{2}$	$\frac{1}{2}$	$-\frac{1}{2}$	10
x	1	0	4	-1	2	80
P	0	0	22	0	10	1000

e Give a reason why this tableau is optimal. **(1)**

f Write down the number of each type of chest that should be made to maximise the profit. State the maximum profit. **(2)**

← **Sections 7.1, 7.2, 7.3**

(E/P) 10 A linear programming problem is defined by the following conditions.

Maximise $P = x + 2y + 3z$

subject to:

$$x + y + 2z \leq 10$$
$$x + 3y + z \leq 15$$
$$2x + y + z \geq 12$$
$$x, y, z \geq 0$$

a Explain why this problem cannot be solved directly using the simplex algorithm. **(1)**

The initial tableau for the two stage simplex method is shown below.

Basic variable	x	y	z	s_1	s_2	s_3	a_1	Value
s_1	1	1	2	1	0	0	0	10
s_2	1	3	1	0	1	0	0	15
a_1	2	1	1	0	0	-1	1	12
P	-1	-2	-3	0	0	0	0	0
I	-2	-1	-1	0	0	1	0	-12

A **b** Rewrite the constraints as equations using the notation given in the tableau. **(2)**

c Explain the purpose of the final row of the tableau and show how it was derived. **(3)**

d Choose a suitable pivot and carry out one iteration. **(4)**

e Explain what your answer to part **d** shows. **(1)**

In the second stage of the two-stage simplex method, the following tableau is produced.

Basic variable	x	y	z	s_1	s_2	s_3	Value
z	0	$\frac{1}{3}$	1	$\frac{2}{3}$	0	$\frac{1}{3}$	$\frac{8}{3}$
s_2	0	$\frac{7}{3}$	0	$-\frac{1}{3}$	1	$\frac{1}{3}$	$\frac{23}{3}$
x	1	$\frac{1}{3}$	0	$-\frac{1}{3}$	0	$-\frac{2}{3}$	$\frac{14}{3}$
P	0	$-\frac{2}{3}$	0	$\frac{5}{3}$	0	$\frac{1}{3}$	$\frac{38}{3}$

f Explain how you know that the solution represented by the tableau is not optimal. **(1)**

g Carry out one more iteration and interpret the result. **(4)**

← **Section 7.4**

(E) 11 Here is a linear programming problem that is to be solved using the two-stage simplex method.

Maximise $P = 2x - y + z$

subject to:

$$x + 2y + 3z \leq 18$$
$$3x + y + z \geq 6$$
$$2x + 5y + z \geq 20$$
$$x, y, z \geq 0$$

a Express the constraints as equations using slack, surplus and artificial variables. **(2)**

b Write a new objective function for the first stage in terms of non-basic variables. **(2)**

A **c** Write the initial tableau for the first stage in standard form. **(3)**

d Complete the first stage and give the basic feasible solution. **(5)**

← **Section 7.4**

E/P **12** The Big-M method is to be used to solve the following linear programming problem.

Maximise $P = x + 3y + 4z$

subject to:

$$3x + 2y + z \leqslant 24$$
$$5x + 3y + 2z \leqslant 60$$
$$x \geqslant 2$$
$$y \geqslant 0$$

a Rewrite the constraints as equations using slack, surplus and artificial variables. **(2)**

b Write a modified objective function in terms of M and non-basic variables. **(2)**

c Construct the initial tableau for the Big-M method. **(3)**

d Explain why after the first iteration x replaces a_1 as a basic variable. **(1)**

e Apply the Big-M method to solve the problem. **(5)**

← **Section 7.5**

E/P **13** A linear programming problem is expressed as:

Minimise $C = 2x - 3y + z$

subject to:

$$4x + 3y + 2z \leqslant 36$$
$$x + 4z \leqslant 52$$
$$x + y \geqslant 10$$
$$x, y, z \geqslant 0$$

a Rewrite the constraints as equations using slack, surplus and artificial variables. **(2)**

b Rewrite the objective function in a form suitable for using the big-M method. **(2)**

A **c** Construct the initial tableau for the big-M method. **(2)**

d Complete the solution and state the values of x, y, z and C. **(6)**

← **Section 7.5**

E **14** The precedence table for activities involved in a small project is shown below.

Activity	Preceding activities
A	–
B	–
C	–
D	B
E	A
F	A
G	B
H	C, D
I	E
J	E
K	F, G, I
L	H, J, K

Draw an activity network to model this project. You should not include any dummies in your network. **(5)**

← **Section 8.1**

E **15** The precedence table for activities involved in manufacturing a toy is shown below.

Activity	Preceding activities
A	–
B	–
C	–
D	A
E	A
F	B
G	B
H	C, D, E, F
I	E
J	E
K	I
L	I
M	G, H, K

a Draw an activity network to model this process using exactly one dummy. **(5)**

b Explain briefly why it is necessary to use a dummy in this case. **(1)**

← **Sections 8.1, 8.2**

(E/P) 16 a Draw an activity network for the project described in this precedence table, using as few dummies as possible. **(5)**

Activity	Must be preceded by:
A	–
B	A
C	A
D	A
E	C
F	C
G	B, D, E, F
H	B, D, E, F
I	F, D
J	G, H, I
K	F, D
L	K

b A different project is represented by the activity network shown below. The duration of each activity is shown in brackets.

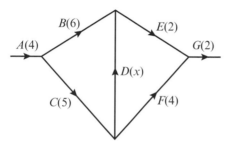

Find the range of values of x that will make D a critical activity. **(2)**

← Sections 8.2, 8.4

(E/P) 17*

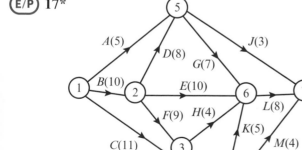

The diagram shows the activity network used to model a building project. The activities are represented by the edges.

The number in brackets on each edge represents the time, in days, to complete the activity.

a Calculate the early time and the late time for each event. **(3)**

b Calculate the total float for each activity. **(2)**

c Hence determine the critical activities. Write down the length of the critical path. **(2)**

Owing to the breakdown of a piece of equipment the time taken to complete activity H increases to 9 days.

d Obtain the new critical path and its length. **(2)**

← Sections 8.3, 8.4, 8.5

(E) 18*

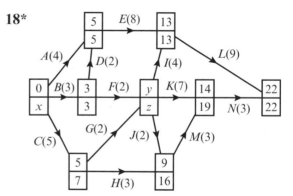

A trainee at a building company is using critical path analysis to help plan a project. The diagram shows the trainee's activity network. Each activity is represented by an arc and the number in brackets on each arc is the duration of the activity, in hours.

a Find the values of x, y and z. **(2)**

b State the total length of the project and list the critical activities. **(2)**

c Calculate the total float time on:
  i activity N
  ii activity H **(2)**

← Sections 8.3, 8.4, 8.5

E **19*** A project is modelled by the activity network shown below. The activities are represented by the arcs. The number in brackets on each arc gives the time, in hours, to complete the activity. The numbers in circles give the event numbers.

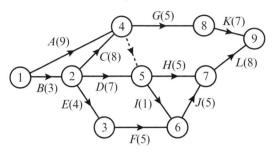

a Explain the purpose of the dotted line from event 4 to event 5. **(1)**

b Calculate the early time and the late time for each event. **(3)**

c Determine the critical activities. **(1)**

d Obtain the total float for each of the non-critical activities. **(2)**

e On a grid, draw a cascade (Gantt) chart, showing the answers to parts **c** and **d**. **(3)**

← Sections 8.2, 8.3, 8.4, 8.5, 8.6

E/P **20*** A building project is modelled by the activity network shown below. The activities are represented by the arcs. The number in brackets on each arc gives the time, in hours, taken to complete the activity.

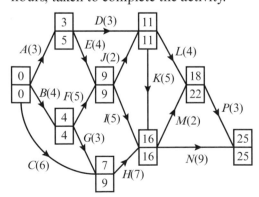

a Determine the critical activities and state the length of the critical path. **(2)**

b State the total float for each non-critical activity. **(2)**

c On a grid, draw a cascade (Gantt) chart for the project. **(4)**

A manager visits the project site 6.5 hours after the start of the project.

d Given that the project is on schedule, which activities must be taking place? **(2)**

← Sections 8.4, 8.5, 8.6

A **21** The table gives information about the activities involved in a software **E/P** development project.

Activity	Immediately preceding activities	Duration (days)	Workers
A	–	3	1
B	–	2	3
C	A	4	2
D	B	3	2
E	C, D	2	1
F	E	4	3
G	D	4	1
H	E, G	3	4
I	F	3	1

a Draw a resource histogram for the project. You may assume that activities start at their earliest possible time, and that once started, an activity is completed without a break. **(5)**

b State the minimum time required to complete the project and the number of workers required. **(1)**

c Use resource levelling to show that the project can be completed using 5 workers without extending the time for the project. **(2)**

d Given that only 4 workers are available, calculate the minimum time by which the project will be delayed. **(2)**

← Sections 8.7

A **22*** This activity network represents a project.
E/P The number in brackets represents the
duration, in days, of that activity.

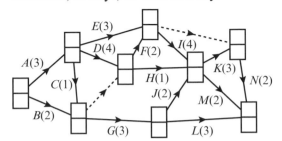

a Copy and complete the diagram to
show the early and late times for each
event. **(3)**

b State the minimum time required to
complete the project. **(1)**

c Describe the critical path. **(1)**

d Draw a resource histogram. Assume
that one worker is assigned to each
activity and that each activity starts at
its earliest time. **(4)**

e State the number of workers used. **(1)**

f Use resource levelling to show that the
project can be completed, using just
two workers, without extending the
time for the project. **(4)**

← Sections 8.3, 8.4, 8.7

E/P **23*** The network below shows the activities
involved in the process of producing a
perfume. The activities are represented
by the arcs. The number in brackets on
each arc gives the time, in hours, taken to
complete the activity.

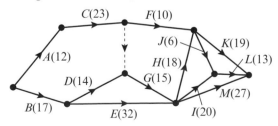

a Calculate the early time and the late
time for each event, showing them on a
diagram. **(3)**

b Hence determine the critical activities.
(1)

A c Calculate the total float time for D. **(1)**
Each activity requires only one person.

d Find a lower bound for the number
of workers needed to complete the
process in the minimum time. **(1)**

Given that there are only three workers
available, and that workers may **not** share
an activity,

e schedule the activities so that the
process is completed in the shortest
time. Use a time line. State the new
shortest time. **(4)**

← Sections 8.3, 8.4, 8.8

E/P **24*** A project is modelled by the activity
network shown below. The activities are
represented by the edges. The number in
brackets on each edge gives the time, in
days, taken to complete the activity.

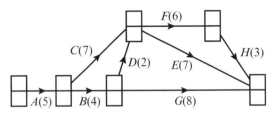

a Copy and complete the diagram to
show the early and late times for each
event. **(3)**

b Hence determine the critical activities
and the length of the critical path. **(2)**

c Obtain the total float for each of the
non-critical activities. **(2)**

d Draw a cascade (Gantt) chart showing
the information obtained in parts **b**
and **c**. **(3)**

Each activity requires one worker.
Only two workers are available.

e Draw up a schedule and find the
minimum time in which the two
workers can complete the project. **(3)**

← Sections 8.3, 8.4, 8.5, 8.6, 8.8

Challenge

1 A linear programming problem is given as:

Maximise $P = x + 5y$

subject to:

$$x + 4y \leqslant 24$$
$$y - x \leqslant 0$$
$$x < 8$$
$$x, y \geqslant 0$$

a Show the feasible region for this problem on a graph, and find the optimal values of x and y.

b Given the further restriction that $x, y \in \mathbb{Z}$, find a solution to the problem by testing values near your answer to part **a**.

c Show that the optimal integer solution does not lie near your answer to part **a**.

d With reference to the gradient of the objective line, describe conditions in which the optimal integer solution may not lie close to the optimal vertex of the feasible region.

← Sections 6.1, 6.2, 6.3, 6.4

2 A linear programming problem is expressed as:

Minimise $C = 2x - 3y + z$

subject to:

$$4x + 3y + 2z \leqslant 36$$
$$x + 4z \leqslant 52$$
$$x + y \geqslant 10$$
$$x, y, z \geqslant 0$$

Choose a suitable method to solve this problem, giving the optimal values of C, x, y and z.

← Section 7.5

Exam-style practice
Further Mathematics
AS Level
Decision 1

Time: 50 minutes
You must have: Mathematical Formulae and Statistical Tables, Calculator

1   **a**   Use the bubble sort to write these numbers in ascending order.

       21   16   11   25   18   15   23   10

You should only show the state of the list at the **end of each pass**.   **(3)**

    b   Show that the bubble sort algorithm has quadratic order.   **(3)**

A computer takes 0.018 seconds to apply the bubble sort to a list of 50 numbers.

    c   Estimate how long it would take the computer to sort a list of 1000 numbers using the bubble sort. Explain why your answer is only an estimate.   **(2)**

2*   The diagram shows the time taken, in minutes, to drive along some sections of road.

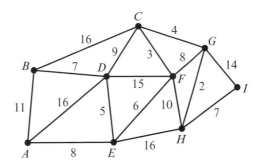

    a   Use Dijkstra's algorithm to find the quickest route from A to I. State the time required.   **(5)**

Roadworks cause a delay of 5 minutes on the section GH.

    b   Find the new quickest route and the corresponding time required.   **(2)**

3   A magazine offers two advertising packages for its classified section:

       Package A: picture area: $20\,cm^2$, price: £40 including up to 25 words

       Package B: picture area: $30\,cm^2$, price: £55 including up to 60 words

In a particular issue, up to $1000\,cm^2$ of picture space is available, and at most 1600 words can be used in total.

The magazine needs to generate a minimum of £1200 of revenue from the advertising sales.

The profit made on each Package A sold is £24.

The profit made on each Package B sold is £32.

Formulate this as a linear programming problem in which the object is to maximise the total profit. Write each constraint in its simplest form.

You should not attempt to solve the problem. **(6)**

4* The diagram shows a network of tunnels that are to be inspected. The number on each arc represents the time required, in minutes, to inspect each tunnel. The inspection route is to start and finish at *B* and every tunnel must be traversed at least once. The total time taken is to be minimised.

The total weight of the network is 144.

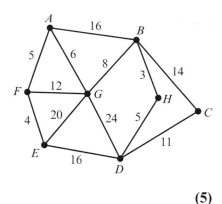

a Use an appropriate algorithm to determine which tunnels need to be traversed twice and state the total time required for the inspection. **(5)**

A new tunnel is added to the network, joining *A* directly to *E*. A manager states that after this addition, only tunnel *FG* should be traversed twice to create a minimum inspection route.

b Explain why the manager is incorrect. **(2)**

Following the introduction of the new tunnel, the duration of the minimum inspection route is found to have increased by 4 minutes.

c Find the length of the new tunnel. **(2)**

5* Here is an activity network for an engineering project. The number in brackets on each arc gives the time in hours to complete the activity.

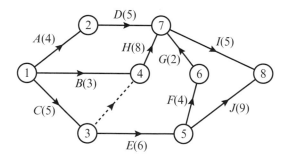

a Draw a precedence table for the project showing the immediately preceding activities. **(3)**
b Explain the purpose of the dotted line between events 3 and 4. **(1)**
c Draw a diagram showing the early time and late times for each event. **(3)**
d Describe the critical path. **(1)**
e Calculate the total float for activity *D*. **(2)**

Exam-style practice
Further Mathematics
A Level
Decision 1

Time: 1 hour and 30 minutes
You must have: Mathematical Formulae and Statistical Tables, Calculator

1* **a** Explain why there is no Hamiltonian cycle for this network that begins AB... **(1)**

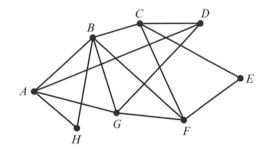

  b Complete the Hamiltonian cycle for the network that begins AH... **(1)**

  c Use the planarity algorithm to determine whether the graph is planar.

   You must make your method clear. **(4)**

2 The following lengths of cable, in metres, are required for a building project.

   42   31   36   18   27   33   41   47   12   24   16

  a Carry out a quicksort to put the lengths in decreasing order of size. **(4)**

  The cable is supplied in reels of length 80 metres.

  b Use the first fit decreasing algorithm to determine the number of reels required. **(3)**

  The quicksort algorithm has order $n \log n$. A computer applies the quicksort to a list of 800 numbers in 0.034 seconds.

  c Estimate the time required for the computer to sort a list of 5000 numbers. Explain why your answer is only an estimate. **(3)**

3* **a** Explain what is meant when a network is described as semi-Eulerian. **(1)**

The diagram represents a road network that must be checked for damage after a storm.

The number on each arc is the length of the corresponding road in miles.

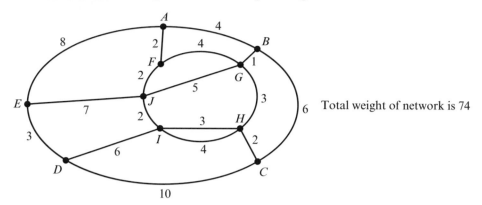

Total weight of network is 74

Each road must be traversed at least once to complete the inspection, starting at E and finishing at B. The total distance travelled is to be kept to a minimum.

b Work out the minimum length of the route. **(6)**

c State which roads need to be traversed twice. **(2)**

d In a change of plan, the route is to be finished at C. Work out how this affects the total distance travelled. **(3)**

4* **a** Name an algorithm that can be used to find the shortest path between two vertices in a network. **(1)**

Floyd's algorithm is to be applied to a network. The initial distance table and route table are shown below.

	A	B	C	D	E
A	–	5	∞	∞	∞
B	5	–	11	16	∞
C	10	11	–	8	6
D	∞	16	8	–	10
E	∞	∞	6	10	–

	A	B	C	D	E
A	A	B	C	D	E
B	A	B	C	D	E
C	A	B	C	D	E
D	A	B	C	D	E
E	A	B	C	D	E

b Name the arc that can only be traversed in one direction. **(1)**

c Use Floyd's algorithm to produce a complete table of least distances. Show the distance and route tables after each iteration. **(7)**

d Find the shortest distance from A to E and state the route used. **(2)**

5 Angie supplies two types of party-pack for children's parties:

A **mini-pack** containing 12 sandwiches, 5 slices of cake, 5 orange drinks and 6 blackcurrant drinks.

A **mega-pack** containing 30 sandwiches, 20 slices of cake, 5 orange drinks and 12 blackcurrant drinks.

Angie can make at most 240 sandwiches.

The number of sandwiches she can make is at most double the number of slices of cake she can make.

She wants to use at least 55 orange drinks and at least 96 blackcurrant drinks.

Assume that Angie makes x mini-packs and y mega-packs.

a Write and simplify 4 inequalities to represent the constraints, other than the non-negativity constraints. **(4)**

b Represent the inequalities on a diagram and label the feasible region R. **(4)**

Angie makes a profit of £8 on each mini-pack and £21 on each mega-pack. She wants to maximise her profit.

c Use vertex testing to work out how many of each pack she should make and state the corresponding profit. **(4)**

6* The diagram shows an activity network for a publishing project. The activities are shown on arcs and the figures in brackets represent the duration of each activity in days.

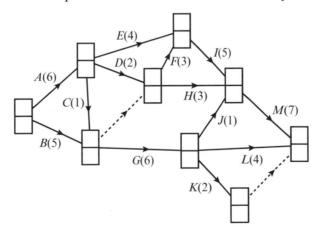

a Complete the diagram by showing the early and late event times. **(2)**

b Draw a resource histogram for the project. Assume that each activity is started as soon as possible and may be completed by one worker without any breaks. State the number of workers required in this instance. **(4)**

c Describe how the project can be completed using just 3 workers without extending the time for completion. **(2)**

7 **a** Describe the purpose of the first stage of the two-stage simplex method. **(2)**

A linear programming problem is described as:

Maximise $P = x + 2y + z$

subject to:

$$2x + y + z \leqslant 50$$
$$x + 3y + z \leqslant 60$$
$$x \geqslant 10$$

The partly completed initial tableau for this problem is

Basic variable	x	y	z	s_1	s_2	s_3	a_1	Value
s_1	2	1	1	1	0	0	0	50
s_2	1	3	1	0	1	0	0	60
a_1	1	0	0	0	0	-1	1	10
P								
I								

In the tableau, s_1 and s_2 are slack variables, s_3 is a surplus variable, a_1 is an artificial variable and I is a modified objective function.

b Complete the tableau. **(3)**

c Find the optimal solution for the first stage and state the value of each of the variables. **(3)**

d Set up the tableau for the second stage. Give the value and position of the pivot. **(2)**

e Complete the solution and state the values of P, x, y and z. **(6)**

CHAPTER 1

Prior knowledge 1

1 a 47   b 24
2 300

Exercise 1A

1 a

A	B
244	125
122	250
61	500
30	1000
15	2000
7	4000
3	8000
1	16 000
Total	30 500

b

A	B
125	244
62	488
31	976
15	1952
7	3904
3	7808
1	15 616
Total	30 500

c

A	B
256	123
128	246
64	492
32	984
16	1968
8	3936
4	7872
2	15 744
1	31 488
Total	31 488

2 a 1 $\frac{a}{b} = \frac{9}{4}$ $\frac{c}{d} = \frac{4}{3}$ $a = 9, b = 4, c = 4, d = 3$

   2 $e = ad = 9 \times 3 = 27$
   3 $f = bc = 4 \times 4 = 16$
   4 Answer is $\frac{27}{16}$

   b It divides the first fraction by the second fraction.

3 a 1, 4, 9, 16, 25, 36, 49, 64, 81, 100
   b The first 10 square numbers.

4 a i

Step	A	r	C	$\lvert r - C \rvert$	s	Print r
1	253	12				
2			21.083			
3				9.083		
4					16.542	
5		16.542				
6 → 2			15.294			
3				1.248		
4					15.918	
5		15.918				
6 → 2			15.894			
3				0.024		
4					15.906	
5		15.906				
6 → 2			15.906			
3 → 7				0		
7						r = 15.906
8 Stop						

ii

Step	A	r	C	$\lvert r - C \rvert$	s	Print r
1	79	10				
2			7.900			
3				2.1		
4					8.950	
5		8.95				
6 → 2			8.827			
3				0.123		
4					8.889	
5		8.889				
6 → 2			8.887			
3 → 7				0.002		
7						Print 8.889

iii

Step	A	r	C	$\lvert r - C \rvert$	s	Print r
1	4275	50				
2			85.500			
3				35.5		
4					67.750	
5		67.75				
6 → 2			63.100			
3				4.65		
4					65.425	
5		65.425				
6 → 2			65.342			
3				0.083		
4					65.384	
5		65.384				
6 → 2			65.383			
3 → 7				0.001		
7						Print 65.384

b Finds the square root of A.

Exercise 1B

1 a

a	b	c	d	d < 0?	d = 0	x
4	−12	9	0	No	Yes	1.5

Equal roots are 1.5

b

a	b	c	d	d < 0?	d = 0?	x_1	x_2
−6	13	5	289	No	No	$-\frac{1}{3}$	$\frac{5}{2}$

Roots are $-\frac{1}{3}$ and $\frac{5}{2}$

c

a	b	c	d	$d < 0$
3	−8	11	−68	Yes

No real roots

2 **a i** Output is 21 **ii** Output is 5
 b It will find the largest number in the list.
 c Box 6 – changed to 'Is $n < 8$?'
3 **a** 1.8041 (4 d.p.)
 b 1.8041 (4 d.p.)
  The sequence produced in part **b** is initially quite different to the sequence produced in part **a** but both sequences converge to the same root.
4 **a i** 13 **ii** 17 **iii** 15
 b The algorithm finds the highest common factor of the two input values.
5 **a**

A	B	$A < B$?	Output
18	7	No	
11	7	No	
4	7	Yes	4

 b Calculates remainder when A is divided by B.
 c 0

Exercise 1C

1 **i a** 16 15 23 18 25 11 19 34
    15 16 18 23 11 19 25 34
    15 16 18 11 19 23 25 34
    15 16 11 18 19 23 25 34
    15 11 16 18 19 23 25 34
    11 15 16 18 19 23 25 34
    11 15 16 18 19 23 25 34
   b 23 16 34 18 25 15 19 11
    23 34 18 25 16 19 15 11
    34 23 25 18 19 16 15 11
    34 25 23 19 18 16 15 11
    34 25 23 19 18 16 15 11

 ii a E N T O R K S W
    E N O R K S T W
    E N O K R S T W
    E N K O R S T W
    E K N O R S T W
    E K N O R S T W
   b N T W O R K S E
    T W O R N S K E
    W T R O S N K E
    W T R S O N K E
    W T S R O N K E
    W T S R O N K E

 iii a A5 D2 A1 B4 C7 C2 B3 D3
    A5 A1 B4 C7 C2 B3 D2 D3
    A1 A5 B4 C2 B3 C7 D2 D3
    A1 A5 B4 B3 C2 C7 D2 D3
    A1 A5 B3 B4 C2 C7 D2 D3
    A1 A5 B3 B4 C2 C7 D2 D3
   b D3 D2 A5 B4 C7 C2 B3 A1
    D3 D2 B4 C7 C2 B3 A5 A1
    D3 D2 C7 C2 B4 B3 A5 A1
    D3 D2 C7 C2 B4 B3 A5 A1

2 Ch St Br Bu Cr Ev Yo
 Ch Br Bu Cr Ev St Yo
 Br Bu Ch Cr Ev St Yo
 Br Bu Ch Cr Ev St Yo
 Bridlington, Burton, Chester, Cranleigh, Evesham, Stafford, York

3 **a** 1
 b One pass is sufficient if the items are already in ascending order.
 c n
 d n passes are needed if the smallest item is at the end of the list.
4 **a** Move through the list comparing each pair of numbers. If the first number of a pair is greater than or equal to the second, make no change. If the first number of a pair is less than the second, swap them.
 b 63 57 55 48 48 72 49 61 39 32
  63 57 55 48 72 49 61 48 39 32
  63 57 55 72 49 61 48 48 39 32
  63 57 72 55 61 49 48 48 39 32
  63 72 57 61 55 49 48 48 39 32
  72 63 61 57 55 49 48 48 39 32
  72 63 61 57 55 49 48 48 39 32

Exercise 1D

1 **a** 2 3 4 5 6 7 8
 b 8 7 6 5 4 3 2
2 **a** 7 11 14 17 18 20 22 25 29 30
 b 30 29 25 22 20 18 17 14 11 7
3 **a** C E H J K L M N P R S
 b C E H J K L M N P R S
4 **a** Amy 93 Greg 91 Janelle 89 Sophie 77
  Dom 77 Lucy 57 Alison 56 Annie 51
  Harry 49 Josh 37 Alex 33 Sam 29
  Myles 19 Hugo 9
 b 93 91 89 77 77 57 56 51 49 37 33 29 19 9
5 **a** $(n-1) + (n-2) + \dots 3 + 2 + 1 = \dfrac{n(n-1)}{2}$
 b The bubble sort would be quicker, for example, if the items are to be put in increasing order and if the only item out of place is the largest.
 c i Bubble sort. Only the 7 is out of place and it will be moved to its final position in the first pass. A second pass is still needed to complete the bubble sort. A total of 11 comparisons is needed for the bubble sort and 14 are needed for the quick sort.
  ii Quick sort. This is the worst case for the bubble sort. The 1 is at the wrong end of the list and only moves one place with each pass.
6 **a** There are 9 names in the list. The 5th name (Mellor) is taken as the pivot. Starting at the beginning of the list, each name is compared with Mellor and placed on the appropriate side to produce two sub lists. The process is repeated for each sub-list with pivot of G on the left and Mi on the right.
 b H S F G (Me) C Mi W A (M is the pivot)
  H F (G) C A Me S (Mi) W (1st pass)
  F (C) A (G)(H) Me Mi S (W) (2nd pass)

Challenge
a Answers will vary.
b Put the Ace of Hearts at the end.

Online Full worked solutions are available in SolutionBank.

Exercise 1E

1 a 5 bins

   b **i** Bin 1: 18 + 4 + 23 + 3
       Bin 2: 8 + 27
       Bin 3: 19 + 26
       Bin 4: 30
       Bin 5: 35
       Bin 6: 32

     ii Bin 1: 35 + 8 + 4 + 3
       Bin 2: 32 + 18
       Bin 3: 30 + 19
       Bin 4: 27 + 23
       Bin 5: 26

    iii Bin 1: 32 + 18
       Bin 2: 27 + 23   } full bins
       Bin 3: 35 + 8 + 4 + 3
       Bin 4: 19 + 26
       Bin 5: 30

2 a Bin 1: A(30) + B(30) + C(30) + D(45) + E(45)
     Bin 2: F(60) + G(60) + H(60)
     Bin 3: I(60) + J(75)
     Bin 4: K(90)
     Bin 5: L(120)
     Bin 6: M(120)

   b Bin 1: M(120) + I(60)
     Bin 2: L(120) + H(60)
     Bin 3: K(90) + J(75)
     Bin 4: G(60) + F(60) + E(45)
     Bin 5: D(45) + C(30) + B(30) + A(30)

   c Lower bound 5 so **b** optimal.

   d Bin 1: M(120)
     Bin 2: L(120)
     Bin 3: A(30) + K(90)
     Bin 4: F(60) + G(60) } full bins
     Bin 5: H(60) + I(60)
     Bin 6: J(75) + E(45)
     Bin 7: B(30) + C(30) + D(45)

3 a e.g. First-fit does not rely on observation.

   b 4

   c Bin 1: A, B, C, D; Bin 2: E, I, J; Bin 3: F, G, H
     Each of the three lanes is full, so solution is optimal.

4 a 4 rolls

   b 5 rolls, 15 m wasted

   c Doesn't always give an optimal solution

   d e.g. Bin 1: A, C, L; Bin 2: B, D, E, F; Bin 3: G, H, I,
     Bin 4: J, K
     No carpet is wasted.

5 a Bin 1: H(25) + A(8)
     Bin 2: G(25)
     Bin 3: F(24) + B(16)
     Bin 4: E(22) + C(17)
     Bin 5: D(21)

   b Lower bound is 4.

   c There are 5 programs over 20 MB. It is not possible
     for any two of these to share a bin. So at least 5
     bins will be needed, so 4 will be insufficient.

Exercise 1F

1 a 1.12 seconds   **b** 5.19 seconds

2 a To pack the kth item requires at most $k - 1$
     comparisons (if every item placed so far is in a
     separate bin). Hence the total number of

comparisons for n items is $\sum_{k=1}^{n}(k-1) = \frac{1}{2}n(n-1)$
which is a quadratic expression.

   b $0.72 \times \left(\dfrac{6200}{400}\right)^2 = 173$ seconds (3 s.f.)

   c The exact run-time will depend on the specific
     lengths of pipe.

3 a If the size of the problem is multiplied by k, then
     the algorithm will take approximately k^2 times as
     long to run.

   b $0.028 \times \left(\dfrac{500}{50}\right)^2 = 0.028 \times 100 = 2.8$ seconds

4 If the runtime of bubble sort is $an^2 + bn + c$ and the
   runtime of first-fit bin-packing is $pn^2 + qn + r$ then the
   combined runtime will be $(an^2 + bn + c) + (pn^2 + qn + r)$
   $= (a + p)n^2 + (b + q)n + (c + r)$ which is a quadratic
   expression, so the combined process has quadratic (n^2)
   order.

5 n^4: The bubble sort (n^2) is applied to a list of n^2 items.

Mixed exercise 1

1

15	2	27	16	1	38
2	15	16	1	27	38
2	15	1	16	27	38
2	1	15	16	27	38
1	2	15	16	27	38

2 a

42	31	25	26	41	22
42	31	26	41	25	22
42	31	41	26	25	22
42	41	31	26	25	22

   b 15

3

8	4	13	(2)	17	9	15
[2]	8	4	13	(17)	9	15
[2]	8	4	(13)	9	15	[17]
[2]	8	(4)	9	[13]	(15)	[17]
[2]	[4]	8	(9)	[13]	[15]	[17]
[2]	[4]	(8)	9	[13]	[15]	[17]

4 a

111	103	77	81	(98)	68	82	115	93
111	(103)	115	[98]	77	81	(68)	82	93
111	(115)	[103]	[98]	77	81	(82)	93	[68]
[115]	(111)	[103]	[98]	(93)	[82]	77	(81)	[68]
[115]	[111]	[103]	[98]	[93]	[82]	81	(77)	[68]

   b **i** Bin 1: 115 + 82
      Bin 2: 111 + 81
      Bin 3: 103 + 93
      Bin 4: 98 + 77
      Bin 5: 68

     ii No room in bin 1 (3 left) or bin 2 (8 left) or bin 3
      (4 left) but room in bin 4.

5 a Rank the times in descending order and use them
     in this order. Number the bins, starting with bin 1
     each time.

b Bin 1: 100
Bin 2: 92
Bin 3: 84 + 30
Bin 4: 75 + 42
Bin 5: 60 + 52
unused tape: 65 minutes

c There is room on tape 2 for one of the 25-minute programmes but no room on any tape for the second programme.

6 a The two 1.2 m lengths cannot be 'made up' to 2 m bins since there are only 3 × 0.4 m lengths. Two of these can be used to make a full bin, 1.2 + 0.4 + 0.4, but the second 1.2 m cannot be made up to 2 m since there is only one remaining 0.4 m length.

b Bin 1: 1.6 + 0.6
Bin 2: 1.4 + 1
Bin 3: 1.2 + 1.2
Bin 4: 1 + 1 + 0.4
Bin 5: 0.6 + 0.6 + 0.6 + 0.4
Bin 6: 0.4

c Bin 1: 1.6 + 0.4 + 0.4
Bin 2: 1.4 + 1
Bin 3: 1.2 + 1.2
Bin 4: 1 + 1 + 0.4
Bin 5: 0.6 + 0.6 + 0.6 + 0.6

7 a Output 4.8
b It selects the number nearest to 5.
c It would select the number furthest from 5.

8 a 1st pass: 2.0 1.3 1.6 0.3 1.3 0.3 0.2 2.0 0.5 0.1
2nd pass: 2.0 1.6 1.3 1.3 0.3 0.3 2.0 0.5 0.2 0.1
3rd pass: 2.0 1.6 1.3 1.3 0.3 2.0 0.5 0.3 0.2 0.1
4th pass: 2.0 1.6 1.3 1.3 2.0 0.5 0.3 0.3 0.2 0.1
5th pass: 2.0 1.6 1.3 2.0 1.3 0.5 0.3 0.3 0.2 0.1
6th pass: 2.0 1.6 2.0 1.3 1.3 0.5 0.3 0.3 0.2 0.1
7th pass: 2.0 2.0 1.6 1.3 1.3 0.5 0.3 0.3 0.2 0.1
8th pass: 2.0 2.0 1.6 1.3 1.3 0.5 0.3 0.3 0.2 0.1

b Bin 1: 2.0; Bin 2: 2.0; Bin 3: 1.6 + 0.3 + 0.1
Bin 4: 1.3 + 0.5 + 0.2; Bin 5: 1.3 + 0.3
5 lengths of pipe needed

c Yes: Total length required is 9.6 m, so lower bound is 4.8, rounded up to 5 lengths of pipe.

9 a M V C A ⒟ B K S
C ⒜ B ▢D▢ M V ⒦ S
▢A▢ C ⒝▢D▢▢K▢ M ⒱ S
▢A▢▢B▢⒞▢D▢▢K▢ M ⒮ V
▢A▢▢B▢▢C▢▢D▢▢K▢ ⒨ ▢S▢ V

b $0.3 \times \dfrac{1000 \log 1000}{100 \log 100} = 0.3 \times \dfrac{1000 \times 3}{100 \times 2} = 4.5$ seconds

Challenge
a i 0.977 seconds **ii** 5191 seconds
b 192
c e.g. If it checks small potential factors first, all even numbers will be factorised very quickly compared to a number which is a product of two large prime factors.

CHAPTER 2

Prior knowledge 2

1 If $a = b + c$, then ABC would be on a straight line, and if $a > b + c$, then sides b and c would not have sufficient combined length to meet at a **single point** A.

2
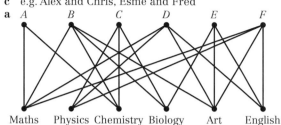

Exercise 2A

1 a i a student   **ii** that a pair of students are friends
b Becky, Dhevan, Esme
c e.g. Alex and Chris, Esme and Fred

2 a
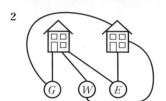

Maths   Physics Chemistry Biology   Art   English

b Maths and art

3 a i Marylebone, Oxford Circus, Victoria
ii Marylebone, Baker St, Green Park, Victoria
b i 11 min
ii 12 min 40 sec
iii 7 min
c e.g. It will take longer to change at some stations than others.

4 a 40 min
b Aberdeen/Cork
c Dublin – it is the airport with the most connections

5 a *PTV*
b Not correct. *PTQRSV* is 25 km

Challenge
a 6 routes   **b** 18 routes

Exercise 2B

1 a For example
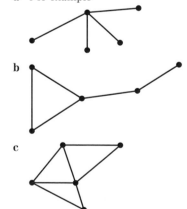

b

c

2 a is not simple. There are two edges connecting *C* with *D*.
b and **c** are simple.
d is not simple. There is a loop attached to *U*.

3 a and **c** are connected.
b is not connected, there is no path from *J* to *G*, for example.
d is not connected, there is no path from *W* to *Z*, for example.

Online Full worked solutions are available in SolutionBank.

4 a Any four of these

F A B D   F E D
F A C B D   F E C B D
F A B C E D   F E C A B D
F A C E D

b e.g. F A C B D E F, D B A F E D

c

Vertex	A	B	C	D	E	F
Degree	3	3	3	2	3	2

d Here are examples. (There are many others.)

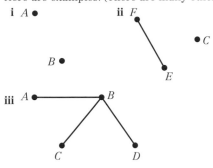

i A•

ii F   •C

B•

E

iii

e Sum of degrees = 3 + 3 + 3 + 2 + 3 + 2 = 16
number of edges = 8
sum of degrees = 2 × number of edges for this graph

5

Vertex	J	K	L	M	N	P	Q	R	S
Degree	1	2	3	1	1	4	2	1	1

Here are some possible subgraphs (there are many others).

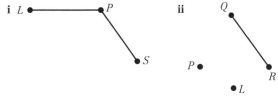

i L•   •P

ii Q

P•   •R

•L

•S

iii J•   •S

Sum of degrees = 1 + 2 + 3 + 1 + 1 + 4 + 2 + 1 + 1 = 16
number of edges = 8
sum of degrees = 2 × number of edges for this graph

6 a For example

4 vertices of degree 2

b For example

3 vertices all even, all of degree 2

c The sum of degrees = 2 × number of edges, so the sum of degrees must be even. Any vertices of odd degree must therefore 'pair up'. So there must be an even number of vertices of odd degree.

7

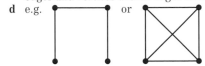

8 a A Hamiltonian cycle is a cycle that includes every vertex.

b PQRSTP, PQSRTP, PTSRQP, PTRSQP

c A subgraph of some graph G is a graph whose edges and vertices are all edges and vertices of G.

d e.g.   or

Exercise 2C

1 a and **b** are trees.
c is not a tree, it is not a connected graph.
d is not a tree, it contains a cycle.

2

i

ii

iii

iv

v

vi

vii

viii

ix

x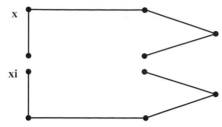

xi

3 BCEF

4 A, C

5 **a** **i** A connected graph with no cycles
  ii A subgraph which includes all vertices and is a tree
  b The graph is not connected.

6 **a**

  b Each vertex will have a degree of $n - 1$
  c 190

Challenge

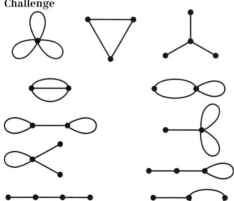

b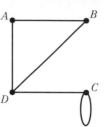

c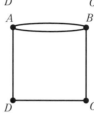

d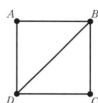

3 **a**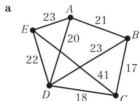

b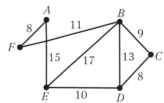

Exercise 2D

1

	A	B	C	D	E	F	G
A	0	1	1	0	2	0	0
B	1	2	1	0	0	0	0
C	1	1	0	2	0	1	0
D	0	0	2	0	0	1	1
E	2	0	0	0	0	1	0
F	0	0	1	1	1	0	1
G	0	0	0	1	0	1	0

2 **a**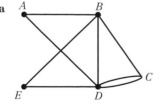

4

	A	B	C	D	E	F
A	—	14	11	—	—	—
B	—	—	10	13	11	—
C	—	—	—	12	—	—
D	—	—	—	—	—	10
E	—	—	—	—	—	7
F	—	—	—	—	9	—

5 **a**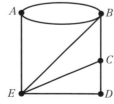

  b E

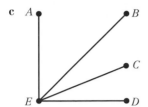

Exercise 2E

1 **a** Starting with *ABC* leaves no return path to *A* to complete the cycle.

b *ABEDFCA* or *ABEFDCA*

c e.g.

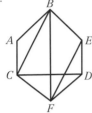

BC (I), *BF*, *CD*, *EF*,
BC (I), *BF* (I), *CD* (O), *EF*
BC (I), *BF* (I), *CD* (O), *EF* (I)
Graph is planar.

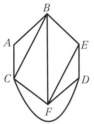

2 **a** A planar graph is one that can be drawn in a plane such that no two edges meet except at a vertex.

b e.g. *AGBHCEDFA*

c

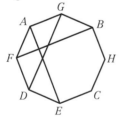

AE (I), *GF*, *GD*
GF and *GD* cross so graph is non-planar.

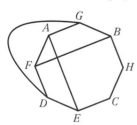

There are still two edges crossing inside the polygon. Neither can be removed without crossing the edge that is already outside.

Similarly, taking *AE* or *BF* outside first, the remaining edges cross inside and neither can be taken outside without crossing the outside edge.

The conclusion is that the original graph is not planar.

3 **a**

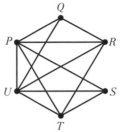

b The connections can be made without crossing any wires.

c

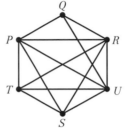

PR (I), *PU*, *PS*, *QU*, *RS*, *RT*, *UT*
PR (I), *PU*, *PS*, *QU* (O), *RS* (I), *RT* (I), *UT*
RT and *RS* must be in the same set, and hence *PS* and *TU* must be in the same set, but they cross, so graph is non-planar.

4

A Hamiltonian cycle is *AWCEBGA*. Redraw the graph inside a hexagon:

Fix edge *AE* and try to take edges *BW* and *CG* outside the cycle so that they don't cross. This is impossible.

Repeat, fixing edges *BW* and *CG*, which will also result in edges crossing. Therefore conclude that the graph is not planar.

Challenge

a The graph has no Hamiltonian cycle.

b Take edges *AC* and *EC* outside. No edges cross.

Mixed exercise 2

1

2 **a**, **e** and **h** are isomorphic.
b and **i** are isomorphic.
c and **g** are isomorphic.
d and **f** are isomorphic.

3 **a**

b **i**

ii

iii

4 **a** A distance matrix gives the weights of edges between pairs of vertices, whereas an adjacency matrix gives the number of edges between pairs of vertices.

b

	A	B	C	D	E	F
A	—	20	18	16	—	—
B	20	—	15	—	—	50
C	18	15	—	10	20	30
D	16	—	10	—	23	—
E	—	—	20	23	—	25
F	—	50	30	—	25	—

c e.g.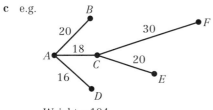

Weight = 104

5 $v - 1$

6 *PQR, PQTR, PTR, PSR*

7 **a** 3
b **i** e.g. *BEAD*   **ii** e.g. *ACEA*
c **i** e.g.

ii e.g.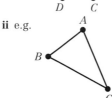

8 **a** The sum of the degrees of the vertices must be even. However, $3 + 1 + 2 = 7$ is odd, so there is no 4-vertex graph whose vertices have these degrees.
b 5

9 **a** *AGBECFDHA*
b

BH (I), *GC*, *DE*
BH (I), *GC* (O), *DE*
BH (I), *GC* (O), *DE* (I)
Graph is planar.

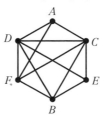

10 **a** e.g. *ACEBFDA*
b Drawing the graph inside a hexagon gives

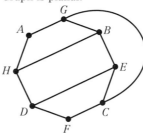

DC (I), *AF, CF, CB, DE, DB*
DC (I), *AF* (O), *CF, CB, DE, DB*
DC (I), *AF* (O), *CF, CB, DE* (I), *DB* (I)
DC (I), *AF* (O), *CF* (O), *CB* (O), *DE* (I), *DB* (I)
Graph is planar.

Online Full worked solutions are available in SolutionBank.

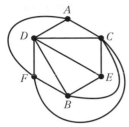

c The graph is not planar. Working should show that two arcs always cross.

Challenge

a $V = 7$, $E = 13$, $R = 7$; $7 + 7 - 13 = 1$

b

$V = 1$, $E = 4$, $R = 4$; $1 + 4 - 4 = 1$

c $E = R$, so $V + R - E = V = 1$

d We will prove the following statement by induction: The relationship $V + R - E = 1$ holds for all connected planar graphs with n vertices'. Basis step: part **c** shows that this statement holds for $n = 1$. Induction step: Assume true for any planar graph with $n - 1$ vertices. Then, given a graph G of n vertices, contract one edge to obtain a graph G' of $n - 1$ vertices. The induction hypothesis implies that G' satisfies the relation, and then part **d** implies that G also satisfies the relation.

CHAPTER 3

Prior knowledge 3

1 a 5

b

	A	B	C	D	E	F	G
A	–	12	15	–	–	4	9
B	12	–	7	14	–	–	–
C	15	7	–	9	5	11	–
D	–	14	9	–	16	–	–
E	–	–	5	16	–	–	2
F	4	–	11	–	–	–	3
G	9	–	–	–	2	3	–

c e.g. *AF* (4), *FG* (3), *GE* (2), *EC* (5), *BC* (7), *CD* (9). Total weight 30.

Exercise 3A

1 a

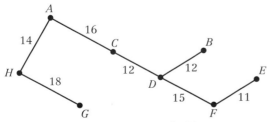

EF, *BD*, *CD*, *AH*, *DF*, *AC*, *GH*; weight 98

b

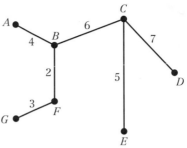

BF, *FG*, *AB*, *CE*, *BC*, *CD*; weight 27

c

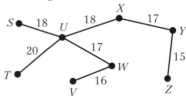

DE, *EF*, *CD*, *EG*, *GJ*, *AB*, *GH*, *BH*; weight 22.4

2 a i A tree is a connected graph with no cycles.
ii A minimum spanning tree is a tree of minimum total weight that connects all of the nodes.

b Writing the arcs in order of their weights: *YZ* (15), *VW* (16), *XY* (17), *UW* (17), *UX* (18), *WX* (18), *SU* (18), *WZ* (18), *UV* (19), *TU* (20), *ST* (22), *TV* (23)
Underlined arcs are in the minimum spanning tree. Total weight = 121

c

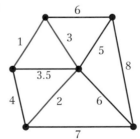

d The minimum spanning tree is not unique. For example, *UX* can be replaced with *WX*.

3 a

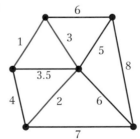

The three shortest edges do not form a cycle.

b

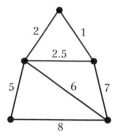

The three shortest edges form a cycle.

4 a

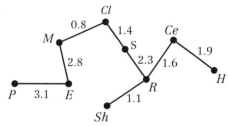

Arcs were added in the order MCl (0.8), ShR (1.1), ClS, CeR, CeH, SR, ME, PE

b 15 km

Exercise 3B

1

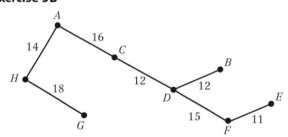

weight: 98

b

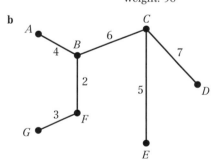

weight: 27

c

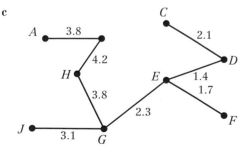

weight: 22.4

2 In Prim's algorithm, the starting point can be any node, whereas Kruskal's algorithm starts from the arc of least weight.
In Prim's algorithm, each new node and arc is added to the existing tree as it builds, whereas in applying Kruskal's algorithm, the arcs are selected according to their weight and may not be connected until the end.

3 a

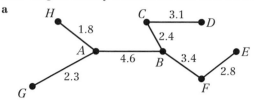

Selection order: AH, AG, AB, BC, CD, BF, EF

b £17 340

4 a

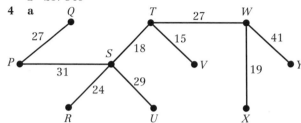

b £231 000

c Replace WX with VX. This is the cheapest way to link X to the spanning tree. The total cost is now £231 000 + £28 000 − £19 000 = £240 000

5 a Prim's algorithm identifies the next node to link to the existing tree. Linking a new node cannot form a cycle.

b One minimum connector is shown below.

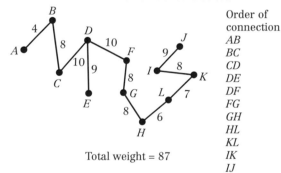

Total weight = 87

Order of connection
AB
BC
CD
DE
DF
FG
GH
HL
KL
IK
IJ

c The minimum connector is not unique; there are three minimum connectors with total weight 87.

Exercise 3C

1 a Arcs in order
AF (9)
FB (14)
AC (20)
AE (25)
DE (26)
weight = 94

b Arcs in order
RS (28)
ST (16)
SU (19)
UV (37)
weight = 100

2 Arcs in order
   BS (49)
   SM (44)
   SN (56)
   NL (37)
   weight = 186

3 **a** cost = €1014
  b

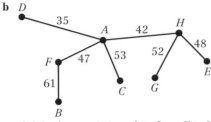

  c **i** It is cheaper to translate from E to H then from H to G at a cost of 48 + 52 = 100 euro rather than 159 euro per 1000 words.
     ii A direct translation is likely to be more accurate then a translation via another language.

4 **a** order of arcs
   XE (26)
   EG (18)
   EH (23)
   HA (25)
   AF (20)
   FB (16)
   AD (22)
   FC (24)
   CI (26)

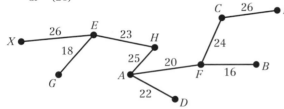

  b 9 oil rigs and 1 depot make 10 nodes.
    24 oil rigs and 1 depot make 25 nodes.
    Estimated time = $0.7 \times \left(\dfrac{25}{10}\right)^3$ = 10.9 seconds (1 d.p.)
  c **i** Any distance less than 26 miles will change the minimum connector as I will link directly to X.
     ii Any distance of 26 miles or more will not change the minimum connector as the shortest way to connect I to the rest of the tree will be to connect to C.

Exercise 3D

1 **a** Shortest route: $S - B - E - H - G - C - F - T$
    Length of shortest route: 20
  b Shortest route: $S - A - B - D - F - H - T$
    Length of shortest route: 15
2 **a** A to Q $A - F - E - I - N - P - Q$ Length 37
  b A to L $A - F - G - M - L$ Length 19
  c M to A $M - G - F - A$ Length 17
  d P to A $P - N - I - E - F - A$ Length 24
3 Shortest route: $A - B - E - C - G - F$ Length 38
4 Shortest route: $S_1 - B - C - F - E - G - T$
    Length of shortest route: 1660
5 **a** 94 − 18 = 76 EH 76 − 24 = 52 CE
    52 − 22 = 30 BC 30 − 30 = 0 AB
    Shortest route A to H: $A - B - C - E - H$
    Length 94
  b Shortest route A to H via G: $A - D - G - H$
    Length 96

c Shortest route A to H not using CE: $A - D - F - E - H$
  Length 95
6 Shortest route: $S - B - C - T$
  Length of shortest route: 10
7 **a** Quickest route is $SCEHT$. Shortest time = 31 min.
  b **i** Route changes to $SABDGT$. New time = 32 min
     ii The driver should change the route to $HEGT$ to save 1 minute. Total travel time 38 min.
  c $0.026 \times \left(\dfrac{40}{10}\right)^2 = 0.026 \times 16 = 0.416$ seconds
  d The time required is not directly proportional to the square of the number of locations – this is just an approximation.

Exercise 3E

1 **a**

Initial tables

	P	Q	R	S
P	–	9	5	12
Q	9	–	3	7
R	5	3	–	∞
S	12	7	∞	–

	P	Q	R	S
P	P	Q	R	S
Q	P	Q	R	S
R	P	Q	R	S
S	P	Q	R	S

1st iteration

	P	Q	R	S
P	–	9	5	12
Q	9	–	3	7
R	5	3	–	17
S	12	7	17	–

	P	Q	R	S
P	P	Q	R	S
Q	P	Q	R	S
R	P	Q	R	P
S	P	Q	P	S

2nd iteration

	P	Q	R	S
P	–	9	5	12
Q	9	–	3	7
R	5	3	–	10
S	12	7	10	–

	P	Q	R	S
P	P	Q	R	S
Q	P	Q	R	S
R	P	Q	R	Q
S	P	Q	Q	S

3rd iteration

	P	Q	R	S
P	–	8	5	12
Q	8	–	3	7
R	5	3	–	10
S	12	7	10	–

	P	Q	R	S
P	P	R	R	S
Q	R	Q	R	S
R	P	Q	R	Q
S	P	Q	Q	S

4th iteration (no change)

	P	Q	R	S
P	–	8	5	12
Q	8	–	3	7
R	5	3	–	10
S	12	7	10	--

	P	Q	R	S
P	P	R	R	S
Q	R	Q	R	S
R	P	Q	R	Q
S	P	Q	Q	S

b The 4th iteration route table shows that the shortest route from R to S is via Q.
It also shows that the shortest route from R to Q is direct and the shortest route from Q to S is direct.
It follows that the shortest route from R to S is RQS.

2

	A	B	C	D
A	–	17	13	8
B	3	–	4	6
C	7	4	–	5
D	8	9	5	–

3

	J	K	L	M
J	–	6	11	4
K	6	–	9	10
L	11	9	–	7
M	4	8	7	–

4 a The distance table is not symmetrical about the leading diagonal.
b $x = 6$, $y = 6$, $z = 23$

5 a The output of Floyd's algorithm gives the shortest distance between every pair of nodes. The output of Dijkstra's algorithm gives the shortest distance from the start node to every other.

b i

	A	B	C	D	E	F	G
A	–	7	10	∞	∞	4	11
B	7	–	2	5	∞	∞	∞
C	10	2	–	4	3	8	∞
D	∞	5	4	–	6	∞	∞
E	∞	∞	3	6	–	7	9
F	4	∞	8	∞	7	–	6
G	11	∞	∞	∞	9	6	–

ii

	A	B	C	D	E	F	G
A	–	7	10	∞	∞	4	11
B	7	–	2	5	∞	11	18
C	10	2	–	4	3	8	21
D	∞	5	4	–	6	∞	∞
E	∞	∞	3	6	–	7	9
F	4	11	8	∞	7	–	6
G	11	18	21	∞	9	6	–

	A	B	C	D	E	F	G
A	A	B	C	D	E	F	G
B	A	B	C	D	E	A	A
C	A	B	C	D	E	F	A
D	A	B	C	D	E	F	G
E	A	B	C	D	E	F	G
F	A	A	C	D	E	F	G
G	A	A	A	D	E	F	G

6 a Jared : Dijkstra's   Amy : Floyd's
b 1st iteration (no change)

	A	B	C	D	E	F
A	–	15	∞	∞	25	∞
B	15	–	22	∞	8	11
C	∞	22	–	14	∞	10
D	∞	∞	14	–	∞	∞
E	25	8	∞	∞	–	21
F	∞	11	10	6	21	–

	A	B	C	D	E	F
A	A	B	C	D	E	F
B	A	B	C	D	E	F
C	A	B	C	D	E	F
D	A	B	C	D	E	F
E	A	B	C	D	E	F
F	A	B	C	D	E	F

2nd iteration

	A	B	C	D	E	F
A	–	15	37	∞	23	26
B	15	–	22	∞	8	11
C	37	22	–	14	30	10
D	∞	∞	14	–	∞	6
E	23	8	30	∞	–	19
F	26	11	10	6	19	–

	A	B	C	D	E	F
A	A	B	B	D	B	B
B	A	B	C	D	E	F
C	B	B	C	D	B	F
D	A	B	C	D	E	F
E	B	B	B	D	E	B
F	B	B	C	D	B	F

7 a $n^2 - 3n + 2$
b cubic
c $0.012 \times \left(\dfrac{100}{30}\right)^3 = 0.44$ sec (2 d.p.)

Mixed exercise 3
1 a i Arcs are labelled with initial letters of the nodes.
CK add to tree
SH add to tree
CE add to tree
EK reject
CH add to tree
HW add to tree
CS reject
HQ add to tree
QS reject
QD add to tree
KS reject
DW reject
EW reject

Online Full worked solutions are available in SolutionBank.

ii *EC*
  CK
  CH
  HS
  HW
  HQ
  QD

b

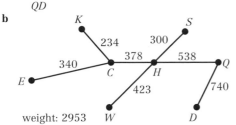

weight: 2953

2 a i *LT*
    MT
    MQ
    NQ
    ST
    QR
    NP

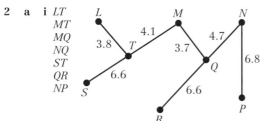

ii *MQ* (3.7) add to tree
  LT (3.8) add to tree
  MT (4.1) add to tree
  NQ (4.7) add to tree
  MN (5.3) reject
  { *ST* (6.6) add to tree
  { *QR* (6.6) add to tree
  NP (6.8) add to tree
  reject remaining arcs

b Start off the tree with *QT* and *PR* then apply
  Kruskal's algorithm. Prim's algorithm requires the
  'growing' tree to be connected at all times. When
  using Kruskal's algorithm the tree can be built from
  non-connected sub-trees.

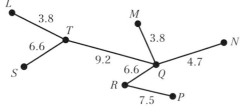

3 a

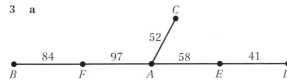

Selection order: *AC, AE, ED, AF, FB*

b length 332 mm

c $0.02 \times \left(\dfrac{240}{80}\right)^3 = 0.54$ sec

d The time required is not directly proportional to n^3
  but this is used as an approximation.

4 a Arcs in order: Entrance 2–office; Entrance 2–
  Entrance 4; Entrance 4–Entrance 3; Office–Entrance
  1

b length 3112 m

5 a

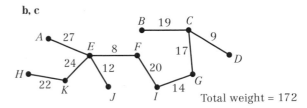

EF CD EJ FJ GI CG DG BC FI HK EK CF JK AE AB AH

b, c

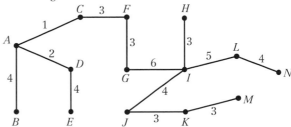

Total weight = 172

d $e = v - 1$

6 a weight = 45 so 4500 m needed

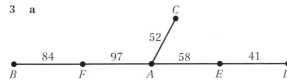

b weight = 47 so 4700 m

7 a Possible paths are $A - H - G - E - I - K$
  and $A - H - J - I - K$
  and $A - B - C - K$
  $A, B, F, H, D, G, J, E, C, I, K$

ii
$44 - 9 = 35$ *IK* $44 - 9 = 35$ *IK*
$35 - 10 = 25$ *EI* $35 - 17 = 18$ *JI*
$25 - 10 = 15$ *GE*   or $18 - 8 = 10$ *HJ*
$15 - 5 = 10$ *HG* $10 - 10 = 0$ *AH*
$10 - 10 = 0$ *AH*
  or
$44 - 12 = 32$ *CK*
$32 - 25 = 7$ *BC*
$7 - 7 = 0$ *AB*

b $A - H - G - E - I - K$ and $A - H - J - I - K$
  and $A - B - C - K$

c The arcs could be roads.
  The nodes could be junctions.
  The number on each arc could be distance in km.
  The network, together with Dijkstra's algorithm,
  could be used to find the shortest route from *A* to *K*.

8 Order of vertex labelling:
  S C B A E D G F T
  Route: $S - C - E - F - T$
  $411 - 101 = 310$ *FT*
  $310 - 123 = 187$ *EF*
  $187 - 85 = 102$ *CE*
  $102 - 102 = 0$ *SC*

287

9 a

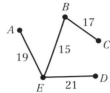

Total weight = 72

b Students' own answer

c Prim's algorithm may be quicker on a graph with a large number of arcs, such as a complete network, as Kruskal's algorithm would require arcs to be sorted by weight.

10 a

1st iteration

	P	Q	R	S			P	Q	R	S
P	–	12	$\propto$	16		P	P	Q	R	S
Q	12	–	15	28		Q	P	Q	R	P
R	$\propto$	15	–	10		R	P	Q	R	S
S	16	20	10	–		S	P	Q	R	S

2nd iteration

	P	Q	R	S			P	Q	R	S
P	–	12	27	16		P	P	Q	Q	S
Q	12	–	15	28		Q	P	Q	R	P
R	27	15	–	10		R	Q	Q	R	S
S	16	20	10	–		S	P	Q	R	S

3rd iteration

	P	Q	R	S			P	Q	R	S
P	–	12	27	16		P	P	Q	Q	S
Q	12	–	15	25		Q	P	Q	R	R
R	27	15	–	10		R	Q	Q	R	S
S	16	20	10	–		S	P	Q	R	S

4th iteration

	P	Q	R	S			P	Q	R	S
P	–	12	26	16		P	P	Q	S	S
Q	12	–	15	25		Q	P	Q	R	R
R	26	15	–	10		R	S	Q	R	S
S	16	20	10	–		S	P	Q	R	S

b From route table: shortest distance from R to P is via S.
Shortest distance from R to S is direct.
Shortest distance from S to P is direct.
So shortest route from R to P is RSP.

11 a Prim's algorithm or Kruskal's algorithm.

b 20 miles

c Dijkstra's algorithm

d Shortest distance = 11 miles
Route: $ADEFG$

e Floyd's algorithm

Challenge

For a network of n vertices, after the rth vertex has been selected you need to compare $(n - r)$ values of $\min(Y)$ with XY, where X is the most recently selected vertex. You then need to choose the smallest value of $\min(Y)$, which

requires a further $(n - r - 1)$ comparisons. The number of comparisons at each step is $(n - 1) + (n - r - 1) = 2n - 2r - 1$, so the total number of comparisons is:

$$\sum_{r=1}^{n-1}(2n - 2r - 1) = 2\sum_{r=1}^{n-1}n - 2\sum_{r=1}^{n-1}r - \sum_{r=1}^{n-1}1$$
$$= 2n(n - 1) - n(n - 1) - (n - 1)$$
$$= n^2 - 2n + 1$$

Which has order n^2.

CHAPTER 4

Prior knowledge 4

1 a 16

b Sum of valencies = 2 × number of arcs

c 4

d From part **b**, the sum of the valencies must be even.

Exercise 4A

1 a

Vertex	A	B	C	D	E	F
Degree	2	3	2	3	3	3

There are 4 nodes with odd degree so the graph is neither Eulerian nor semi-Eulerian.

b

Vertex	G	H	I	J	K
Degree	3	4	3	2	4

There are precisely 2 nodes of odd degree (G and I) so the graph is semi-Eulerian.

A possible route starting at G and finishing at I is:
$G - H - K - I - J - K - G - H - I$.

c

Vertex	L	M	N	P	Q	R
Degree	2	4	2	4	2	4

All vertices have even degree, so the graph is Eulerian. A possible route starting and finishing at L is:
$L - M - N - P - M - R - P - Q - R - L$.

2 a i

Vertex	A	B	C	D	E	F	G	H
Degree	4	2	4	2	2	4	2	2

ii

Vertex	A	B	C	D	E	F	G
Degree	4	4	2	4	2	4	4

b i A possible route is:
$A - B - C - A - F - C - E - G - H - F - D - A$.

ii A possible route is:
$A - C - F - A - B - E - G - B - D - G - F - D - A$.

3 a i

Vertex	R	S	T	U	V	W
Degree	2	2	3	3	2	2

Precisely 2 vertices of odd valency (T and U) so semi-Eulerian.

ii

Vertex	H	I	J	K	L	M	N
Degree	2	4	3	2	3	4	4

Precisely two nodes of odd degree (J and L) so semi-Eulerian.

b i A possible route starting of T and finishing at U is:
$T - R - S - U - W - V - T - U$.

ii A possible route starts at J and finishes at L:
$J - K - L - M - J - I - M - N - I - H - N - L$.

Online Full worked solutions are available in SolutionBank.

4 a There are an odd number of odd nodes.
b i $x = 1$
ii Semi-Eulerian since there are two odd nodes.
c Numerous possible answers e.g.:

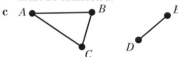

5 a Not connected. There are no connections from A, B or C to D or E.
b Neither. To be Eulerian or semi-Eulerian the graph must be connected.
c

6 The graph is Eulerian since all of the nodes are even.
7 a n must be odd.
b

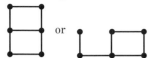

8 The example given in the question **1a** is a counterexample. $ABEFCDA$ is a Hamiltonian cycle, but the graph is not Eulerian.

Challenge

1 a

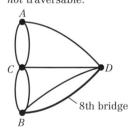

Vertex	A	B	C	D
Degree	3	3	5	3

There are more than two odd nodes, so the graph is *not* traversable.

b

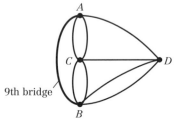

We will start at A and finish at C so these still need to have odd degree. We can only have two odd degrees so B and D must have even degrees (see table).
We need to change the degree of B and of D. So we build a bridge from B to D.

Vertex	A	B	C	D
Degree with 7 bridges	odd	*odd*	odd	*odd*
Degree wanted	odd	*even*	odd	*even*

c

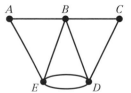

We will start at B and finish at C so these vertices need to be the two vertices with odd degree. We need A and D to have even degree (see table). We need to change the degree of node A and of node B.
So we build a bridge from A to B.

Vertex	A	B	C	D
Degree with 8 bridges	*odd*	*even*	odd	even
Degree wanted	*even*	*odd*	odd	even

d

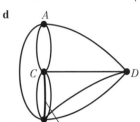

All vertices now need to have even degree.
This means we need to change the degrees of nodes B and C.
So the 10th bridge needs to be built from B to C.

Vertex	A	B	C	D
Degree with 9 bridges	even	*odd*	*odd*	even
Degree wanted	even	*even*	*even*	even

Exercise 4B

1 a All degrees are even, so the network is traversable and can return to its start.

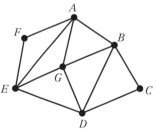

A possible route is:
$A - B - C - D - B - G - D - E - G - A - E - F - A$.
length of route = weight of network
$= 285$

b The degrees of D and E are odd; the rest are even.
We must repeat the shortest path between D and E, which is the direct path DE.
We add this extra arc to the diagram.

A possible route is:
$A - B - C - D - E - D - B - E - A$.
length of route = weight of network + arc DE
$= 61 + 11$
$= 72$

c The degrees of *B* and *E* are odd; the rest are even.
We must repeat the shortest path from *B* to *E*.
By inspection this is *BCE*, length 260.
We add these extra arcs to the diagram.

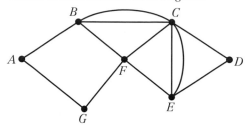

A possible route is:
$A - B - C - D - E - C - B - F - C - E - F - G - A$.
length of route = weight of network + *BCE*
$= 1055 + 260$
$= 1315$

d The degrees of *B* and *G* are odd; the rest are even.
We must repeat the shortest path from *B* to *G*.
By inspection this is *BDHG*, length 183.
We add these arcs to the diagram.

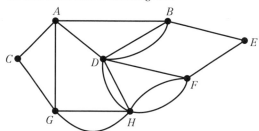

A possible route is:
$A - B - E - F - D - B - D - A - G - H - F - H - D - H - G - C - A$.
length of route = weight of network + *BDHG*
$= 995 + 183$
$= 1178$

2 a Odd degrees at *B*, *D*, *E* and *F*.
Considering all possible pairings and their weights.
$BD + EF = 130 + 85 = 215 \leftarrow$ least weight
$BE + DF = 110 + 178 = 288$
$BF + DE = 125 + 93 = 218$
We need to repeat arcs *BD* and *EF*.
The length of the shortest route
$= $ weight of network + 215
$= 908 + 215$
$= 1123$
Adding *BD* and *EF* to the diagram gives:

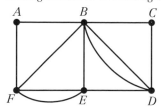

A possible route is:
$A - B - C - D - B - E - D - B - F - E - F - A$.

b Odd degrees at *C*, *D*, *E* and *G*
Considering all possible pairings and their weights
$CD + EG = 130 + 75 = 205$
$CE + DG = 157 + 92 = 249$
$CG + DE = 82 + 120 = 202 \leftarrow$ least weight
We need to repeat arcs *CG* and *DE*.

The length of the shortest route
$= $ weight of network + 202
$= 938 + 202$
$= 1140$
Adding *CG* and *DE* to the diagram gives:

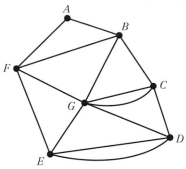

A possible route is:
$A - B - C - G - D - C - G - E - D - E - F - G - B - F - A$.

3 a The odd nodes are *B*, *D*, *G* and *I*.
The minimum path lengths for each pairing are:
$BD + GI = 22 + 30 = 52$
$BG + DI = 37 + 42 = 79$
$BI + DG = 32 + 18 = 50$
The arcs to be traversed twice are *BI* and *DG*.
Adding *BI* and *DG* to the diagram gives:

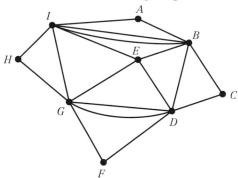

A possible route is
$A - B - C - D - B - I - E - B - I - G - D - F - G - E - D - G - H - I - A$.

b Minimum time required $= 326 + 50 = 376$ ms

4 a Odd degrees at *B*, *D*, *E* and *F*.
Considering all possible pairings and their weight
$BD + EF = 250 + 200 = 450 \leftarrow$ least weight
$BE + DF = 350 + 380 = 730$
$BF + DE = 300 + 180 = 480$
We need to repeat arcs *BD* and *EF*.
Adding these to the diagram gives:

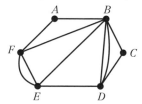

A possible route is:
$A - B - C - D - B - D - E - F - B - E - F - A$.
length $= 1910 + 450 = 2360$ m.

Online Full worked solutions are available in SolutionBank.

b We will still have two odd valencies.
We need to select the pair that gives the least path.
From part **a** our six choices are
BD (250), *EF* (200), *BE* (350), *DF* (380), *BF* (300) and
DE (180).
The shortest is *DE* (180) so we choose to repeat this.
It is the other two vertices (*B* and *F*) that will be our
start and finish.
For example, start at *B*, finish at *F*
length of route = 1910 + 180 = 2090 m

5 a Each arc must be traversed twice, whereas in the
standard problem each arc need only be visited once.
This has the same effect as doubling up all the edges
The length of the route = 2 × weight of network
= 2 × 89 = 178 km

b Odd nodes are *C, D, E, G*.
Considering all possible pairings.
CD + *EG* = 13 + 5 = 18
CE + *DG* = 15 + 3 = 18
CG + *DE* = 10 + 7 = 17 ← least weight
We need to repeat arcs *CG* and *DE*.
Length of route = 89 + 17 = 106 km

c As *EG* can be omitted the valencies of *E* and *G*
become even, therefore the only odd nodes are *C*
and *D*. Shortest route from *C* to *D* is 13.
Length of route = 84 + 13 = 97 km

Exercise 4C

1 There are 6 odd nodes *B, C, D, E, G* and *H*.
B and *G* remain as odd nodes.
The minimum path lengths for the pairings are:
CD + *EH* = 3 + 14 = 17
CE + *DH* = 4 + 13 = 17
CH + *DE* = 10 + 1 = 11 ← least weight
The edges that must be traversed twice are *CH* and *DE*.
Length of route = 96 + 11 = 107 km

2 a There are 6 odd nodes *B, C, D, E, F* and *G*.
Starting at *C* so always remains odd
Case (i): Finishing at *E*
The pairings are:
BD + *FG* = 21 + 7 = 28
BF + *DG* = 16 + 19 = 35
BG + *DF* = 9 + 12 = 21 ← least weight
Case (ii): Finishing at *G*
The pairings are:
BD + *EF* = 21 + 8 = 29
BE + *DF* = 24 + 12 = 36
BF + *DE* = 16 + 11 = 27
The edges that must be traversed twice are *BG*
and *DF*.

b The length of the route is 106 + 21 = 127 km

c Case (i): Finishing at *E*   *FG* = 7
Case (ii): Finishing at *G*   *EF* = 8
Total length of minimum route is
$106 + 7 + BD = 127 + \frac{1}{2}BD$
$\frac{1}{2}BD = 127 - 113 = 14 \Rightarrow BD = 28$

3 a *AG, GB, CD*
b 630 + 2220 = 2850 m
c 570 + 2160 = 2730 m

4 a *PQWVRUTS*, 286 m
b Since the path represents the shortest route from
P to *S*, passing through every other vertex along
the way, the shortest route between *any* pair of
vertices must be contained within it. In particular,
the shortest route between any pair of odd vertices
must be contained within it.

c Using the answer to part **b**, the minimum pairing is
found by using the path found in part **a**. The pipes
to be traversed twice are *PQ, WV, UT* and *TS*.
Length of route = 853 + 48 + 35 + 22 + 31 = 989 m

Mixed exercise 4

1 a Eulerian – all nodes even
b Neither – more than 2 odd nodes

2 e.g.

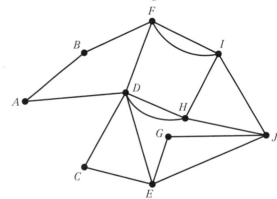

Any not connected graph with 6 even nodes

3 a *x* = 3
b The orders of the vertices are 29, 21, 17 and 3
The graph is neither Eulerian not semi-Eulerian
since it has more than 2 odd vertices.

4 a

Vertex	A	B	C	D	E	F	G	H	I	J
Degree	2	2	2	5	4	3	2	3	3	4

b *DF* + *HI* = 19 + 36 = 55
DH + *FI* = 22 + 27 = 49 ← least weight
DI + *HF* = 46 + 41 = 87
Repeat *DH* and *FI*
Add these to the network to get

A possible route is
*A – B – F – I – J – G – E – J – H – D – F – I – H – D
– C – E – D – A*.

c Length = 725 + 49 = 774

5 a The odd vertices are *Q, R, T* and *V*.
QR + *TV* = 104 + 189 = 293
QT + *RV* = 171 + 115 = 286
QV + *RT* = 163 + 123 = 286
The postman can either repeat *QT* and *RV* or *QV*
and *RT*.

b The total length of the route is 1890 + 286 = 2176 m
c Only *QV* now needs to be repeated.
Total length = 1890 – 123 + 163 = 1930 m
The route is now 246 m shorter.

6 a 15 m. The route is *GFD*.
b The odd vertices are *G, B, C* and *D*.
GB + *CD* = 16 + 3 = 19 ← least weight
GC + *BD* = 18 + 10 = 28
GD + *BC* = 15 + 7 = 22
GA, AB and *CD* should be traversed twice.
Total length = 118 + 19 = 137 m
c *x* + 10 = 19, so *x* = 9

7 a

Vertex	A	B	C	D	E	F	G	H	I
Degree	2	3	4	3	4	2	6	3	3

Odd valencies at B, D, H and I.

b Considering all possible complete
pairings and their weight
$BD + HI = 7.2 + 3.4 = 10.6$
$BH + DI = 7.6 + 4 = 11.6$
$BI + DH = 5.6 + 4.3 = 9.9 \leftarrow$ least weight

> Shortest routes: BD is BED; BH is $BEGH$, DI is DGI;
> BI is BEI, DH is DGH.

Repeat BE, EI and DG, GH.
Adding these arcs to the network gives

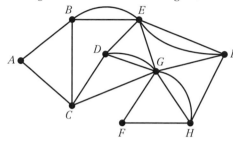

A possible route is:
$A - B - E - I - H - G - I - E - B - C - D - G -$
$D - E - G - H - F - G - C - A$.

c Length $= 51.4 + 9.9 = 61.3$ km

d If BD is included B and D now have even valency.
Only H and I have odd valency.
So the shortest path from H to I needs to be repeated.
Length of new route $= 51.4 + BD +$ path from H to I
$= 51.4 + 6.4 + 3.4$
$= 61.2$ km
This is (slightly) shorter than the previous route so
choose to grit BD since it saves 0.1 km.

8 a Odd nodes are B, C, E and H
Consider possible pairings and weight
$BC + EH = 68 + 150 = 218$
$BE + CH = 95 + 73 = 168 \leftarrow$ least weight
$BH + CE = 141 + 85 = 226$
Repeat BD, DE and CH
A possible route is:
$A - B - C - H - E - G - H - C - D - B - D - E - A -$
$F - E - D - A$
Total length $= 1011 + 168 = 1179$ m

b If the salesman can start at B and end at C those
nodes can stay odd, therefore only route to be
travelled twice is EH at 150m.
Total length $= 1011 + 150 = 1161$ m
This would decrease the total distance the salesman
has to walk.

9 a Route inspection algorithm:
Identify any vertices with odd valency.
Consider all possible complete pairings of these
vertices.
Select the complete pairing that has the least sum.
Add a repeat of the arcs indicated by this pairing to
the network.

b Odd vertices B, D, F, H
Considering all complete pairings
$BD + FH = 14 + 15 = 29$
$BF + DH = 10 + 26 = 36$
$BH + DF = 12 + 16 = 28 \leftarrow$ least weight
Repeat BH and DF.
Adding these arcs to the network gives

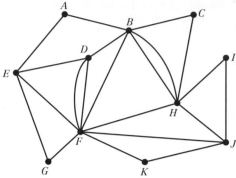

A possible route is:
$A - B - H - C - B - H - I - J - H - F - J - K -$
$F - B - D - F - D - E - F - G - E - A$.

c Length of route $= 249 + 28 = 277$

d i We will still have to repeat the shortest path
between a pair of the odd nodes.
We will choose the pair that requires the shortest
path.
The shortest path of the six is BF (10).
We will use D and H as the start and finish nodes.
ii 259

e Each edge, having two ends, contributes two to the
sum of valencies for the network.
Therefore the sum $= 2 \times$ number of edges
The sum is even so any odd valencies must occur
in pairs.

10 a Odd nodes are A, B, D, E, F and G
Starting at B so can leave as odd
Case (i): Land at D
$AE + FG = 19 + 10 = 29$
$AF + EG = 7 + 22 = 29$
$AG + EF = 6 + 12 = 18 \leftarrow$ least weight
Case (ii) Land at F
$AD + EG = 26 + 22 = 48$
$AE + DG = 19 + 20 = 39$
$AG + ED = 6 + 14 = 20$
Better to use landing strip at D.

b $168 + 18 = 186$ miles

c Odd notes unchanged.
Case (i): Land at D
$AE + FG = 40 + 13 = 53$
$AF + EG = 7 + 34 = 41$
$AG + EF = 6 + 47 = 53$
Case (ii) Land at F
$AD + EG = 26 + 34 = 60$
$AE + DG = 40 + 20 = 60$
$AG + ED = 6 + 14 = 20 \leftarrow$ least weight
Now better to land at F.
$168 - (10 + 25 + 12) + 20 = 141$ miles

Challenge

a Shortest time from A to B is 30 minutes.
Route $A - C - D - G - H - E - F - B$
or $A - C - D - G - F - B$

b $143 + AC + DG + HE + FB$
$= 143 + 2 + 5 + 4 + 8 = 162$ minutes

CHAPTER 5

Prior knowledge 5

1 a

	A	B	C	D	E	F	G
A	–	22	–	–	15	–	8
B	22	–	18	–	–	–	21
C	–	18	–	30	–	11	–
D	–	–	30	–	19	–	–
E	15	–	–	19	–	15	–
F	–	–	11	–	15	–	25
G	8	21	–	–	–	25	–

b i Using the matrix form of Prim's algorithm:

	1	6	5	7	3	4	2
	A	B	C	D	E	F	G
A	–	22	–	–	15	–	8
B	22	–	⑱	–	–	–	21
C	–	18	–	30	–	⑪	–
D	–	–	30	–	⑲	–	–
E	⑮	–	–	19	–	15	–
F	–	–	11	–	⑮	–	25
G	⑧	21	–	–	–	25	–

ii Using Kruskal's algorithm:

8	AG
11	CF
15	AE and EF
18	BC
19	DE

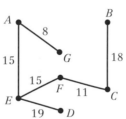

Exercise 5A

1 a

	A	B	C	D	E
A	–	7	10	9	5
B	7	–	3	11	12
C	10	3	–	8	12
D	9	11	8	–	4
E	5	12	12	4	–

AC – the shortest route is ABC length 10
AD – the shortest route is AED length 9
BD – the shortest route is BCD length 11
BE – the shortest route is BAE length 12

b

	A	B	C	D	E
A	–	5	12	7	9
B	5	–	7	2	4
C	12	7	–	5	7
D	7	2	5	–	2
E	9	4	7	2	–

AC – the shortest route is ABDC length 12
AE – the shortest route is ABDE length 9
BC – the shortest route is BDC length 7
BE – the shortest route is BDE length 4
CE – the shortest route is CDE length 7

c

	A	B	C	D	E	F
A	–	10	18	13	15	18
B	10	–	8	3	5	8
C	18	8	–	5	3	10
D	13	3	5	–	2	5
E	15	5	3	2	–	7
F	18	8	10	5	7	–

AC – the shortest route is ABDEC length 18
AF – the shortest route is ABDF length 18
BC – the shortest route is BDEC length 8
BE – the shortest route is BDE length 5
BF – the shortest route is BDF length 8
CD – the shortest route is CED length 5
CF – the shortest route is CEDF length 10
EF – the shortest route is EDF length 7

d

	A	B	C	D	E	F	G
A	–	10	9	10	17	20	20
B	10	–	3	20	11	10	20
C	9	3	–	19	8	13	23
D	10	20	19	–	27	20	10
E	17	11	8	27	–	8	18
F	20	10	13	20	8	–	10
G	20	20	23	10	18	10	–

AC – the shortest route is ABF length 20
AG – the shortest route is ADG length 20
BE – the shortest route is BCE length 11
CF – the shortest rotue is CBF length 13
CG – the shortest route is CBFG length 23
DE – the shortest route is DACE length 27

e

	A	B	C	D	E	F
A	–	12	21	20	7	26
B	12	–	17	18	5	24
C	21	15	–	11	20	17
D	16	4	11	–	9	6
E	7	5	22	13	–	19
F	27	25	21	32	20	–

e.g. AD – the shortest route is AED length 20
DA – the shortest route is DBEA length 16
DF – the shortest route is DF length 6
FD – the shortest route is FCD length 32

2 a In the classical problem, each vertex must be visited exactly once. In the practical problem, each vertex must be visited at least once.

b i ABECEDEBA (20)
 ii ABCDEA (26)

3 a PQRTSUQP (39)

b Consider removing the arcs RS and TU, since they have the greatest weight. Also disregard PQ for the moment – it must be included (twice) in order to access P. All vertices are then even which makes the remaining network Eulerian. The minimum length route is then made up of the length of the Eulerian circuit plus 2 × PQ. This will be the same regardless of the starting point.

c Every route must visit Q twice.

4 Students' own answers, e.g.

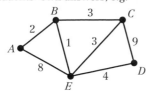

Challenge

1 Since the network is complete, every vertex is accessible from every other vertex. From any start vertex, there must exist a route that visits every other vertex directly in turn. There must also exist a direct route back to the start vertex. The solution is given by the route of this type for which the total distance is a minimum.

2 The number of *distinct* Hamiltonian cycles is $\dfrac{(n-1)!}{2}$
The order of the algorithm is $(n-1)!$
The value of $(n-1)!$ grows rapidly as n increases showing that the algorithm is no use for large n.

Exercise 5B

1 a

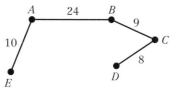

Initial upper bound = 102

b DE shortcut route length 79
c $ABCDEA$ length 79

2 a

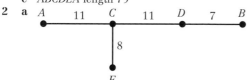

b 74
c BE as shortcut or AB as shortcut (other answers possible)
d **i** Using BE: $ACEBDCA$ length 62
ii Using AB: $ACECDBA$ length 58

3 a

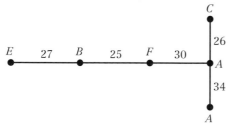

Initial upper bound = 284
b Many possibilities, e.g. DE, EC, DF and EC
c DE gives $ACAFBEDA$ length 231
EC gives $ADAFBECA$ length 217
DF and EC gives $ACEBFDA$ length 190

4 a

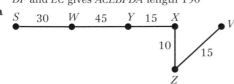

Initial upper bound = 230
b For example use VS
c Route $SWYXZVS$ length 190

Exercise 5C

1 a 45
b The lower bound represents a tour so it must be optimal.
2 a 50 (deleting A)   49 (deleting B)
b Better lower bound is 50 since it is the highest.
3 a 183 (deleting A)   177 (deleting B)
b Better lower bound is 183 since it is the highest.
c $183 <$ optimal value $\leqslant 190$
4 a 170 (deleting S)   145 (deleting V)
b Better lower bound is 170 because it is higher.
c $170 <$ optimal value $\leqslant 190$

Exercise 5D

1 a $D_7B_{12}C_8E_{14}A_{19}D = 60$
b $E_8C_{11}A_{13}B_7D_{14}E = 53$ or $E_8C_{11}D_7B_{13}A_{14}E = 53$
c The better upper bound is 53 since this is lower.
2 a $Z_{10}X_{15}Y_{40}V_{55}W_{30}S_{70}Z = 220$
b $X_{10}Z_{15}V_{40}Y_{45}W_{30}S_{55}X = 195$
$V_{15}Z_{10}X_{15}Y_{45}W_{30}S_{75}V = 190$
c The better upper bound is 190 because it is lower.

3 Applied to n vertices the algorithm has cubic order.
$0.27 \times \left(\dfrac{20}{12}\right)^3 = 1.25$ seconds

4 a 1200 minutes
b $U_{120}S_{150}R_{120}V_{150}T_{180}W_270U = 990$
and
$U_{120}T_{150}V_{120}R_{150}S_{240}W_{270}U = 1050$
c $V_{120}R_{150}S_{120}U_{120}T_{180}W_{300}V = 990$ and
$V_{120}R_{150}U_{120}S_{210}T_{180}W_{300}V = 1080$ and
$V_{120}R_{150}U_{120}T_{180}W_{240}S_{210}V = 1020$
d The better upper bound is 990 because it is lower.

5 a Nearest neighbour route from B has length $648 + x$
Nearest neighbour route from A has length $639 + x$
Total length $= 1287 + 2x = 1419, \Rightarrow x = 66$
b An upper bound for the optimal length $= 639 + 66$
$= 705$ miles

6 a AE – the shortest route is ACE length 9 minutes
EA – the shortest route is $ECBA$ length 14 minutes
b $A_3C_2B_7D_4E_{14}A = 30$ minutes
c In the original network: $ACBCDECBA$. Not Hamiltonian.
d e.g. $ADECBA$ has total length 25 minutes

Mixed exercise 5

1 a Either Kruskal's: EF, DE, CD, BD, AC, EG
or Prim's (e.g.): AC, CD, DE, EF, BD, EG
b 7004
c e.g. use AB and DG
Route $ACDEFEGDBA$ length 6005

2 a

	A	B	C	D	E
A	–	7	13	4	3
B	7	–	17	7	10
C	13	17	–	10	13
D	4	7	10	–	5
E	3	10	13	5	–

b $A_3E_5D_7B_{17}C_{13}A = 45$
c $AEDBDCA$ (BC is not on the original network)
d $0.85 \times \left(\dfrac{12}{5}\right)^3 = 11.75$ seconds

e The time taken to complete the algorithm is not proportional to the cube of the number of hostels. However, the calculation in part **d** is based on this as an approximation.

3 a *SC SF FA AB CD DE*

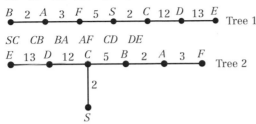

b 74

c Tree 1: use *BE* as short cut
Route: *SCDEBAFS* length 56
Tree 2: use *EF* as short cut
Route: *SCBAFEDCS* length 53

d $C_2S_5F_3A_2B_{17}D_{13}E_{21}C = 63$
$D_{12}C_2S_5F_3A_2B_{19}E_{13}D = 56$

e The better upper bound is 53 since it is smaller

f Route *SCBAFEDCS*

g 44

4 a In the classical problem each vertex must be visited exactly once before returning to the start.
In the practical problem each vertex must be visited at least once before returning to the start.

b

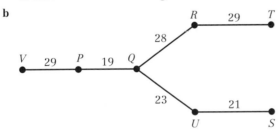

Order: *PQ QU US QR* $\begin{Bmatrix} TR \\ VP \end{Bmatrix}$

c Use *VT* and *QS* as shortcuts giving a length of 213
Route *PQUSQRTVP*

d $P_{19}Q_{23}U_{21}S_{51}R_{29}T_{37}V_{29}P = 209$

e 186

f 186 < optimal value ≤ 209

5 a

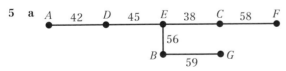

Order: *AD DE EC EB CF BG*

b 596

c The minimum connector has been doubled and each arc in it repeated

d Use *AE* and *GF* as shortcuts
Route: *ADEBGFCEA* length 427

e 352 km

f The lower bound will give an optimal solution if it is a tour.
If the minimum spanning tree has no 'branches' so the two end vertices have valency 1, and all other vertices have valency 2, then if the two least arcs are incident on the two vertices of valency 1 an optimal solution cannot be found.

6 a

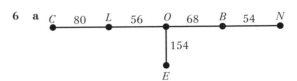

Order: *LO OB BN LC OE*

b i 824 miles
ii Use *NC* as a shortcut
Route: *LOEOBNCL* length 653

c 573

7 a 110 + *x* = 140, so *x* = 30

b 151 miles
$$B \overset{16}{-} A \overset{12}{-} E \overset{22}{-} C \overset{23}{-} G \overset{30}{-} F \overset{30}{-} D \overset{18}{-} B$$

c 81 + 19 + 20 = 120 miles

d 120 < optimal length ≤ 140

8 a $D_4B_5E_7A_{21}C_{17}F_{32}D = 86$ miles
$E_5B_{12}A_{20}D_6F_{21}C_{20}E = 84$ miles

b The cycle starting at *E* as it has a shorter total length.

c In the original network: *EBEAEDFCDBE*.

Challenge

For *n* vertices: The number of distance to compare from some start vertex is
$$(n-1) + (n-2) + \ldots + 2 + 1 = \left(\frac{n-1}{2}\right)n$$
The process must be repeated for each vertex. This gives the total number of comparisons as $\frac{(n-1)n^2}{2}$. Finally, Select the tour of shortest length (*n* − 1 comparisons) to give $\frac{(n-1)n^2}{2} + n - 1 = \frac{(n-1)(n^2+2)}{2}$, which is cubic order.

Review exercise 1

1 a

a	b	c	d	e	f	$f = 0$?
645	255	2.53	2	510	135	No
255	135	1.89	1	135	120	No
135	120	1.13	1	120	15	No
120	15	8	8	120	0	Yes

answer is 15

b The first row would be
255 645 0.40 0 0 255 No
but the second row would then be the same as the first row in the table above. So in effect this new first line would just be an additional row at the start of the solution.

c Finds the highest common factor of *a* and *b*.

2 a Total length = 390 cm, so 4 planks are needed.
b 6 planks   **c** 5 planks
d There are 5 lengths over 50 cm, so none of these can be paired together. Therefore minimum of 5 lengths are required.

3 a
80	55	84	25	34	25	75	17	5	3
80	84	55	34	25	75	25	17	5	3
84	80	55	34	75	25	25	17	5	3
84	80	55	75	34	25	25	17	5	3
84	80	75	55	34	25	25	17	5	3

b 403 ÷ 100 = 4.03 so 5 bins are needed.

c Bin 1: 84 + 5 + 3
Bin 2: 80 + 17
Bin 3: 75 + 25
Bin 4: 55 + 34
Bin 5: 25

4 a For example, 45 37 18 [46] 56 79 90 81 51

b For example, 56 45 79 46 37 90 81 51 18

c 0.576 seconds (3 s.f.)

5 a For example,

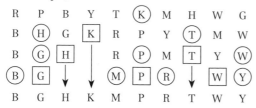

list in order

b $0.024 \times \dfrac{2000 \times \log 2000}{1200 \times \log 1200} = 0.043$ seconds (3 d.p.)

6 a Since the graph is simple, there are no loops, so each of the degree-5 vertices must be joined to each of the other vertices. This means that each of the other vertices has degree at least 2.

b e.g.

7 a

b 4 **c** 5

8 a A Hamiltonian cycle is a cycle that includes every vertex.

b *ABDCFGEA*

c

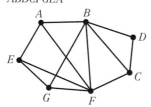

BG(I), *BF*, *BC*, *AF*, *EF*
BG(I), *BF*, *BC*, *AF*(O), *EF*(O)

BG(I), *BF*(I), *BC*, *AF*(O), *EF*(O)

BG(I), *BF*(I), *BC*(I), *AF*(O), *EF*(O)

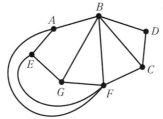

Graph is planar.

d

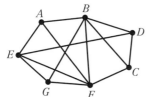

BG(I), *BF*, *BC*, *AF*, *EF*, *ED*
ED and *EF* cross so graph is non-planar.

9 a *GC*, *FD*, *GF*, reject *CD*, *ED*, reject *EF*, *BC*, *AG*, reject *AB*.

b

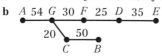

cost = (20 + 25 + 30 + 35 + 50 + 54) × £1000
= £214 000

10 a i Method:
- Start at *A* and use this to start the tree.
- Choose the shortest edge that connects a vertex already in the tree to a vertex not yet in the tree. Add it to the tree.
- Continue adding edges until all vertices are in

the tree. $AF, FC \begin{Bmatrix} FB \\ \text{or} \\ BC \end{Bmatrix}, FD, EB$

ii

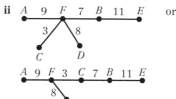

or

iii The tree is not unique, there are 2 of them (see above).

b i number of edges = 7 – 1 = 6

ii number of vertices = *n* + 1

11 a

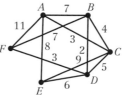

b *BD*, $\begin{Bmatrix} AC \\ DF \end{Bmatrix}$, *BC*, reject *CD*, *DE*; length 18 km

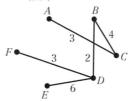

c *DB*, *DF*, *BC*, *CA*, *DE*

12 a In Prim's algorithm, the starting point can be any node, whereas Kruskal's algorithm starts from the arc of least weight.
In Prim's algorithm, each new node and arc is added to the existing tree as it builds, whereas in applying Kruskal's algorithm, the arcs are selected according to their weight and may not be connected until the end.

b i *GH*, *GI*, *HF*, *FD*, *DA*, *AB*, *AC*, *DE*

ii *GH*, *AB*, *AC*, *AD*, reject *BD*, *DF*, *GI*, reject *BC*, *FH*, reject *DG*, *DE*

c weight is 76

13 a Route: $S - A - C - G - T$, length: 82

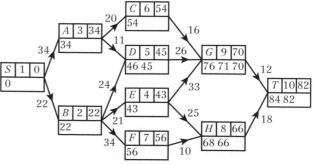

b For example $82 - 12 = 70$ GT
                      $70 - 16 = 54$ CG
                      $54 - 20 = 34$ AC
                      $34 - 34 = 0$ SA

c Shortest route from S to H + HT $S - B - F - H - T$
  length: 84

14 a

```
A 1 0    4    B 3 4    6    C 7/8 10    6    D 12 6
0             4            10              16
1             2            8               4
E 2 1    4    F 4 5    5    G 8/7 10    2    H 10/9 12
1             5            10              12
5             7            3               1
I 5 6    2    J 6 8    4    K 9/10 12    1   L 11 13
6             12 8         12              13
```

For example

13 – 1 = 12 HL or	13 – 1 = 12 KL
12 – 2 = 10 GH	12 – 4 = 8 JK
10 – 5 = 5 FG	8 – 2 = 6 IJ
5 – 4 = 1 EF	6 – 5 = 1 EI
1 – 1 = 0 AE	1 – 1 = 0 AE

Shortest path is $\left\{ \begin{matrix} A - E - F - G - H - L \\ A - E - I - J - K - L \end{matrix} \right\}$ length 13

b State the other path.

15 a Initial tables

	A	B	C	D			A	B	C	D
A	–	10	5	∞		A	A	B	C	D
B	10	–	3	7		B	A	B	C	D
C	5	3	–	11		C	A	B	C	D
D	1	7	11	–		D	A	B	C	D

First iteration

	A	B	C	D			A	B	C	D
A	–	10	5	∞		A	A	B	C	D
B	10	–	3	7		B	A	B	C	D
C	5	3	–	11		C	A	B	C	D
D	1	7	6	–		D	A	B	A	D

Second iteration

	A	B	C	D			A	B	C	D
A	–	10	5	17		A	A	B	C	B
B	10	–	3	7		B	A	B	C	D
C	5	3	–	10		C	A	B	C	B
D	1	7	6	–		D	A	B	A	D

Third iteration

	A	B	C	D			A	B	C	D
A	–	8	5	15		A	A	C	C	C
B	8	–	3	7		B	C	B	C	D
C	5	3	–	10		C	A	B	C	B
D	1	7	6	–		D	A	B	A	D

Fourth iteration

	A	B	C	D			A	B	C	D
A	–	8	5	15		A	A	C	C	C
B	8	–	3	7		B	C	B	C	D
C	5	3	–	10		C	A	B	C	B
D	1	7	6	–		D	A	B	A	D

b 15 miles
c The final route table shows that the shortest route from A to D is via C. The shortest route from C to D is via B. So, the shortest route from A to D is $ACBD$.
d i $BCADB$ **ii** 30 miles   **iii** $BCACBDB$
  iv e.g. $BDACB$ has total length 16 miles.

16 $x = CD + DA = 7 + 10 = 17$
$y = AB + BC = 5 + 3 = 8$
$z = AB + BD = 5 + 6 = 11$

17 a i Shortest path through A is $18 + y$ or 26 both of which are greater than 17.
Shortest path through C is 23, which is $\geqslant 17$. So shortest path cannot go through A or C.
  ii Shortest path must go through B.
$S - B - D - T = 13 + x$
          $13 + x = 17$
               $x = 4$
b If $y = 0$ shortest path is $S - A - D - T = 18$
If $y = 5$ shortest path is $S - C - D - T = 23$
so range is 18 to 23.
c For example, a person seeking the quickest route from home to work through a city. The arcs are the roads that may be chosen, the number the time, in minutes, to journey along that road. The nodes represent junctions.

18 a Odd vertices are B_1, B_2, E, G
$B_1 B_2 + EG = 65 + 18 = 83$
$B_1 E + B_2 G = 41 + 42 = 83$
$B_1 G + B_2 E = 26 + 30 = 56$ ← least weight
Repeat $B_1 D, DG, B_2 A, AE$
Route: For example, $F - A - E - A - B_2 - A - C - E - F - G - D - H - G - D - B_1 - D - F$
(All correct routes have 17 letters in their 'word')
length = $129 + 56 = 185$ km
b Now only the route between E and G needs repeating so repeat $EF + FG = 18$
length of new route = $129 + 18 = 147$ km

19 a All arcs are to be traversed twice, this is, in effect, repeating each arc. So all valencies are even
b E.g. $A - B - D - G - F - G - D - C - E - A - E - C - A - F - E - F - B - F - A - B - D - C - A$
(all correct routes will have 23 letters in their name)
length = $2 \times 6 = 12$ km.

20 a

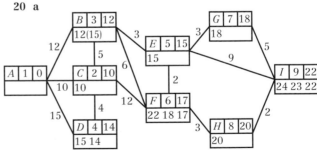

Shortest route is $A - B - E - F - H - I$, length 22 km
b i Odd vertices are A and I (only), so we need to repeat the shortest route from A to I. This was found in **a**.
So repeat AB, BE, EF, FH, HI.
  ii For example $A - B - C - A - D - C - E - H - I$ $- H - E - F - I - G - F - E - B - F - B - A$ (20 letters in route)
  iii $91 + 22 = 113$ km XX

21

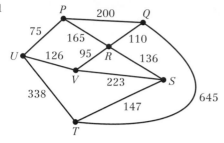

Total length = 2260 m
Odd nodes P, Q, S, T, U, V
T and P remain odd.
QS + UV = 246 + 126 = 372 ← least weight
QU + SV = 275 + 223 = 498
QV + SU = 205 + 349 = 554
QS and UV gives the shortest pairing.
  a Roads to be traversed twice: QR, RS, UV
  b Shortest route is 2260 + 372 = 2632 m

22 Odd valencies are at A, B, C, D, F, G
Route starts at A and finishes at G so these can remain
odd.
Remaining odd vertices are B, C, D, F:
BC + DF = 0.8 + 1.7 = 2.5
BD + CF = 1.3 + 2.3 = 3.6
BF + CD = 1.5 + 0.7 = 2.2 ← least weight
Repeat BF and CD
  a length = 9.5 + 2.2 = 11.7 km
  b A – B – C – A – G – B – D – C – D – E – F – B – F –
    G (14 letters in route)
  c Repeating AC and BF = 2.1
    Minimum distance = 11.6 km
    The engineer is correct. His new route is 0.1 km
    shorter.

23 a In the *practical* T.S.P. each vertex must be visited *at least once*.
    In the *classical* T.S.P. each vertex must be visited
    exactly once.

  b AB, DF, DE, (reject EF) $\begin{Bmatrix} FG \\ AC \end{Bmatrix}$, EH $\begin{Bmatrix} DC \\ \text{or} \\ BE \end{Bmatrix}$

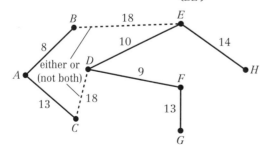

  c 170 km
  d e.g. when CD is part of tree
    Use GH (saving 26) and BD (saving 19) giving a new
    U.B. of 125 km
    Tour ABDEHGFDCA
    e.g. when BE is part of tree
    Use CG (saving 40) giving a new U.B. of 130 km
    Tour ABEHEDFGCA

24 a

	A	B	C	D	E	F
A	0	20	30	32	12	15
B	20	0	10	ⓐ25	ⓐ32	16
C	30	10	0	15	ⓐ35	19
D	32	ⓐ25	15	0	20	ⓐ34
E	12	ⓐ32	ⓐ35	20	0	16
F	15	16	19	ⓐ34	16	0

  b 101 km tour AEFBCDA
  c In the original network AD is not a direct path.
    The tour becomes AEFBCDEA
  d e.g. $\left.\begin{array}{l} BCDEAFB \\ CBFAEDC \\ DCBFAED \\ EAFBCDE \\ FAEDCBF \end{array}\right\}$ length 88

25 a i 714
       ii 552 (ACBDEC)
   b 472
   c 472 < solution ⩽ 552
26 a 45
   b i AEFCDBA – length 49
      ii Choose a tour that does not use AB
         e.g. DB(6), BC(10), CF(8), FE(7), EA(7)
         Complete with AD(8), DBCFEAD
         Total length 46
27 a Order of arcs: AB, BC, CF, FD, FE

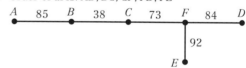

  b i 744
     ii e.g. AD saves 105 giving 639
         or   AE saves 180 giving 564
              AF saves 96 giving 648
              DE saves 66 giving 678
   c 498
28 a Order of arcs: AC, AD, DF, CE, EB

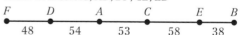

  b i 502
     ii Finding a shortcut to below 360, e.g. FB leaves 351
   c M.S.T. is DF, CE, EB, FB length 244
     The 2 shortest arcs are AC (53) and AD (54) giving
     a total of 351
   d The optimal solution is 351 and is
     A – C – E – B – F – D – A
   e

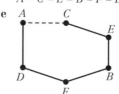

Online Full worked solutions are available in SolutionBank.

29 a Order of arcs: *AB*, *AF*, *FG*, *EG*, *DG*, *CG*

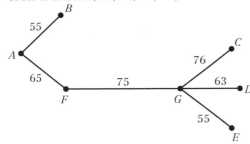

b 778 km

c e.g. *AC* as a shortcut gives *ABCGDGEGFA* with length 632 km

d

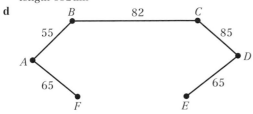

e 470 km

f Route *ABCDEGFA*, length 482 km

Challenge

1 a Let *G* be any finite simple graph with more than one vertex and with number of vertices = $n \geq 2$. The maximal degree of any vertex in *G* is $\leq n - 1$. Also, if our graph *G* is not connected, then the maximal degree is $< n - 1$.

Case 1: Assume that *G* is connected. We cannot have a vertex of degree 0 in *G*, so the set of vertex degrees is a subset of $S = \{1, 2, \ldots, n - 1\}$. Since the graph *G* has *n* vertices and there are $n - 1$ possible degree options, there must be two vertices of the same degree in *G*.

Case 2: Assume that *G* is not connected. *G* has no vertex of degree $n - 1$, so the set of vertex degrees is a subset of $S' = \{0, 1, 2, \ldots, n - 2\}$. Again we have *n* vertices and $n - 1$ possible degree options, so there must be two vertices of the same degree in *G*.

b i Possible sets are:
Blue: *ABD*, *ACD*
Red: *BCF*, *DEF*

ii For K_6 any vertex will have a valency of 5, an edge to each of the other points.

With 5 lines there must be at least three of one colour so there are four points connected by the same colour.

For the three lines connecting the new points, if one is the original colour then a set of three vertices is made with the original colour. If none are the original colour then the three vertices make a set of three themselves.

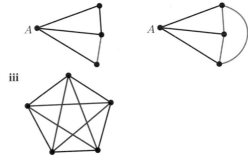

iii

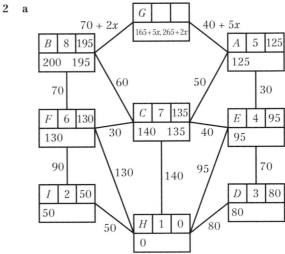

2 a

Via *A*   *H – E – A – G*   length $165 + 5x$
Via *B*   *H – E – C – B – G*   length $265 + 2x$

b $165 + 5x = 265 + 2x \Rightarrow x = 33\frac{1}{3}$
So range is $0 \leq x \leq 33\frac{1}{3}$

3 a $9\frac{1}{2}x - 26$

b The only vertices of odd order are *B* and *C*, we have to repeat the shortest path between *B* and *C*.
If $x \geq 9$ the shortest path is *BC* (direct)
Weight of network + *BC* = 100
$\left(9\frac{1}{2}x - 26\right) + x = 100 \Rightarrow x = 12$
If $x < 9$ the shortest path is *BAC* of length $2x - 9$
$\left(9\frac{1}{2}x - 26\right) + 2x - 9 = 100 \Rightarrow x = 11\frac{17}{23} \not< 9$
so inconsistent.

c $x = 12$

4 a Minimum spanning tree = 751
Initial upper bound = $2 \times 751 = 1502$
Taking shortcut *AH* saves $120 + 131 - 144 = 107$
Tour length = 1395
Tour route: *ABACAEHFGDGFHA*

b Delete *G* and use Prim's algorithm starting at *A*
RMST = 672
Lower bound by deleting $G = 672 + 144 + 155 = 971$
Route is not a tour, so does not give an optimal solution.

CHAPTER 6

Prior knowledge 6

1 $y > 2, x + y \leqslant 7, y \leqslant x + 1$

2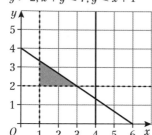

Exercise 6A

1 Number of boxes of gold assortment = x Number of boxes of silver assortment = y
Objective: maximise $P = 80x + 60y$
Constraints
- time to make chocolate, $30x + 20y \leqslant 300 \times 60$ which simplifies to $3x + 2y \leqslant 1800$ •——— All units now in minutes.
- time to wrap and pack $12x + 15y \leqslant 200 \times 60$ which simplifies to $4x + 5y \leqslant 4000$
- 'At *least* twice as many silver as gold' $2x \leqslant y$
- non-negativity $x, y \geqslant 0$
In summary: maximise $P = 80x + 60y$
subject to
$3x + 2y \leqslant 1800$
$4x + 5y \leqslant 4000$
$\qquad 2x \leqslant y$
$\qquad x, y \geqslant 0$

2 Number of type $A = x$ Number of type $B = y$
Objective: minimise $C = 6x + 10y$
Constraints
- Display must be at least 30 m long $x + 1.5y \geqslant 30$ which simplifies to $2x + 3y \geqslant 60$
- 'At least twice as many x as y' $2y \leqslant x$
- At *least* six type B $y \geqslant 6$
- non-negativity $x, y \geqslant 0$
In summary: minimise $C = 6x + 10y$
subject to:
$2x + 3y \geqslant 60$
$\qquad 2y \leqslant x$
$\qquad y \geqslant 6$
$\qquad x, y \geqslant 0$

3 Number of games of Cludopoly = x Number of games of Trivscrab = y
Objective: maximise $P = 1.5x + 2.5y$
Constraints
- First machine: $5x + 8y \leqslant 10 \times 60$ which simplifies to $5x + 8y \leqslant 600$ ——— All units now in minutes.
- Second machine: $8x + 4y \leqslant 10 \times 60$ which simplifies to $2x + y \leqslant 150$
- 'At *most* 3 times as many x as y' $3y \geqslant x$
- non-negativity $x, y \geqslant 0$
In summary: maximise $P = 1.5x + 2.5y$
subject to:
$5x + 8y \leqslant 600$
$\quad 2x + y \leqslant 150$
$\qquad 3y \geqslant x$
$\qquad x, y \geqslant 0$

4 Number of type 1 bookcases = x Number of type 2 bookcases = y
Objective: maximise $S = 40x + 60y$
Constraints
- budget $150x + 250y \leqslant 3000$ which simplifies to $3x + 5y \leqslant 60$
- floor space $15x + 12y \leqslant 240$ which simplifies to $5x + 4y \leqslant 80$
- 'At most $\frac{1}{3}$ of all bookcases to be type 2' $y \leqslant \frac{1}{3}(x + y)$ which simplifies to $2y \leqslant x$
- At least 8 type 1 $x \geqslant 8$
- non-negativity $x, y \geqslant 0$

   Online Full worked solutions are available in SolutionBank.

In summary: maximise $S = 40x + 60y$
subject to:
$3x + 5y \leqslant 60$
$5x + 4y \leqslant 80$
$\quad\quad 2y \leqslant x$
$\quad\quad\quad x \geqslant 8$
$\quad x, y \geqslant 0$

5 Let x = number of kg of indoor feed and y = number of kg of outdoor feed
   Objective: maximise $P = 7x + 6y$
   Constraints
   - Amount of A $10x + 20y \leqslant 5 \times 1000$ which simplifies to $x + 2y \leqslant 500$
   - Amount of B $20x + 10y \leqslant 5 \times 1000$ which simplifies to $2x + y \leqslant 500$ •————— All units now in grams.
   - Amount of C $20x + 20y \leqslant 6 \times 1000$ which simplifies to $x + y \leqslant 300$
   - 'At *most* 3 times as much y as x' $y \leqslant 3x$
   - At least 50 kg of x $x \geqslant 50$
   - non-negativity $y \geqslant 0$ ($x \geqslant 0$ is unnecessary because of the previous constraint)
   In summary: maximise $P = 7x + 6y$
   subject to
   $x + 2y \leqslant 500$
   $2x + y \leqslant 500$
   $\quad x + y \leqslant 300$
   $\quad\quad\quad y \leqslant 3x$
   $\quad\quad\quad x \geqslant 50$
   $\quad\quad\quad y \geqslant 0$

6 Number of A smoothies = x Number of B smoothies = y Number of C smoothies = z
   Objective: maximise $P = 60x + 65y + 55z$
   Constraints
   - oranges $x + \frac{1}{2}y + 2z \leqslant 50$ which simplifies to $2x + y + 4z \leqslant 100$
   - raspberries $10x + 40y + 15z \leqslant 1000$ which simplifies to $2x + 8y + 3z \leqslant 200$
   - kiwi fruit $2x + 3y + z \leqslant 100$
   - apples $2x + \frac{1}{2}y + 2z \leqslant 60$ which simplifies to $4x + y + 4z \leqslant 120$
   - non-negativity $x, y, z \geqslant 0$
   In summary: maximise $P = 60x + 65y + 55z$
   subject to:
   $\quad 2x + y + 4z \leqslant 100$
   $2x + 8y + 3z \leqslant 200$
   $\quad 2x + 3y + z \leqslant 100$
   $\quad 4x + y + 4z \leqslant 120$
   $\quad\quad\quad x, y, z \geqslant 0$

7 Let number of hours of work for factory $R = x$ Let number of hours of work for factory $S = y$
   Objective: minimise $C = 300x + 400y$
   Constraints
   - milk $1000x + 800y \geqslant 20\,000$ which simplifies to $5x + 4y \geqslant 100$
   - yoghurt $200x + 300y \geqslant 6000$ which simplifies to $2x + 3y \geqslant 60$
   - 'At *least* $\frac{1}{3}$ of total time for R' $x \geqslant \frac{1}{3}(x + y)$ which simplifies to $2x \geqslant y$
   - 'At *least* $\frac{1}{3}$ of total time for S' $y \geqslant \frac{1}{3}(x + y)$ which simplifies to $2y \geqslant x$
   - non-negativity $x, y \geqslant 0$
   In summary: minimise $C = 300x + 400y$
   subject to:
   $5x + 4y \geqslant 100$
   $2x + 3y \geqslant 60$
   $\quad\quad 2x \geqslant y$
   $\quad\quad 2y \geqslant x$
   $\quad x, y \geqslant 0$

Exercise 6B

1 a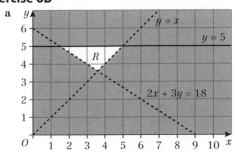

b
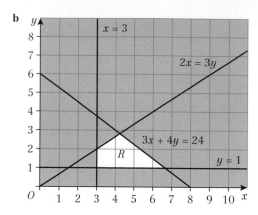

c
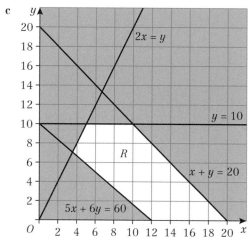

d
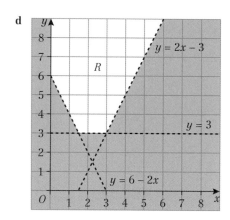

2
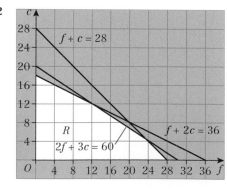

3
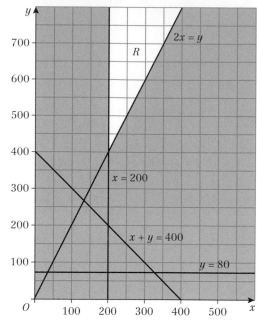

4 Let x represent the number of type A and y represent the number of type B.

$$x \geqslant 200, \frac{y}{9} \leqslant x \leqslant \frac{2y}{3}, x + y \leqslant 3000, y \geqslant 0$$

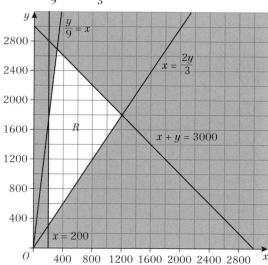

5 a $y \leqslant \dfrac{x}{2} + 10 \qquad y \geqslant \dfrac{x}{4} \qquad x \geqslant 0$

b
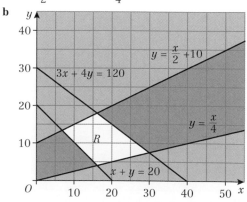

Online Full worked solutions are available in SolutionBank.

Exercise 6C

1 **a** Need intersection of $4x + y = 1400$ and $3x + 2y = 1200$ ———— Objective line passes through $(200, 0)$ and $(0, 400)$.
$(320, 120)$ $m = 760$

   b $(0, 400)$ $N = 1600$ ———— Objective line passes through $(400, 0)$ and $(100, 0)$.

   c Need intersection of $x + 3y = 1200$ and $3x + 2y = 1200$
$\left(171\frac{3}{7}, 342\frac{6}{7}\right)$ $P = 514\frac{2}{7}$

   d $(350, 0)$ $Q = 2100$

2 **a** $(0, 90)$ $E = 90$

   b Need intersection of $6y = x$ and $3x + 7y = 420$
$(100.8, 16.8)$ $F = 168$

   c Need intersection of $9x + 10y = 900$ and $3x + 7y = 420$ ———— Objective line passes through $(80, 0)$ and $(0, 60)$.
$\left(63\frac{7}{11}, 32\frac{8}{11}\right)$ $G = 321\frac{9}{11}$

   d Same intersection as in **b** $(100.8, 16.8)$ $H = 201.6$ ———— Objective line passes through $(120, 0)$ and $(0, 20)$.

3 **a** Need intersection of $3x + y = 60$ and $5y = 3x$
$\left(16\frac{2}{3}, 10\right)$ $J = 56\frac{2}{3}$

   b Need intersection of $y = 4x$ and $9x + 5y = 450$
$\left(15\frac{15}{29}, 62\frac{2}{29}\right)$ $K = 77\frac{17}{29}$

   c Need intersection of $3x + y = 60$ and $y = 4x$ ———— Objective line passes through $(10, 0)$ and $(0, 60)$.
$\left(8\frac{4}{7}, 34\frac{2}{7}\right)$ $L = 85\frac{5}{7}$

   d Need intersection of $9x + 5y = 450$ and $5y = 3x$ ———— Objective line passes through $(40, 0)$ and $(0, 80)$.
$(37.5, 22.5)$ $m = 97.5$

4 **a** C **b** A **c** B **d** D **e** C **f** A **g** B **h** D **i** C **j** D

5

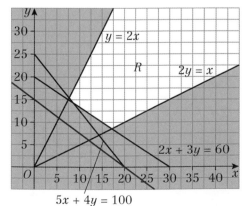

Objective line passes through $(0, 350)$ and $(300, 0)$.
Maximum point is $(200, 100)$. $P_{max} = 2000$

6

Objective line passes through $(0, 15)$ and $(20, 0)$.
Intersection of $2x + 3y = 60$ $5x + 4y = 100$
$\left(8\frac{4}{7}, 14\frac{2}{7}\right)$ value $= 8285\frac{5}{7}$

7 a

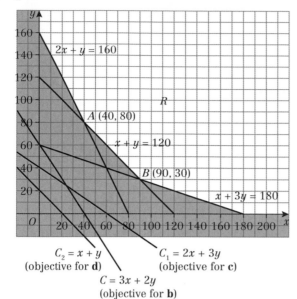

$C_2 = x + y$
(objective for **d**)

$C_1 = 2x + 3y$
(objective for **c**)

$C = 3x + 2y$
(objective for **b**)

b

Vertices	$C = 3x + 2y$
(0, 160)	320
(40, 80)	280
(90, 30)	330
(180, 0)	540

so minimum is (40, 80) value of $C = 280$

c (90, 30) $C_1 = 270$

d C_2 is parallel to $x + y = 120$ so all points from A to B are optimal points.

8 a

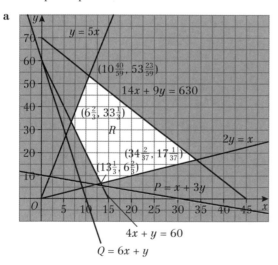

$4x + y = 60$

$Q = 6x + y$

b i $\left(13\frac{1}{3}, 6\frac{2}{3}\right)$ $P = 33\frac{1}{3}$

ii $\left(34\frac{2}{37}, 17\frac{1}{37}\right)$ $Q = 221\frac{13}{37}$

9 a

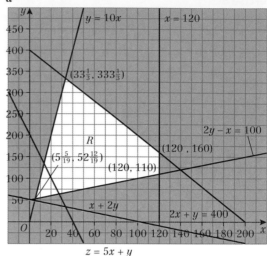

$z = 5x + y$

b i (120, 160) $z = 760$

ii $\left(5\frac{5}{19}, 52\frac{12}{19}\right)$ $z = 78\frac{18}{19}$

c Optimal point $\left(33\frac{1}{3}, 333\frac{1}{3}\right)$ optimal value of $x + 2y = 700$

Challenge

a

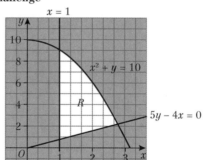

b Gradient of objective line $= -3$

Max value of P occurs at the point where the objective line is a tangent to the curve $x^2 + y = 10$

Gradient of curve is $\dfrac{dy}{dx} = -2x$

$-2x = -3 \Rightarrow x = 1.5, \ y = 10 - 1.5^2 = 7.75$

Max $P = 3 \times 1.5 + 7.75 = 12.25$

Exercise 6D

1 a

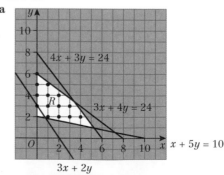

$3x + 2y$

Maximum integer value
(5, 1) $3x + 2y = 17$

Online Full worked solutions are available in SolutionBank.

b

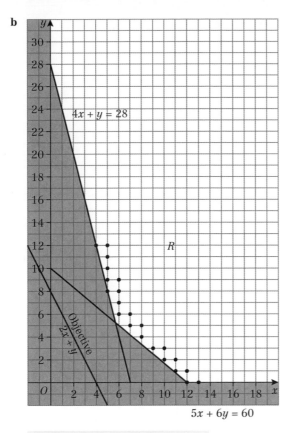

$5x + 6y = 60$

Minimum integer value (6, 5) $2x + y = 17$

c

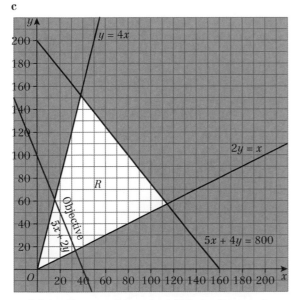

Solving $2y = x$ and $5x + 4y = 800$ simultaneously
gives $\left(114\frac{2}{7}, 57\frac{1}{7}\right)$

Test integer values nearby.

Points	$2y \geqslant x$	$5x + 4y \leqslant 800$	$5x + 2y$
(114, 57)	✓	✓	684
(114, 58)	✓	✗	—
(115, 57)	✗	✗	—
(115, 58)	✓	✗	—

So optimal point is (114, 57)
Value 684

d

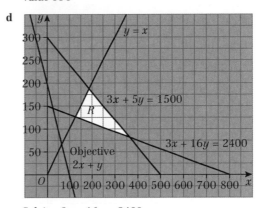

Solving $3x + 16y = 2400$
$3x + 5y = 1500$
simultaneously gives $\left(363\frac{7}{11}, 81\frac{9}{11}\right)$
Taking integer point

Points	$3x + 5y \leqslant 1500$	$3x + 16y \geqslant 2400$
(363, 81)	✓	✗
(363, 82)	✓	✓
(364, 81)	✓	✗
(364, 82)	✗	✓

So optimal integer point is (363, 82)
Value 808

2 The table shows points in R on, or close to, a vertex
with integer coordinate.

Points	$P = 5x + 3y$
(10, 0)	50
(10, 30)	140
(12, 34)	162
(13, 34)	167
(30, 0)	150

Maximum value of $P = 167$ when $x = 13$, $y = 34$

3

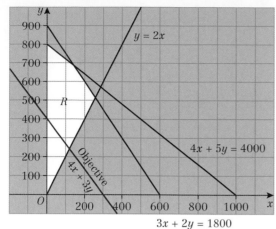

$$3x + 2y = 1800$$

Intersection of $4x + 5y = 4000$ and $3x + 2y = 1800$

gives $\left(142\frac{6}{7}, 685\frac{5}{7}\right)$

Testing nearby integer points

Points	$4x + 5y \leqslant 4000$	$3x + 2y \leqslant 1800$	$80x + 60y$
(142, 685)	✓	✓	52 460
(142, 686)	✓	✓	52 520
(143, 685)	✓	✓	52 540
(143, 686)	✗	✗	—

So maximum integer solution is 52 540 pennies at (143, 685).

4 a $2x + 3y \geqslant 60$
$2y \leqslant x$
$y \geqslant 6$
$x, y \geqslant 0$

b, c

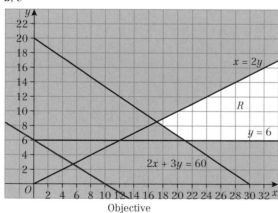

Objective
$6x + 10y$

Intersection of $2x + 3y = 60$ and $y = 6$ (21, 6)
cost $C = 6x + 10y$
so minimum cost $= 6 \times 21 + 60$
$= £186$

5 a $5x + 8y \leqslant 600$
$2x + y \leqslant 150$
$3y \geqslant x$
$x, y \geqslant 0$

b, c

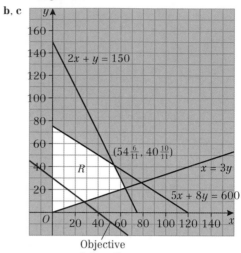

Objective
$1.5x + 2y$

Intersection of $5x + 8y = 600$
$2x + y = 150$ giving $\left(54\frac{6}{11}, 40\frac{10}{11}\right)$

Points	$5x + 8y \leqslant 600$	$2x + y \leqslant 150$	$1.5x + 2y$
(54, 40)	✓	✓	161
(54, 41)	✓	✓	163
(55, 40)	✓	✓	162.5
(55, 41)	✗	✗	—

So maximum value is 163 at (54, 41).

6 a Let x be the number of type 1 bookcases purchased.
Let y be the number of type 2 bookcases purchased.
Maximise: $S = 40x + 60y$
subject to:
$3x + 5y \leqslant 60$
$5x + 4y \leqslant 80$
$2y \leqslant x$
$x \geqslant 8$
$x, y \geqslant 0$

b

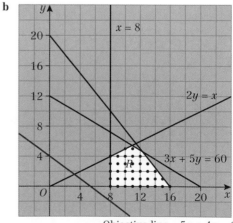

Objective line $5x + 4y = 80$
$40x + 60y$

Online Full worked solutions are available in SolutionBank.

Using Method 1 from Example 13 shows you that the optimal integer solution is (11, 5) giving 740 m of shelving.

Using Method 2 gives you the following solution: Intersection of $3x + 5y = 60$

$5x + 4y = 80$ giving $\left(12\frac{4}{13}, 4\frac{8}{13}\right)$

Points	$3x + 5y \leq 60$	$5x + 4y \leq 80$	$40x + 60y$
(12, 4)	✓	✓	720
(12, 5)	✗	✓	
(13, 4)	✓	✗	
(13, 5)	✗	✗	

Maximum value is 720 at (12, 4).

In this instance, the solution produced by Method 2 is actually incorrect, but it requires a very particular set of circumstances to create this discrepancy. It is generally safe to assume that a solution found using Method 2 will be correct, but do check your graph to see whether there could be an alternative optimal integer solution.

Mixed exercise 6

1 a *flour*: $200x + 200y \leq 2800$ so $x + y \leq 14$
      fruit: $125x + 50y \leq 1000$ so $5x + 2y \leq 40$
   b Cooking time $50x + 30y \leq 480$ so $5x + 3y \leq 48$
   c

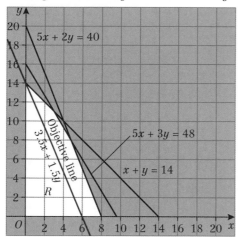

   d $P = 3.5x + 1.5y$
   e Maximum integer solution is £28.50 at (6, 5) so Mr Baker should bake 6 cakes and 5 fruit loaves.
   f P_{max} = £28.50

2 a $0.08x + 0.08y \leq 6.4$ so $x + y \leq 80$
   b cost: $6x + 4.8y \leq 420$ $5x + 4y \leq 350$
   c Display $30x \leq 2 \times 20y$ $3x \leq 4y$
   d

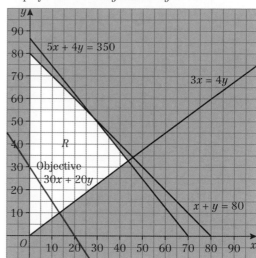

   e Maximum integer solution is 1950 at (43, 33) so the librarian should buy 43 CD storage units and 33 DVD storage units.

3 a i   Total number of people
            $54x + 24y \geq 432$ so $9x + 4y \geq 72$
      ii number of adults $x + y \leq 12$
      iii number of large coaches $x \leq 7$
   b

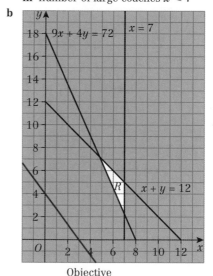

Objective
$84(4x + 3y)$

   c minimise $C = 336x + 252y$
                   $= 84(4x + 3y)$
   d Objective line passes through (0, 4) (3, 0)
   e Minimum integer solution is 3108 at (7, 3) so the headteacher should hire 7 large coaches and 3 small coaches for a cost of £3108.

4 a $4x + 5y \leq 47$
                 $y \geq 2x - 8$
         $4y - x - 18 \leq 0$
             $x, y \geq 0$
   b Solving simultaneous equations $y = 2x - 8$
                                        $4x + 5y = 47$
                                        $6\frac{3}{14}, 4\frac{3}{7}$

c i For example where x and y
    • types of car to be hired
    • number of people etc

ii (6, 4)

5 a $2\frac{1}{2}x + 3y \leq 300$ $[5x + 6y \leq 600]$
    $5x + 2y \leq 400$
    $2y \leq 150$ $[y \leq 75]$

b Maximise $P = 2x + 4y$

c

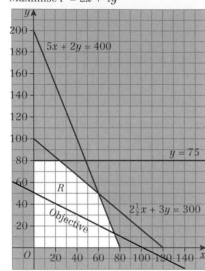

Maximum integer solution is $P = 30$ at (30, 75) so the company should make 30 plates and 75 mugs for a profit of £360.

d The optimal point is at the intersection of $y = 75$ and $2\frac{1}{2}x + 3y = 300$.
So the constraint $5x + 2y \leq 400$ is not at its limit.
At (30, 75) $5x + 2y = 300$ so 100 minutes are unused.

Challenge

a (60, 30, 35)

b (60, 30, 35) From solving $x + y + 2z = 160$, $x - z = 25$, $y + 2z = 100$
(40, 70, 15) From solving $x - z = 25$, $y + 2z = 100$, where $z = 15$
(60, 70, 15) From solving $x + y + 2z = 160$, $y + 2z = 100$, where $z = 15$
(40, 90, 15) From solving $x + y + 2z = 160$, $x - z = 25$, where $z = 15$

c

Vertex	P
(60, 30, 35)	175
(40, 70, 15)	275
(60, 70, 15)	315
(40, 90, 15)	335

Optimum value of P is 335 when $x = 40$, $y = 90$ and $z = 15$

CHAPTER 7

Prior knowledge 7

1 $x = 10$, $y = -3$

2 24

3 a $2x + 3y - z = 5$ (1)
    $3y + 2z = 0$ (2)
    (1) × 2: $4x + 6y - 2z = 10$ (3)
    (2) + (3): $4x + 9y = 10$ as required.

b $x = 3\frac{2}{5}$, $y = -\frac{2}{5}$, $z = \frac{3}{5}$

Exercise 7A

1 Let x_1, x_2 and x_3 be the numbers of round, square and rectangular boxes respectively.
Maximise $P = 12x_1 + 10x_2 + 11x_3$
subject to: $4x_1 + 2x_2 + 3x_3 + r = 360$
            $2x_1 + 3x_2 + 3x_3 + s = 360$
            $x_1, x_2, x_3, r, s \geq 0$

2 Let x_A, x_B, x_C and x_D be the numbers of types A, B, C and D backpacks made.
Maximise $P = 8x_A + 7x_B + 6x_C + 9x_D$
subject to: $2.5x_A + 3x_B + 2x_C + 4x_D + r = 1400$
           $10x_A + 12x_B + 8x_C + 15x_D + s = 9000$
           $5x_A + 7x_B + 4x_C + 9x_D + t = 4800$
           $x_A + x_B + x_C + x_D + u = 500$
           $x_A, x_B, x_C, x_D, r, s, t, u \geq 0$

3 Let x_A, x_C and x_S be the numbers of adults, children and senior members.
Maximise $P = 40x_A + 10x_C + 20x_S$
subject to: $x_A + x_C + x_S + r = 100$
           $-x_A + x_C - x_S + s = 0$
           $-2x_A + x_C + x_S + t = 0$
           $x_A, x_C, x_S, r, s, t \geq 0$

4 Let x_r, x_f and x_m be the numbers of batches of rock cakes, fairy cakes and muffins made.
Maximise $N = 10x_r + 18x_f + 12x_m$
Subject to: $220x_r + 100x_f + 250x_m + r = 3000$
           $100x_r + 100x_f + 50x_m + s = 2000$
           $50x_r + 100x_f + 75x_m + t = 1500$
           $x_r, x_f, x_m, r, s, t \geq 0$

5 Let x_s, x_m and x_l be the numbers of small, medium and large boxes.
Minimise $C = 0.3x_s + 0.5x_m + 0.8x_l$
subject to: $x_s + 3x_m + 7x_l \geq 280$
           $3x_s + 8x_m + 18x_l \geq 600$
           $-x_s + x_m + x_l \leq 0$
           $-x_m + 2x_l \leq 0$
           $x_s, x_m, x_l \geq 0$

Exercise 7B

1 $P = 38$ when $x = 0$ $y = 3$ $z = 5$ $r = 0$ $s = 0$

2 $P = 260$ when $x = 0$ $y = 40$ $z = 10$ $r = 0$ $s = 0$

3 $P = 11$ when $x = 2$ $y = 1$ $z = 0$ $r = 0$ $s = 0$ $t = 1$

4 $P = 105.5$ when $x_1 = \frac{77}{4}$, $x_2 = 0$, $x_3 = \frac{57}{4}$, $x_4 = 0$, $r = 33$, $s = 0$, $t = \frac{109}{4}$, $u = 0$

5 $P = -336$ when $x = 0$, $y = 0$, $z = \frac{21}{2}$, $r = 420$, $s = 84$, $t = 0$, $u = 16$

6 $P = -2$ when $x = 0$, $y = 1$, $z = 0$, $r = 0$, $s = 4$, $t = 1$

7 *For Q1*

b $P + \frac{8}{3}x + r + \frac{4}{3}s = 38$
    $\frac{1}{2}x + y + \frac{1}{2}r = 3$
    $\frac{7}{6}x + z - \frac{1}{2}r + \frac{1}{3}s = 5$

c $P = 38 - \frac{8}{3}x - r - \frac{4}{3}s$
so increasing x, r or s would decrease P.

For Q2

b $P + \frac{1}{2}x + 2r + \frac{3}{2}s = 260$
    $\frac{1}{4}x + y + \frac{1}{2}r - \frac{1}{4}s = 40$
    $\frac{1}{4}x + z + \frac{1}{4}s = 10$

c $P = 260 - \frac{1}{2}x - 2r - \frac{3}{2}s$, so increasing x, r or s would decrease P.

Online Full worked solutions are available in SolutionBank.

For Q3

b $P + 8z + \frac{4}{5}r + \frac{3}{5}s = 11$

$x - 5z + \frac{3}{5}r - \frac{4}{5}s = 2$

$y + 5z - \frac{1}{5}r + \frac{3}{5}s = 1$

$15z - r + 2s + t = 1$

c $P = 11 - 8z - \frac{4}{5}r - \frac{3}{5}s$, so increasing z, r or s would decrease P.

For Q4

b $P + \frac{11}{2}x_2 + \frac{1}{2}x_4 + \frac{1}{2}s + u = \frac{211}{2}$

$4x_2 - 3x_4 + r - 2s + u = 33$

$-\frac{1}{4}x_2 + x_3 + \frac{5}{4}x_4 + \frac{3}{4}s - \frac{1}{2}u = \frac{57}{4}$

$\frac{11}{4}x_2 - \frac{3}{4}x_4 - \frac{5}{4}s + t + \frac{1}{2}u = \frac{109}{4}$

$x_1 + \frac{3}{4}x_2 + \frac{1}{4}x_4 - \frac{1}{4}s + \frac{1}{2}u = \frac{77}{4}$

c $P = \frac{211}{2} - \frac{11}{2}x_2 - \frac{1}{2}x_4 - \frac{1}{2}s - u$, so increasing x_2, x_4, s or u would decrease P.

For Q5

b $P - 5x - 12y - 2t = -336$

$-\frac{1}{2}x + \frac{3}{2}y + r - \frac{3}{2}t = 420$

$\frac{3}{2}x - \frac{7}{2}y + s - \frac{3}{2}t = 84$

$\frac{1}{16}x + \frac{3}{16}y + z + \frac{1}{16}t = \frac{21}{2}$

c $P = -366 + 5x + 12y + 2t$, so increasing x, y or t would decrease P.

For Q6

b $P - 6x - 4z - r = -2$

$2x + y + \frac{1}{2}z + \frac{1}{2}r = 1$

$-6x - 2r + s = 4$

$-3x + \frac{5}{2}z - \frac{3}{2}r + t = 1$

c $P = -2 + 6x + 4z + r$, so increasing x, z or r would increase P.

8 a $x = 0$, y $= 0$, $z = \frac{3}{2}$

b $3x + 4y + r + P = 3$

9 a $P - x + 2y - 2z = 0$

b

	x	y	z	r	s	t	Value	
r	2	$\frac{1}{2}$	1	$\frac{1}{2}$	0	0	1	R1 ÷ 2
s	−1	$\frac{3}{2}$	0	$-\frac{1}{2}$	1	0	7	R2 − R1
t	−4	$\frac{5}{2}$	0	$-\frac{3}{2}$	0	1	1	R3 − 3R1
p	3	3	0	1	0	0	2	R4 + 2R1

c $P = 2$ when $x = 0$, $y = 0$, $z = 1$, $r = 0$, $s = 7$, $t = 1$

Exercise 7C

1 a $P = 39$ when $x = 13$, $y = 0$

b $P = 148$ when $x = 0$, $y = 9$, $z = 5$

2 a $2x + 3y + z \leqslant 80$
$4x + 2y + 3z \leqslant 140$
$3x + 4y + 2x \leqslant 96$

b $x \geqslant 0$, $y \geqslant 0$, $z \geqslant 0$

c The most negative value in the objective row is −6, so the pivot is in the z column. The smallest positive θ value is 46, which is in the s row. So the pivot value is 2 in row s, column z

d

Basic variable	x	y	z	r	s	t	Value
r	$\frac{3}{8}$	0	0	1	$\frac{1}{4}$	$-\frac{7}{8}$	31
z	$\frac{5}{4}$	0	1	0	$\frac{1}{2}$	$-\frac{1}{4}$	46
y	$\frac{1}{8}$	1	0	0	$-\frac{1}{4}$	$\frac{3}{8}$	1
P	$\frac{21}{2}$	0	0	0	3	$\frac{7}{2}$	756

This is an optimal solution. Examining the profit equation, increasing x, s, or t would reduce profit, so this solution is optimal.

$$P = 756 - \frac{21}{2}x - 3s - \frac{7}{2}t$$

e $x = 0$, $y = 1$, $z = 46$

Challenge

$k \geqslant \frac{8}{5}$

Exercise 7D

1 a Max $P = 30$ when $x = 15$, $y = 0$

b Min $C = -2$ when $x = \frac{32}{5}$, $y = \frac{14}{5}$, $z = 0$

c Max $P = 24$ when $x = 0$, $y = 0$, $z = 12$

2 Applying the first stage of the two-stage simplex method produces the following tableau after one iteration.

	x_1	x_2	s_1	s_2	a_1	Value
x_2	$\frac{5}{6}$	1	$\frac{1}{6}$	0	0	$\frac{25}{2}$
a_1	$-\frac{1}{3}$	0	$-\frac{2}{3}$	−1	1	2
I	$\frac{1}{3}$	0	$\frac{2}{3}$	1	0	−2

There are no negative values in the bottom row, so the maximum value of I is $-2 \neq 0$ and there is no feasible solution.

3 a i Surplus variables represent the amount by which the actual quantity exceeds the minimum possible value of that quantity.

ii Artificial variables are added to $\geqslant$ constraints so that $s \geqslant 0$ and a basic feasible solution can be obtained.

b There are no negative values in the bottom row of the tableau so it is optimal, and $I = 0$, so there is a basic feasible solution.

c (Initial tableau interpreted from question)

Basic variable	x	y	z	s_1	s_2	s_3	Value
s_1	0	0	2	1	0	1	2
x	1	0	$\frac{1}{2}$	0	$\frac{-1}{2}$	$\frac{-3}{2}$	$\frac{29}{2}$
y	0	1	$\frac{-1}{2}$	0	$\frac{1}{2}$	−1	$\frac{11}{2}$
P	0	0	0	0	−1	−2	13

(1st iteration)

Basic variable	x	y	z	s_1	s_2	s_3	Value
s_3	0	0	2	1	0	1	2
x	1	0	$\frac{7}{2}$	$\frac{3}{2}$	$\frac{-1}{2}$	0	$\frac{35}{2}$
y	0	1	$\frac{3}{2}$	1	$\frac{1}{2}$	0	$\frac{15}{2}$
P	0	0	4	2	−1	0	17

(2nd iteration)

Basic variable	x	y	z	s_1	s_2	s_3	Value
s_3	0	0	2	1	0	1	2
x	1	1	5	$\frac{5}{2}$	0	0	25
s_2	0	2	3	2	1	0	15
P	0	2	7	4	0	0	32

$P = 32$ when $x = 25$, $s_2 = 15$, $s_3 = 2$, $y = z = s_1 = 0$

Exercise 7E

1   a $a_1 \neq 0$

    b   The pivot is the 2 in the z column. The most negative value in the P row is in the z column and the smallest θ value is given by $\frac{6}{2}$.

    c   In the next iteration, a_1 remains as a basic variable $\neq 0$ so this cannot represent a feasible solution.

    d

	x	y	z	s_1	s_2	a_1	Value	Row operation
z	$\frac{1}{2}$	$\frac{1}{2}$	1	$\frac{1}{2}$	0	0	3	R1 ÷ 2
s_2	$\frac{3}{2}$	$-\frac{7}{2}$	0	$-\frac{1}{2}$	-1	1	2	R2 − R1
P	0	$3(1 + M)$	0	$1 + M$	$-M$	0	$6 - 9M$	R3 + R1(2 + 2M)

2   a $x + 3y + z + s_1 = 100$
      $3x - y + s_2 = 52$
      $x - s_3 + a_1 = 20$

    b

	x	y	z	s_1	s_2	s_3	a_1	Value
s_1	1	3	1	1	0	0	0	100
s_2	3	-1	0	0	1	0	0	52
a_1	1	0	0	0	0	-1	1	20
P	$-(4 + M)$	-2	1	0	0	M	0	$-20M$

    c M is an arbitrarily large number.

    d   The pivot is the 3 in the x column.

3   a   The constraints include a mixture of $\leqslant$ and $\geqslant$ signs.

    b $3x - y + s_1 = 110$
      $x + 2y - s_2 + a_1 = 45$

    c   Minimise $C = 4x + 3y$
      Maximise $D = -4x - 3y - Ma_1$
      $D = -4x - 3y - M(45 - x - 2y + s_2)$
      $D + x(4 - M) + y(3 - 2M) + Ms_2 = -45M$

	x	y	s_1	s_2	a_1	Value
s_1	3	-1	1	0	0	110
a_1	1	2	0	-1	1	45
D	$(4 - M)$	$(3 - 2M)$	0	M	0	$-45M$

4   a $x + y + z + s_1 = 20$
      $3x + y + 2z - s_2 + a_1 = 24$

    b $P = 3x + 5y - z - Ma_1$
      $a_1 = 24 - 3x - y - 2z + s_2$
      $P - (3 + 3M)x - (5 + M)y - (2M - 1)z + Ms_2 = -24M$

    c

Basic variable	x	y	z	s_1	s_2	a_1	Value
s_1	1	1	1	1	0	0	20
a_1	3	1	2	0	-1	1	24
P	$-(3 + 3M)$	$-(5 + M)$	$-(2M - 1)$	0	M	0	$-24M$

    d

Basic variable	x	y	z	s_1	s_2	a_1	Value
s_1	0	$\frac{2}{3}$	$\frac{1}{3}$	1	$\frac{1}{3}$	$\frac{-1}{3}$	12
x	1	$\frac{1}{3}$	$\frac{2}{3}$	0	$\frac{-1}{3}$	$\frac{1}{3}$	8
P	0	-4	3	0	-1	$(1 + M)$	24

Basic variable	x	y	z	s_1	s_2	a_1	Value
y	0	1	$\frac{1}{2}$	$\frac{3}{2}$	$\frac{1}{2}$	$\frac{-1}{2}$	18
x	1	0	$\frac{1}{2}$	$\frac{-1}{2}$	$\frac{-1}{2}$	$\frac{1}{2}$	2
P	0	0	5	6	1	$M - 1$	96

$x = 2$, $y = 18$, $z = s_1 = s_2 = a_1 = 0$

Mixed exercise 7

1   a   There are no negative numbers in the profit row.

    b $P + \frac{3}{2}x + \frac{3}{4}r = 840$
      so $P = 840 - \frac{3}{2}x - \frac{3}{4}r$
      Increasing x or r would decrease P.

    c   i   Maximum profit = £840

       ii   Optimum number of $A = 0$, $B = 56$ and $C = 75$

2   a   Maximise $P = 14x + 20y + 30z$
      subject to: $5x + 8y + 10z + r = 25\,000$
                $5x + 6y + 15z + s = 36\,000$
      where r and s are slack variables $x, y, z, r, s \geqslant 0$

    b

b.v.	x	y	z	r	s	Value
r	$1\frac{2}{3}$	④	0	1	$-\frac{2}{3}$	1000
z	$\frac{1}{3}$	$\frac{2}{5}$	1	0	$\frac{1}{15}$	2400
P	-4	-8	0	0	2	72\,000

b.v.	x	y	z	r	s	Value	Row operation
y	⑤⁄₁₂	1	0	$\frac{1}{4}$	$-\frac{1}{6}$	250	R1 ÷ 4
z	$\frac{1}{6}$	0	1	$-\frac{1}{10}$	$\frac{2}{15}$	2300	R2 − $\frac{2}{5}$R1
P	$-\frac{2}{3}$	0	0	2	$\frac{2}{3}$	74\,000	R3 + 8R1

b.v.	x	y	z	r	s	Value	Row operation
x	1	$2\frac{2}{5}$	0	$\frac{3}{5}$	$-\frac{2}{5}$	600	R1 ÷ $\frac{5}{12}$
z	0	$-\frac{2}{5}$	1	$-\frac{1}{5}$	$\frac{1}{5}$	2200	R2 − $\frac{1}{6}$R1
P	0	$1\frac{3}{5}$	0	$2\frac{2}{5}$	$\frac{2}{5}$	74\,400	R3 + $\frac{2}{3}$R1

    c   i $x = 600$ $y = 0$ $z = 2200$

       ii   Profit is £74\,400

       iii   The solution is optimal since there are no negative numbers in the profit row.

3   a $\frac{1}{5}(x + y + z) \geqslant y \Rightarrow -x + 4y - z \leqslant 0$
      $60x + 100y + 160z \leqslant 2000 \Rightarrow 3x + 5y + 8z \leqslant 100$
      $x \geqslant 0$ $y \geqslant 0$ $z \geqslant 0$

    b $S = 2x + 4y + 6z$

    c   There are three variables.

    d   Writing inequalities as equations:
      $-x + 4y - z + r = 0$
      $3x + 5y + 8z + t = 100$
      $S - 2x - 4y - 6z = 0$
      Enter coefficients to get initial tableau.

Online   Full worked solutions are available in SolutionBank.

e Pivot is 8 in initial tableau.

b.v.	x	y	z	r	t	Value	Row operation
y	$-\frac{5}{8}$	$\frac{37}{8}$	0	1	$\frac{1}{8}$	$\frac{25}{2}$	R1 + R2
z	$\frac{3}{8}$	$\frac{5}{8}$	1	0	$\frac{1}{8}$	$\frac{25}{2}$	R2 ÷ 8
S	$\frac{1}{4}$	$-\frac{1}{4}$	0	0	$\frac{3}{4}$	75	R3 + 6R2

f

b.v.	x	y	z	r	t	Value	Row operation
y	$-\frac{5}{37}$	1	0	$\frac{8}{37}$	$\frac{1}{37}$	$2\frac{26}{37}$	R1 ÷ $4\frac{5}{8}$
z	$\frac{17}{37}$	0	1	$-\frac{5}{37}$	$\frac{4}{37}$	$10\frac{30}{37}$	R2 $-\frac{5}{8}$R1
S	$\frac{8}{37}$	0	0	$\frac{2}{37}$	$\frac{28}{37}$	$75\frac{25}{37}$	R3 $+\frac{1}{4}$R1

g There are no negative numbers in the objective row.

h 0 small, 2 medium and 11 large tables (seating 74) at a cost of £1960

4 a $0.05x + 0.08y \leqslant 20 \Rightarrow 5x + 8y \leqslant 2000$

$\frac{1}{5}x + \frac{2}{15}y \leqslant 48 \Rightarrow 3x + 2y \leqslant 720$

b

b.v.	x	y	r	s	Value
r	5	⑧	1	0	2000
s	3	2	0	1	720
P	-1.5	-1.75	0	0	0

c Optimal solution $x = 125\frac{5}{7}$ $y = 171\frac{3}{7}$

Integer solutions needed, so point testing gives $x = 126$ $y = 171$

d The first point is A if y is increased first

The second point is C

5 a $54x + 72y + 36z \leqslant 1800 \Rightarrow 3x + 4y + 2z \leqslant 100$

$60x + 36y + 48z \leqslant 1500 \Rightarrow 5x + 3y + 4z \leqslant 125$

b $P = 12x + 24y + 20z$

c

b.v.	x	y	z	r	s	Value
r	3	④	2	1	0	100
s	5	3	4	0	1	125
P	-12	-24	-20	0	0	0

d

b.v.	x	y	z	r	s	Value	Row operation
y	$\frac{3}{4}$	1	$\frac{1}{2}$	$\frac{1}{4}$	0	25	R1 ÷ 4
s	$\frac{11}{4}$	0	$\frac{5}{2}$	$-\frac{3}{4}$	1	50	R2 − 3R1
P	6	0	-8	6	0	600	R3 + 24R1

b.v.	x	y	z	r	s	Value	Row operation
y	$\frac{1}{5}$	1	0	$\frac{2}{5}$	$-\frac{1}{5}$	15	R1 ÷ $\frac{1}{2}$R2
z	$\frac{11}{10}$	0	1	$-\frac{3}{10}$	$\frac{2}{5}$	20	R2 − $\frac{5}{2}$
P	$\frac{74}{5}$	0	0	$\frac{18}{5}$	$\frac{16}{5}$	760	R3 + 8R2

e There are no negative numbers in the profit row.

f Type A = 0 Type B = 15 Type C = 20

Profit = £760

6 a Maximise $P = 14x + 12y + 13z$

subject to:

Carving $2x + 2.5y + 1.5z \leqslant 8 \Rightarrow 4x + 5y + 3z \leqslant 16$

Sanding $25x + 20y + 30z \leqslant 120$
$\Rightarrow 5x + 4y + 6z \leqslant 24$

b r and s are numbers which indicate the slack time.

$P - 14x - 12y - 13z = 0$
$4x + 5y + 3z + r = 16$
$5x + 4y + 6z + s = 24$

Use these equations to create the initial tableau.

c

b.v.	x	y	z	r	s	Value	Row operation
x	1	$\frac{5}{4}$	$\frac{3}{4}$	$\frac{1}{4}$	0	4	R1 ÷ 4
s	0	$-\frac{9}{4}$	$\frac{9}{4}$	$-\frac{5}{4}$	1	4	R2 − 5R1
P	0	$\frac{11}{2}$	$-\frac{5}{2}$	$\frac{7}{2}$	0	56	R3 + 14R1

d From a zero stock situation we increase the number of lions to 4. We are increasing the profit from 0 to £56.

7 a $3x + 2y + s_1 = 15$
$2x + 5y + s_2 = 20$
$y - s_3 + a_1 = 2$

b Maximise $I = -a_1 = y - s_3 - 2$

c

	x	y	s_1	s_2	s_3	a_1	Value
s_1	3	2	1	0	0	0	15
s_2	2	5	0	1	0	0	20
a_1	0	1	0	0	-1	1	2
P	-1	-3	0	0	0	0	0
I	0	-1	0	0	1	0	-2

d

	x	y	s_1	s_2	s_3	a_1	Value
s_1	3	0	1	0	2	-2	11
s_2	2	0	0	1	5	-5	10
y	0	1	0	0	-1	1	2
P	-1	0	0	0	-3	3	6
I	0	0	0	0	0	1	0

$I = 0$, $a_1 = 0$ as required for the first stage. A basic feasible solution for the second stage is $P = 6$ when $s_1 = 11$, $s_2 = 10$, $y = 2$.

e

	x	y	s_1	s_2	s_3	Value
s_1	3	0	1	0	2	11
s_2	2	0	0	1	5	10
y	0	1	0	0	-1	2
P	-1	0	0	0	-3	6

8 a The x column contains the most negative value in the P row.

The θ values are $\frac{70}{1} = 70$, $\frac{180}{1} = 180$, $\frac{150}{2} = 75$

The pivot is 1 in the x column and the a_1 row.

b

	x	y	z	s_1	s_2	s_3	a_1	Value	Row operation
s_1	0	0	1	1	2	-2	-2	10	R1 − 2R3
y	0	1	-1	0	1	-1	-1	110	R2 − R3
x	1	0	0	0	0	1	1	70	
P	0	0	1	0	M	$M+6$	$2M+1$	300	R4 + R3

$P = 300$ when $x = 70$, $y = 110$, $z = 0$, $s_1 = 10$,
$s_2 = s_3 = a_1 = 0$
Optimal since values in bottom row all positive.

9 a Not feasible since $a_1 = 8 \neq 0$
  b The pivot is the first 2 in the x column. x column has the most negative P row and 2 gives the smallest positive θ value.
  c

	x	y	z	s_1	s_2	s_3	a_1	Value	Row operation
s_1	0	$-\frac{1}{2}$	$\frac{7}{2}$	1	$-\frac{1}{2}$	0	0	$\frac{3}{2}$	R1 − R2
x	1	$\frac{3}{2}$	$\frac{1}{2}$	0	$\frac{1}{2}$	0	0	$\frac{5}{2}$	R2 ÷ 2
a_1	0	−4	−1	0	−1	−1	1	3	R3 − 2R2
P	0	$\frac{5}{2} + 4M$	$-\frac{7}{2} + M$	0	$\frac{1}{2} + M$	M	0	$\frac{5}{2} - 3M$	R4 + (1 + 2M)R2

10 a The basic simplex method can only be used if all the non-negativity constraints are $\leqslant$. As $z \geqslant 200$ we have to use non-basic variables.
  b $x = 600$, $y = 0$, $z = 200$
  c M represents an arbitrarily large number
  d $P - 100x - (80 + 4M)y - (60 + 6M)z + Ms_2 + Ms_4$
  $= -1200M$

Basic variable	x	y	z	s_1	s_2	s_3	s_4	a_1	a_2	Value
s_1	1	1	1	1	0	0	0	0	0	800
s_3	2	2	1	0	0	1	0	0	0	1200
a_1	0	0	1	0	−1	0	0	1	0	200
a_1	0	4	5	0	0	0	−1	0	1	1000
P	−100	$-(80 + 4M)$	$-(60 + 6M)$	0	M	0	M	0	0	$-1200M$

  e The most negative value in the objective row is $-(60 + 6M)$ in the z column
  The smallest positive θ value for the z column is 200 in either the a_1 or a_2 row, so either the 1 or 5 can be used as the pivot value
  Divide all elements in the chosen pivot row by the pivot value
  Use the pivot row to eliminate the pivot's variable from all other rows, such that the pivot column now only contains 1s and 0s
  Repeat until there are no negative values in the objective row

Challenge
$P = 28.5$ when $x = \frac{9}{2}$, $y = \frac{1}{4}$, $z = 7$

CHAPTER 8

Prior knowledge 8

1 Route: $AC\ CE\ ED\ DB$, length: 20

Exercise 8A

1 One possible solution is

Activity	Depends on
A	—
B	A
C	B
D	C
E	D
F	E
G	F

Another possible solution is

Activity	Depends on
A	—
B	A
C	—
D	B, C
E	D
F	E
G	F

2 a

Activity	Depends on
A	—
B	A
C	A
D	A
E	B
F	B
G	C, E
H	D
I	F, G
J	H, I

  b

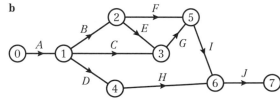

3

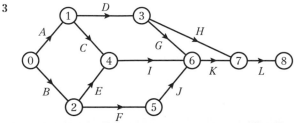

Online Full worked solutions are available in SolutionBank.

4

Activity	Depends on
A	—
B	A
C	A
D	B
E	C
F	E
G	C
H	D, F
I	G
J	G
K	I
L	J

Exercise 8B

1

Activity	Depends on
A	—
B	—
C	A
D	A
E	C
F	B, C, D
G	B, C, D
H	E, F

2 a The dummy shows that activity F depends on activities B and D, whereas activity G only depends on activity B.

b

Activity	Depends on
A	—
B	—
C	A
D	A
E	C
F	B, D
G	B

3

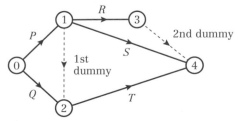

4 a

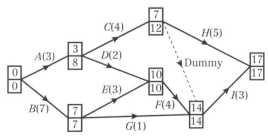

b 1st dummy.
S depends on P only.
T depends on P and Q.
2nd dummy.
So that S and R don't share a start and end event.

5

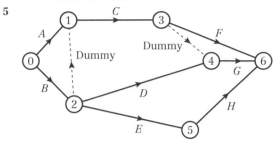

Exercise 8C

1 x is the largest of $7 + 5 = 12$, $5 + 8 = 13$ and $9 + 5 = 14$.
$x = 14$

2 $w = 11$, $x = 16$,
$y = 11$, $z = 9$

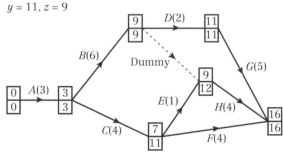

3

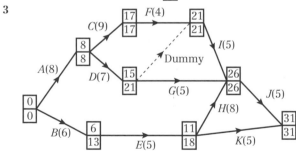

Exercise 8D

1 $x = 3$, $y = 17$, $z = 17$

2 a The critical activities are B, E, H, J and N.
b I, even though it connects too critical events the duration of I can be increased by up to 14 hours without affecting the total time.

3 a

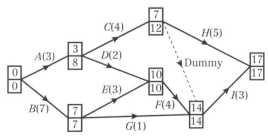

b $7 + 1 \neq 14$
c The critical path is $B - E - F - I$.

4 a

Activity	Depends on
A	—
B	—
C	—
D	A
E	B
F	C
G	D
H	D
I	E, F
J	E, F
K	G
L	H, I
M	J

b

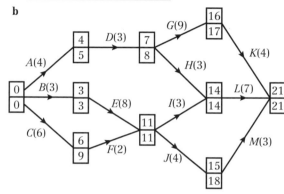

c Critical path: $B - E - I - L$ – length 21 days

Exercise 8E

1

Activity	Total float
A	0
B	$10 - 3 - 0 = 7$
C	$15 - 8 - 6 = 1$
D	0
E	$14 - 4 - 3 = 7$
F	$20 - 5 - 14 = 1$
G	0
H	$22 - 7 - 7 = 7$
I	$28 - 8 - 19 = 1$
J	$22 - 2 - 19 = 1$
K	$29 - 1 - 27 = 1$
L	0

2 a $a = 10$ $b = 19$ $x = 19 - 10 = 9$
Total float $= 3 = 15 - y - a$
$$y = 15 - 3 - 10$$
$$y = 2$$
b Minimum value of $c = 10 + 2 = 12$
c Maximum value of total float of $R = 19 - 4 - 12 = 3$

3 a $x = 3, y = 10, z = 17$
b A, D, I and L
c Total float is 5 days

Exercise 8F

1

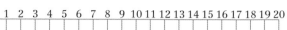
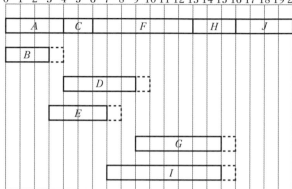

2 **a** $w = 26$ $x = 29$ $y = 34$ $z = 26$

   b Critical activities: B, E, H, M, O

   c Total float for $G = 26 - 5 - 13 = 8$

Total float for $N = 39 - 5 - 29 = 5$

   d 0 1 2 3 4 5 6 7 8 9 10 11 12 13 14 15 16 17 18 19 20 21 22 23 24 25 26 27 28 29 30 31 32 33 34 35 36 37 38 39

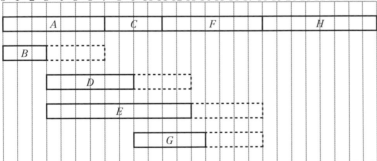

Exercise 8G

1 **a** A, E **b** G, H **c** F, H

2 **a** C, D **b** E, G

3 **a** 0 1 2 3 4 5 6 7 8 9 10 11 12 13 14 15 16 17 18 19 20 21 22 23 24 25 26

B, D and E may be happening at midday on day 5.

   b Only A must be happening at midday on day 7.

4 **a** $x = 15, y = 25, z = 27$

   b B, F, H, K, M

   c A max float 7

C max float 7

d

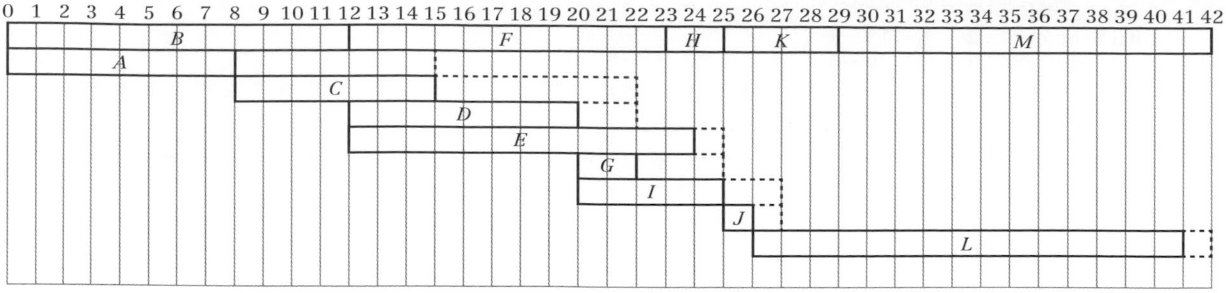

e Yes, activity *I* can start on day 22 and the project can still be completed on time.

Exercise 8H

1 a

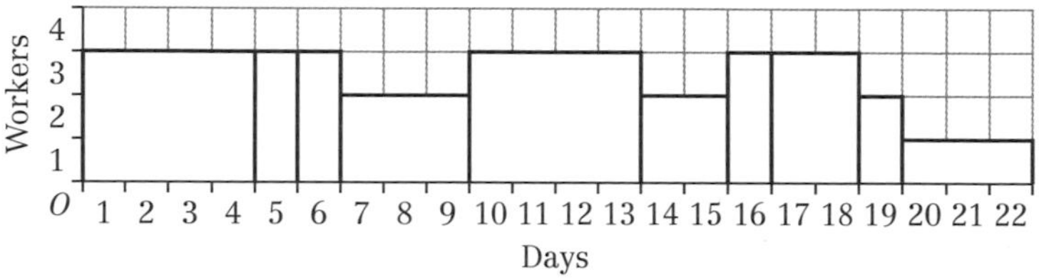

b 3 workers

2 a

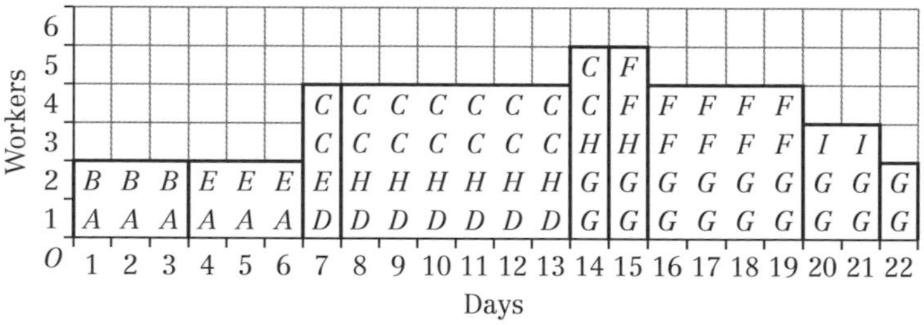

b 5 workers

3 a

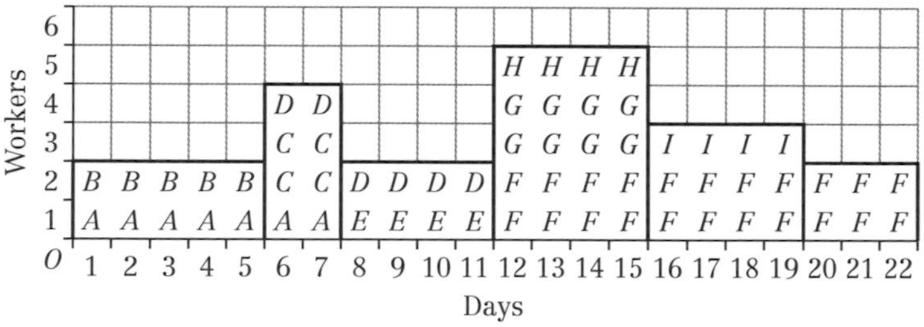

5 workers are needed.

b Delay starting Activity *H* until day 16, project does not need to be extended.

4 a

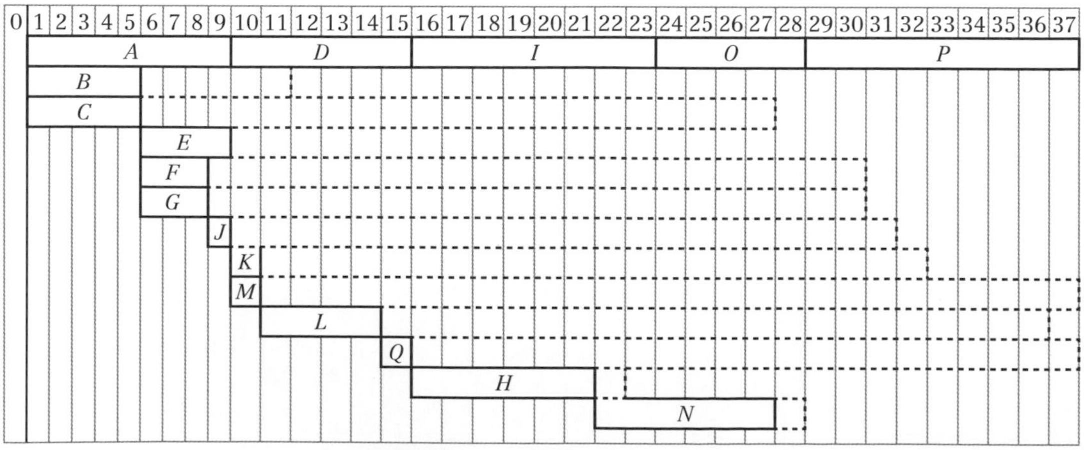

Online Full worked solutions are available in SolutionBank.

b

```
Workers
 6                         M
 5                   G G G J M                         N N
 4   C C C C C F F F J K           H H H H H H N N N N N N
 3   B B B B B E E E K L L L L L I I I I I I I I N N N N     P P P P P P P P P
 2   A A A A A A A A A D D D D D D I I I I I I I I I O O O O O P P P P P P P P P
 1   A A A A A A A A A D D D D D D I I I I I I I I I O O O O O P P P P P P P P P
 O 1 2 3 4 5 6 7 8 9 10 11 12 13 14 15 16 17 18 19 20 21 22 23 24 25 26 27 28 29 30 31 32 33 34 35 36 37
```

c Total float on activity *N* is only 1 day, so delaying it by 2 days would cause the project to be delayed.

d 49 hours

Exercise 8I

1 **a** $\frac{64}{22} = 2.9\ldots$ so, lower bound = 3

b 2 hours is less than the total float for activity *B* (3 hours).

c Activities *J* and *H*

d

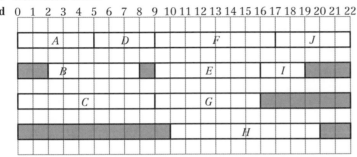

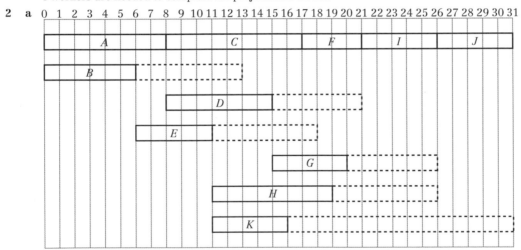

4 workers are needed to complete the project in 22 hours.

2 **a**

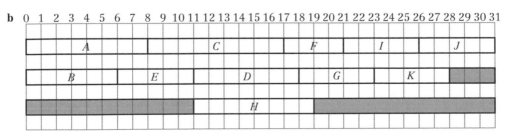

b

3 workers are needed to complete the project in the critical time.

3 The minimum time to complete the project using two workers is 25 days, using the following schedule diagram.

0 1 2 3 4 5 6 7 8 9 10 11 12 13 14 15 16 17 18 19 20 21 22 23 24 25

	A		C			F						I					J		
	B		E			D				G			H				

Mixed exercise 8

1 a Activity D depends on activities A and C, whereas activity E depends only on activity A.
This shows that a dummy is required.
Activity J depends on activities G and I, whereas activity H depends only on activity G.
This shows that a second dummy is required.

b

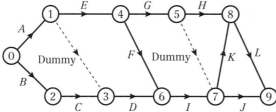

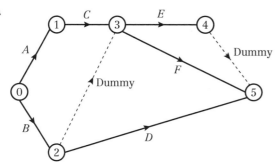

2 a

b Dummy 1 is needed to show dependency.
E and F depend on C and B, but D depends on B only.
Dummy 2 is need so that each activity can be uniquely represented in terms of its event.

3 a

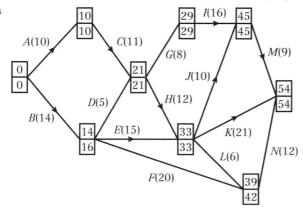

b There are two critical paths:
$A - C - G - I - M$ and $A - C - H - K$
The critical activities are A, C, G, H, I, K
c Total float on D is $21 - 5 - 14 = 2$
Total float on F is $42 - 20 - 14 = 8$

Online Full worked solutions are available in SolutionBank.

d

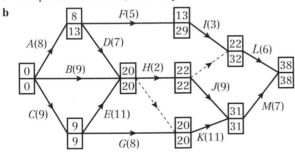

e Day 15: C
Day 25: G, H, E, F

4 a J depends on H alone, but L depends on H and I

b

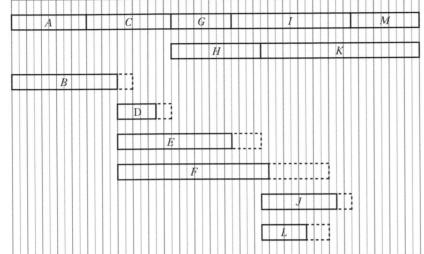

c Total float on $D = 20 - 7 - 8 = 5$
Total float on $E = 20 - 11 - 9 = 0$
Total float on $F = 29 - 5 - 8 = 16$

d

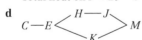

e $\frac{95}{38} = 2.5$ so 3 workers

f For example

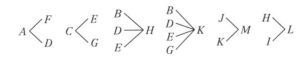

5 a

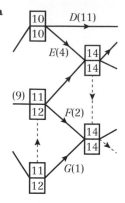

b The critical paths are: $A - E - H - K$ and $A - E - L$

c A critical path is a continuous path from the source node to the sink node such that a delay in any activity results in a corresponding delay in the whole project.

d $\dfrac{\text{Sum of all of the activity times}}{\text{critical time of the project}} = \dfrac{110}{30}$

Lower bound for number of workers is 4.

e D, H, I, J, L

f The answers to part **e** show that at least 5 workers are needed on day 20 in order to complete the project in the minimum time.

g

| 0 | 1 | 2 | 3 | 4 | 5 | 6 | 7 | 8 | 9 | 10 | 11 | 12 | 13 | 14 | 15 | 16 | 17 | 18 | 19 | 20 | 21 | 22 | 23 | 24 | 25 | 26 | 27 | 28 | 29 | 30 |

(Gantt chart showing activities A, B, C, D, E, F, G, H, I, J, K, L)

6 a 28 days

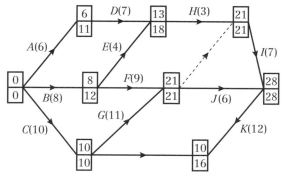

b 3 workers

c

| 0 | 1 | 2 | 3 | 4 | 5 | 6 | 7 | 8 | 9 | 10 | 11 | 12 | 13 | 14 | 15 | 16 | 17 | 18 | 19 | 20 | 21 | 22 | 23 | 24 | 25 | 26 | 27 | 28 |

(Gantt chart showing activities C, A, B, D, E, F, K, H, G, I, J)

d

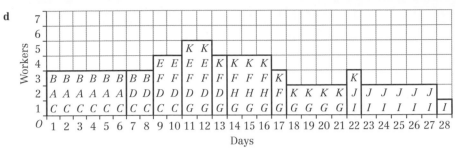

e

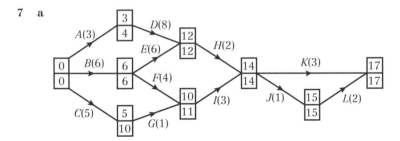

7 a

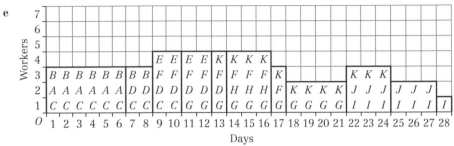

b $B - E - H - K$ and $B - E - H - J - L$

c

[Gantt chart with Days 1–18 on horizontal axis and values 1–15 on vertical axis]

d

	1	2	3	4	5	6	7	8	9	10	11	12	13	14	15	16	17	18	19	20	21	22	23	24
9																								
8	A	A	A									F							H	H				
7	A	A	A			D	D	D	D	D	F	F	F	F					H	H				
6	A	A	A	D	D	D	D	D	D	D	G	F	F	F					H	H			L	L
5	A	A	A	D	D	D	C	C	C	C	C	G	E	E	E	E	E	E	I	I	I	J	L	L
4	B	B	B	B	B	B	C	C	C	C	C	G	E	E	E	E	E	E	I	I	I	J	L	L
3	B	B	B	B	B	B	C	C	C	C	C	G	E	E	E	E	E	E	I	I	I	J	L	L
2	B	B	B	B	B	B	C	C	C	C	C	G	E	E	E	E	E	E	I	I	I	K	K	K
1	B	B	B	B	B	B	C	C	C	C	C	G	E	E	E	E	E	E	I	I	I	K	K	K

O 1 2 3 4 5 6 7 8 9 10 11 12 13 14 15 16 17 18 19 20 21 22 23 24

Challenge

a A, D, F critical. Total floats are B: 2, C: 8, E: 8

b **i** Reduce F by 3 days and D by 1 day, total cost £650.
    ii Reduce F by 3 days, D by 4 days and B by two days. Total cost £1450.

c 27 days. Activity E has already been reduced by 2 days, and critical path is now ACE. No further reduction possible on critical path so no further reduction in total project time possible.

d $P = 100y_B + 200y_D + 400y_E + 150y_F$

e $x_1 \geqslant 12$
    $x_2 + y_B \geqslant 25$
    $x_3 - x_1 + y_D \geqslant 15$
    $x_4 - x_1 \geqslant 8$
    $x_5 - x_4 + y_E \geqslant 9$
    $x_5 - x_3 + y_F \geqslant 10$
    $x_3 - x_2 \geqslant 0$
    $y_B \leqslant 8$
    $y_D \leqslant 10$
    $y_E \leqslant 2$
    $y_F \leqslant 3$
    $x_5 \leqslant 28$
    $x_1, x_2, x_3, x_4, x_5, y_B, y_D, y_E, y_F \geqslant 0$

Review exercise 2

1 a Chemical A: $5x + y \geqslant 10$
    Chemical B: $2x + 2y \geqslant 12$, which simplifies to $x + y \geqslant 6$
    Chemical C: $\frac{1}{2}x + 2y \geqslant 6$, which simplifies to $x + 4y \geqslant 12$
    $x \geqslant 0, y \geqslant 0$

b

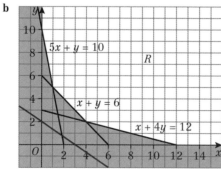

Objective line
$2x + 3y$

c $T = 2x + 3y$

d $(4, 2)$ $T = 14$

2 a Maximise $P = 300x + 500y$

b Finishing $3.5x + 4y \leqslant 56 \Rightarrow 7x + 8y \leqslant 112$ (o.e.)
    Packaging $2x + 4y \leqslant 40 \Rightarrow x + 2y \leqslant 20$ (o.e.)

Online Full worked solutions are available in SolutionBank.

c

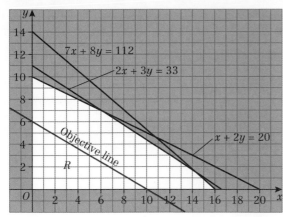

d For example, *Point testing*
- test all corner points in feasible region
- find profit at each and select point yielding maximum.
  Profit line
- Draw profit lines.
- Select point on profit line furthest from the origin.

e Using a correct, complete method.
make 6 Oxford and 7 York, profit = £5300

f Finishing time required = 49 hours, so reduce finishing time by 7 hours.

3 a Objective: maximise $P = 30x + 40y$ (or $P = 0.3x + 0.4y$)
subject to:
$$x + y \geqslant 200$$
$$x + y \leqslant 500$$
$$x \geqslant \frac{20}{100}(x + y) \Rightarrow 4x \geqslant y$$
$$x \leqslant \frac{40}{100}(x + y) \Rightarrow 3x \leqslant 2y$$
$$x, y \geqslant 0$$

b

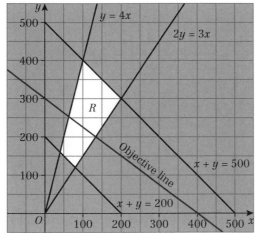

c Visible use of objective line method – objective line drawn or vertex testing
Vertex testing
$(40, 160) \rightarrow 7600$
$(80, 120) \rightarrow 7200$
$(100, 400) \rightarrow 19\,000$
$(200, 300) \rightarrow 18\,000$
Intersection of $y = 4x$ and $x + y = 500$
Maximum profit of £190 at $(100, 400)$, so 'Decide' should make 100 Badge 1 and 400 Badge 2.

4 a

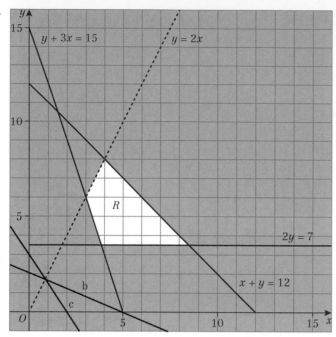

b Visible use of objective line method – objective line drawn or vertex testing.
Minimum cost is £25$\frac{1}{6}$ at ($3\frac{5}{6}$, $3\frac{1}{2}$) so Becky should make $3\frac{5}{6}$ kg of bird feeder food and $3\frac{1}{2}$ kg of bird table food.

c Visible use of objective line method – objective line drawn, or vertex testing.
Maximum profit is £32 at (8, 4) so Becky should make 8 kg of bird feeder food and 4 kg of bird table food.

5 a Objective: maximise $P = 0.4x + 0.2y$ $\quad$ ($P = 40x + 20y$)
subject to:
$x \leqslant 6.5$
$y \leqslant 8$
$x + y \leqslant 12$
$y \leqslant 4x$
$x, y \geqslant 0$

b Visible use of objective line method – objective line drawn [e.g. from (2, 0) to (0, 4)] or all 5 points tested.
vertex testing
[(0, 0) → 0; (2, 8) → 2.4; (4, 8) → 3.2; (6.5, 5.5) → 3.7; (6.5, 0) → 2.6]
Optimal point is (6.5, 5.5) ⇒ 6500 type X and 5500 type Y

c $P = 0.4(6500) + 0.2(5500) = £3700$

6 a Maximise $\quad P = 50x + 80y + 60z$
subject to $\quad x + y + 2z \leqslant 30$
$\quad\quad\quad\quad\quad x + 2y + z \leqslant 40$
$\quad\quad\quad\quad\quad 3x + 2y + z \leqslant 50$
where $\quad\quad x, y, z \geqslant 0$

c The solution found after one iteration has a slack of 10 units of black per day.

d i

b.v.	x	y	z	r	s	t	Value	
z	$\frac{1}{3}$	0	1	$\frac{2}{3}$	$-\frac{1}{3}$	0	$6\frac{2}{3}$	R1 ÷ $\frac{3}{2}$
y	$\frac{1}{3}$	1	0	$-\frac{1}{3}$	$\frac{2}{3}$	0	$16\frac{2}{3}$	R2 − $\frac{1}{2}$R1
t	2	0	0	0	−1	1	10	R3 – no change
P	$-3\frac{1}{3}$	0	0	$13\frac{1}{3}$	$33\frac{1}{3}$	0	$1733\frac{1}{3}$	R4 + 20R1

ii Not optimal as there is a negative value in the profit row.

iii $x = 0 \quad y = 16\frac{2}{3} \quad z = 6\frac{2}{3}$
$P = £1733.33 \quad r = 0 \quad s = 0 \quad t = 10$

7 a Objective: Maximise $P = 4x + 5y + 3z$
subject to $\quad 3x + 2y + 4z \leqslant 35$
$\quad\quad\quad\quad\quad x + 3y + 2z \leqslant 20$
$\quad\quad\quad\quad\quad 2x + 4y + 3z \leqslant 24$
where $\quad\quad x, y, z \geqslant 0$

b $P = 47\frac{1}{4}$, $x = 11\frac{1}{2}$ $y = \frac{1}{4}$, $z = 0$, $r = 0$, $s = 7\frac{3}{4}$, $t = 0$

Online $\quad$ Full worked solutions are available in SolutionBank.

c There is some slack $(7\frac{3}{4})$ on s, so *do not* increase blending; therefore increase processing and packing which are both at their limit at present.

8 a $x + 2y + 4z \leqslant 24$

b i $x + 2y + 4z + s = 24$

  ii $s(\geqslant 0)$ is the slack time on the machine in hours.

c 1 euro

d Profit = 31 euros $y = 7$ $z = 2.5$ $x = r = s = 0$

e Cannot make $\frac{1}{2}$ a lamp.

f e.g. (0, 10, 0) or (0, 6, 3) or (1, 7, 2) checks in **both** inequalities

9 a $2.5x + 10y + 15z \leqslant 300 \Rightarrow x + 4y + 6z \leqslant 120$

  $10x + 20y + 50z \leqslant 1000 \Rightarrow x + 2y + 5z \leqslant 100$

b $P = 10x + 20y + 28z$

c

b.v.	x	y	z	r	s	Value
r	1	4	6	1	0	120
s	1	2	5	0	1	100
P	−10	−20	−28	0	0	0

d First iteration

b.v.	x	y	z	r	s	Value	Row operation
y	$\frac{1}{4}$	1	$\frac{3}{2}$	$\frac{1}{4}$	0	30	R1 ÷ 4
s	$\frac{1}{2}$	0	2	$-\frac{1}{2}$	1	40	R2 − 2R1
P	−5	0	2	5	0	600	R3 + 20R1

Second iteration

b.v.	x	y	z	r	s	Value	Row operation
y	0	1	$\frac{1}{2}$	$\frac{1}{2}$	$-\frac{1}{2}$	10	R1 − $\frac{1}{2}$R2
x	1	0	4	−1	2	80	2R2
P	0	0	22	0	10	1000	R3 + 5R2

e This tableau is optimal as there are no negative numbers in the profit line.

f small 80, medium 10, large 0, profit £1000

10 a The constraints include a mixture of $\leqslant$ and $\geqslant$ variables.

b $x + y + 2z + s_1 = 10$

  $x + 3y + z + s_2 = 15$

  $2x + y + z - s_3 + a_1 = 12$

c The final row represents the new objectives function needed for the first stage of the two-stage simplex method.

d

b.v.	x	y	z	s_1	s_2	s_3	a_1	Value
s_1	0	0.5	1.5	1	0	0.5	−0.5	4
s_2	0	2.5	0.5	0	1	0.5	−0.5	9
x	1	0.5	0.5	0	0	−0.5	0.5	6
P	0	−1.5	−2.5	0	0	−0.5	0.5	6
I	0	0	0	0	0	0	1	0

e There are no negative values in the bottom row, so the optimal value of I is 0 when $a_1 = 0$.

f There is a negative value in the bottom row.

g

b.v.	x	y	z	s_1	s_2	s_3	Value
z	0	0	1	$\frac{5}{7}$	$-\frac{1}{7}$	$\frac{2}{7}$	$\frac{11}{7}$
y	0	1	0	$-\frac{1}{7}$	$\frac{3}{7}$	$\frac{1}{7}$	$\frac{23}{7}$
x	1	0	0	$-\frac{2}{7}$	$-\frac{1}{7}$	$-\frac{5}{7}$	$\frac{25}{7}$
P	0	0	0	$\frac{11}{7}$	$\frac{2}{7}$	$\frac{3}{7}$	$\frac{104}{7}$

The maximum value of P is $\frac{104}{7} = 14\frac{6}{7}$ which occurs when $x = \frac{25}{7}, y = \frac{23}{7}, z = \frac{11}{7}, s_1 = s_2 = s_3 = 0$

11 a $x + 2y + 3z + s_1 = 18$
$3x + y + z - s_2 + a_1 = 6$
$2x + 5y + z - s_3 + a_2 = 20$

b New objective is maximise $I = -(a_1 + a_2)$
$-a_1 = 3x + y + z - s_2 - 6$
$-a_2 = 2x + 5y + z - s_3 - 20$
In terms of non-basic variables, the new objective is maximise $I = 5x + 6y + 2z - s_2 - s_3 - 26$

c

b.v.	x	y	z	s_1	s_2	s_3	a_1	a_2	Value
s_1	1	2	3	1	0	0	0	0	18
a_1	3	1	1	0	−1	0	1	0	6
a_2	2	5	1	0	0	−1	0	1	20
P	−2	1	−1	0	0	0	0	0	0
I	−5	−6	−2	0	1	1	0	0	−26

d 1st iteration y enters the basic

b.v.	x	y	z	s_1	s_2	s_3	a_1	a_2	Value
s_1	$\frac{1}{5}$	0	$\frac{13}{5}$	1	0	$\frac{2}{5}$	0	$-\frac{2}{5}$	10
a_1	$\frac{13}{5}$	0	$\frac{4}{5}$	0	−1	$\frac{1}{5}$	1	$-\frac{1}{5}$	2
y	$\frac{2}{5}$	1	$\frac{1}{5}$	0	0	$-\frac{1}{5}$	0	$\frac{1}{5}$	4
P	$-\frac{12}{5}$	0	$-\frac{6}{5}$	0	0	$\frac{1}{5}$	0	$-\frac{1}{5}$	−4
I	$-\frac{13}{5}$	0	$-\frac{4}{5}$	0	1	$-\frac{1}{5}$	0	$\frac{6}{5}$	−2

2nd iteration x enters the basics

b.v.	x	y	z	s_1	s_2	s_3	a_1	a_2	Value
s_1	0	0	$\frac{33}{13}$	1	$\frac{1}{13}$	$\frac{5}{13}$	$-\frac{1}{13}$	$-\frac{5}{13}$	$\frac{128}{13}$
x	1	0	$\frac{4}{13}$	0	$-\frac{5}{13}$	$\frac{1}{13}$	$\frac{5}{13}$	$-\frac{1}{13}$	$\frac{10}{13}$
y	0	1	$\frac{1}{13}$	0	$\frac{2}{13}$	$-\frac{3}{13}$	$-\frac{2}{13}$	$\frac{3}{13}$	$\frac{48}{13}$
P	0	0	$-\frac{6}{13}$	0	$-\frac{12}{13}$	$\frac{5}{13}$	$\frac{12}{13}$	$-\frac{5}{13}$	$\frac{28}{13}$
I	0	0	0	0	0	0	1	1	0

Basic feasible solution is $x = \frac{10}{13}$, $y = \frac{48}{13}$, $z = 0$, $s_1 = \frac{128}{13}$, $s_2 = s_3 = a_1 = a_2 = 0$

12 a $3x + 2y + z + s_1 = 24$
$5x + 3y + 2z + s_2 = 60$
$x - s_3 + a_1 = 2$

b $P = x + 3y + 4z - Ma_1$
$= x + 3y + 4z - M(2 - x + s_3)$
$P = x(1 + M) + 3y + 4z - 2M - Ms_3$
$P - (1 + M)x - 3y - 4z + Ms_3 = -2M$

c

b.v.	x	y	z	s_1	s_2	s_3	a_1	Value
s_1	3	2	1	1	0	0	0	24
s_2	5	3	2	0	1	0	0	60
a_1	1	0	0	0	0	−1	1	2
P	−(1 + M)	−3	−4	0	0	M	0	−2M

d The most negative value in the P row is in the x-column so, in the first iteration, x enters the basic variables.

e

b.v.	x	y	z	s_1	s_2	s_3	a_1	Value
s_1	0	2	1	1	0	3	−3	18
s_2	0	3	2	0	1	5	−5	50
x	1	0	0	0	0	−1	1	2
P	0	−3	−4	0	0	−1	(1 + M)	2

Online Full worked solutions are available in SolutionBank.

b.v.	x	y	z	s_1	s_2	s_3	a_1	Value
z	0	2	1	1	0	3	-3	18
s_2	0	-1	0	-2	1	-1	1	14
x	1	0	0	0	0	-1	1	2
P	0	5	0	4	0	11	$(M-11)$	74

All entries in the P row are non-negative so the tableau represents the optimal solution.

$x = 2$, $y = 0$, $z = 18$, $s_1 = 0$, $s_2 = 14$, $s_3 = 0$, $a_1 = 0$

13 a $4x + 3y + 2z + s_1 = 36$

$x + 4z + s_2 = 52$

$x + y - s_3 + a_1 = 10$

b Maximise $P = -2x + 3y - z - Ma_1$

$\qquad = -2x + 3y - z - M(10 - x - y + s_3)$

$\qquad = x(M - 2) + y(M + 3) - z - 10M - Ms_3$

Rearranging gives

$P - (M - 2)x - (M + 3)y + z + Ms_3 = -10M$

c

b.v.	x	y	z	s_1	s_2	s_3	a_1	Value
s_1	4	3	2	1	0	0	0	36
s_2	1	0	4	0	1	0	0	52
a_1	1	1	0	0	0	-1	1	10
P	$-(M-2)$	$-(M+3)$	1	0	0	M	0	$-10M$

d

b.v.	x	y	z	s_1	s_2	s_3	a_1	Value
s_1	1	0	2	1	0	3	0	6
s_2	1	0	4	0	1	0	0	52
y	1	1	0	0	0	-1	1	10
P	5	0	1	0	0	-3	0	30

b.v.	x	y	z	s_1	s_2	s_3	a_1	Value
s_3	$\frac{1}{3}$	0	$\frac{2}{3}$	$\frac{1}{3}$	0	1	0	2
s_2	1	0	4	0	1	0	0	52
y	$\frac{4}{3}$	1	$\frac{2}{3}$	$\frac{1}{3}$	0	0	1	12
P	6	0	3	1	0	0	0	36

Maximum value of $P = 36$

Minimum value of $C = -36$

This occurs when $x = 0$, $y = 12$, $z = 0$, $s_1 = 0$, $s_2 = 52$, $s_3 = 2$

14

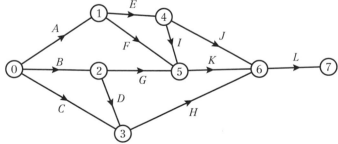

15 a

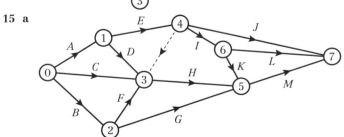

b For example I and J depend only on E. H depends on C, D, E and F.

16 a

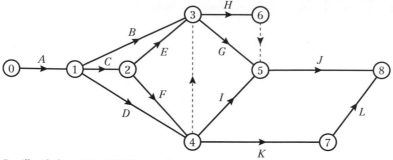

b D will only be critical if it lies on the longest path

Path A to G	Length
$A - B - E - G$	14
$A - C - F - G$	15
$A - C - D == E - G$	$13 + x$

So we need $13 + x$ to be the longest, or equal longest
$13 + x \geqslant 15$
$x \geqslant 2$

17 a

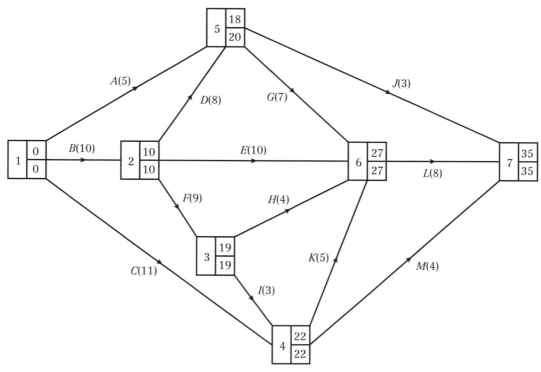

b Total float on $A = 20 - 0 - 5 = 15$ Total float on $H = 27 - 19 - 4 = 4$
Total float on $B = 10 - 0 - 10 = 0$ Total float on $I = 22 - 19 - 3 = 0$
Total float on $C = 22 - 0 - 11 = 11$ Total float on $J = 35 - 18 - 3 = 14$
Total float on $D = 20 - 10 - 8 = 2$ Total float on $K = 27 - 22 - 5 = 0$
Total float on $E = 27 - 10 - 10 = 7$ Total float on $L = 35 - 27 - 8 = 0$
Total float on $F = 19 - 10 - 9 = 0$ Total float on $M = 35 - 22 - 4 = 9$
Total float on $G = 27 - 18 - 7 = 2$

c Critical activities: B, F, I, K and L length of critical path is 35 days
d New critical path is B, F, H, L length of new critical path is 36 days

18 a $x = 0$
$y = 7$ [latest out of $(3 + 2)$ and $(5 + 2)$]
$z = 9$ [earliest out of $(13 - 4)$ and $(19 - 7)$ and $(16 - 2)$]
b Length is 22
Critical activities: B, D, E and L

c i Total float on $N = 22 - 14 - 3 = 5$
ii Total float on $H = 16 - 5 - 3 = 8$

19 a For example, it shows dependence but it is not an activity. G depends on A and C only but H and I depend on A, C and D.

b

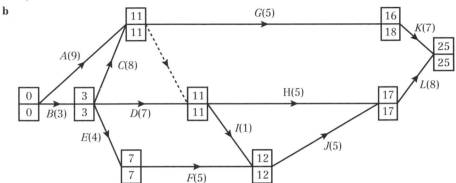

c

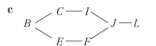

so B, C, E, F, I, J and L

d Total float on $A = 11 - 0 - 9 = 2$
Total float on $D = 11 - 3 - 7 = 1$
Total float on $G = 18 - 11 - 5 = 2$
Total float on $H = 17 - 11 - 5 = 1$
Total float on $K = 25 - 16 - 7 = 2$

e

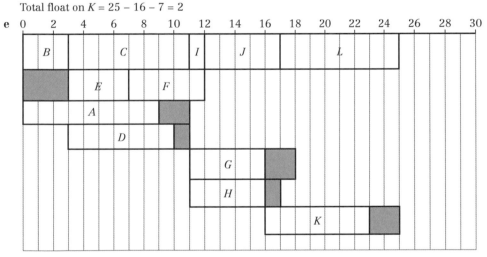

20 a Critical activities are B, F, J, K and N, length of critical path is 25 hours, I is not critical.
b Total float on $A = 5 - 0 - 3 = 2$
Total float on $C = 9 - 0 - 6 = 3$
Total float on $D = 11 - 3 - 3 = 5$
Total float on $E = 9 - 3 - 4 = 2$
Total float on $G = 9 - 4 - 3 = 2$
Total float on $H = 16 - 7 - 7 = 2$
Total float on $I = 16 - 9 - 5 = 2$
Total float on $L = 22 - 11 - 4 = 7$
Total float on $M = 22 - 16 - 2 = 4$
Total float on $P = 25 - 18 - 3 = 4$

c

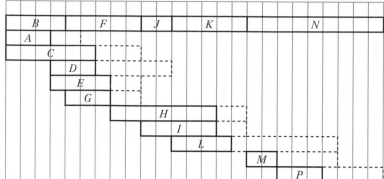

d *F*, *E* and *G*

21 a

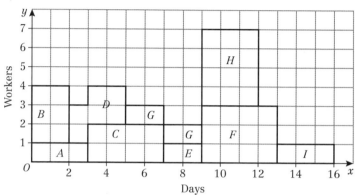

b 16 days, 7 workers
c Delay the start of *H* until time 13
d 3 days

22 a

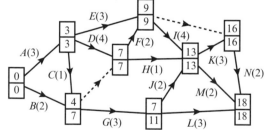

b 18 days
c *ADFIKN*

d

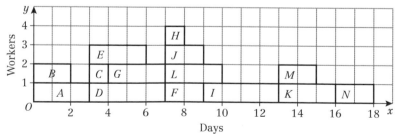

e 4 workers
f e.g. delay the start times of: *E* to time 4
                                   G to time 7
                                   H to time 12
                                   J to time 10
                                   L to time 13
                                   M to time 16

Online Full worked solutions are available in SolutionBank.

23 a

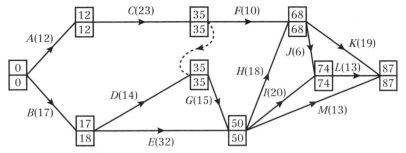

b A, C, E, H, J, K and L

c Total float = $35 - 17 - 14 = 4$

d Either $226 \div 87 = 2.6$ (1 d.p.) so at least 3 workers needed
or 69 hours into the project activities J, K, I and M must be happening so at least 4 workers will be needed.

e

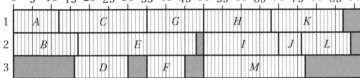

New shortest time is 89 hours.

24 a

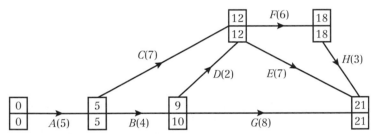

b Critical activities: A, C, F and H; length of critical path = 21

c Total float on $B = 10 - 5 - 4 = 1$ Total float on $E = 21 - 12 - 7 = 2$
Total float on $D = 12 - 9 - 2 = 1$ Total float on $G = 21 - 9 - 8 = 4$

d

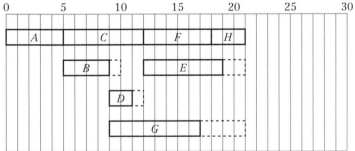

e For example

Minimum time for 2 workers is 24 days.

Challenge

1 a

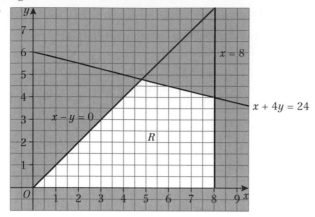

Optimal values $x = y = \frac{24}{5}$

b $x = 4, y = 4 \Rightarrow P = 24$

$x = 5, y = 4 \Rightarrow P = 25$

c $x = 7, y = 4 \Rightarrow P = 27$

d If the gradient of the objective line is similar to the gradient of a constraint that runs through the optimal vertex, then the optimal integer solution may not lie close to the optimal vertex

2 Number of variables > 2 so must use simplex method not a graphical method.

$y = 12, x = z = 0$

$C = -36$

Exam-style practice (AS Level)

1 a

21	16	11	25	18	15	23	10
16	11	21	18	15	23	10	25
11	16	18	15	21	10	23	25
11	16	15	18	10	21	23	25
11	15	16	10	18	21	23	25
11	15	10	16	18	21	23	25
11	10	15	16	18	21	23	25
10	11	15	16	18	21	23	25
10	11	15	16	18	21	23	25

b For a list of n items:

The 1st pass requires $(n - 1)$ comparisons.

The 2nd pass required $(n - 2)$ comparison.

The process continues until the final pass, which requires 1 comparison.

The total number of comparisons is

$1 + 2 + 3 + \ldots + (n - 2) + (n - 1)$

This is an AP with sum $\frac{n}{2}(n - 1)$

Thus the bubble sort has quadratic order.

c $0.018 \times \left(\frac{1000}{50}\right)^2 = 7.2$ seconds

This is an estimate only because the time taken is not directly proportional to n^2. This is used as an approximation.

Online Full worked solutions are available in SolutionBank.

2 a

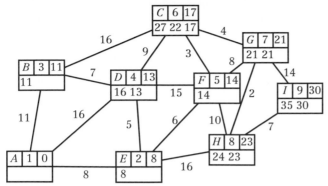

Quickest route: *AEFCGHI*   Time required: 30 minutes

b New quickest route: *AEHI* (or *AEFHI*)   Time required: 31 minutes

3 Maximise $24x + 32y$

Subject to: $2x + 3y \leqslant 100$
    $8x + 11y \geqslant 240$
    $5x + 12y \leqslant 320$
    $x \geqslant 0, y \geqslant 0$

4 a Odd nodes *A*, *E*, *F*, *G*

Pairings: $AE + FG = 9 + 11 = 20$
    $AF + EG = 5 + 16 = 21$
    $AG + EF = 6 + 4 = 10$

The sections that need to be traversed twice are *AG* and *EF*.

Total time required = $144 + 10 = 154$ minutes

b $FG = 12$, Route via *A* = 11, this is shorter so repeat *FA* and *AG*

c $158 - 144 - 11 = 3$

5 a

Activity	Immediately preceding activities
A	–
B	–
C	–
D	*A*
E	*C*
F	*E*
G	*F*
H	*B, C*
I	*D, G, H*
J	*E*

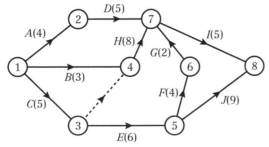

b The dotted lines represent a dummy activity showing that activity *H* depends on activities *B* and *C* whereas activity *E* depends on activity *C* only.

c

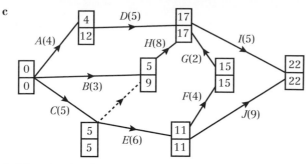

d The critical path is *CEFGI*.

e Total float for activity *D* is $17 - 5 - 4 = 8$ hours

Exam-style practice (A Level)

1 a Starting with *AB* means that *H* cannot be included without repeating either *B* or *A*.

b *AHBCEFGDA*

c

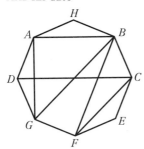

AB(I), *AG*, *BF*, *BG*, *CD*, *CF*
AB(I), *AG*(I), *BF*, *BG*, *CD*, *CF*
AB(I), *AG*(I), *BF*(I), *BG*(I), *CD*(O), *CF*
AB(I), *AG*(I), *BF*(I), *BG*(I), *CD*(O), *CF*(I)
The graph is planar.

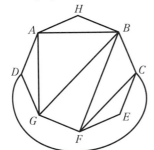

2 a

42	31	36	18	27	33	41	47	12	24	16
42	36	41	47	[33]	31	18	27	12	24	16
42	47	[41]	36	[33]	31	18	27	24	16	[12]
[47]	42	[41]	[36]	[33]	31	[27]	18	24	16	[12]
[47]	[42]	[41]	[36]	[33]	[31]	[27]	[24]	18	16	[12]
[47]	[42]	[41]	[36]	[33]	[31]	[27]	[24]	18	[16]	[12]

All of the numbers have now been selected as pivots, so the list is in order.

b
```
                  18
33    31    27    24    12
47    42    41    36    16
```
5 reels are required.

c $0.034 \times \dfrac{5000 \log 5000}{800 \log 800} = 0.27$ seconds

This is only an estimate because the time taken is only approximately proportional to $n \log n$

Online Full worked solutions are available in SolutionBank.

3 a A network is semi-Eulerian if it has exactly two odd nodes.

b The odd nodes are A, B, C, D, E, F

Possible pairings:
$AC + DF = 10 + 10 = 20$
$AD + CF = 11 + 9 = 20$
$AF + CD = 2 + 10 = 12$

Minimum length = $72 + 12 = 84$ miles

c AF, CD

d Possible pairings:
$AB + DF = 4 + 10 = 14$
$AD + BF = 11 + 5 = 16$
$AF + BD = 2 + 13 = 15$

The route is extended by $14 - 12 = 2$ miles

4 a Dijkstra's or Floyd's Algorithm

b Arc CA

c 1st iteration (no change)

–	5	∞	∞	∞
5	–	11	16	∞
10	11	–	8	6
∞	16	8	–	10
∞	∞	6	10	–

A	B	C	D	E
A	B	C	D	E
A	B	C	D	E
A	B	C	D	E
A	B	C	D	E

2nd iteration

–	5	16	21	∞
5	–	11	16	∞
10	11	–	8	6
21	16	8	–	10
∞	∞	6	10	–

A	B	B	B	E
A	B	C	D	E
A	B	C	D	E
B	B	C	D	E
A	B	C	D	E

3rd iteration

–	5	16	21	22
5	–	11	16	17
10	11	–	8	6
18	16	8	–	10
16	17	6	10	–

A	B	B	B	C
A	B	C	D	C
A	B	C	D	E
C	B	C	D	E
C	C	C	D	E

4th iteration (no change)

–	5	16	21	22
5	–	11	16	17
10	11	–	8	6
18	16	8	–	10
16	17	6	10	–

A	B	B	B	C
A	B	C	D	C
A	B	C	D	E
C	B	C	D	E
C	C	C	D	E

5th iteration (no change)

–	5	16	21	22
5	–	11	16	17
10	11	–	8	6
18	16	8	–	10
16	17	6	10	–

A	B	B	B	C
A	B	C	D	C
A	B	C	D	E
C	B	C	D	E
C	C	C	D	E

d 22, $ABCE$

5 a $12x + 30y \leq 240$,   i.e. $2x + 5y \leq 40$

$12x + 30y \leq 2(5x + 20y)$, i.e. $x \leq 5y$ or $y \geq \dfrac{x}{5}$

$5x + 5y \geq 55$,   i.e. $x + y \geq 11$

$6x + 12y \geq 96$,   i.e. $x + 2y \geq 16$

b

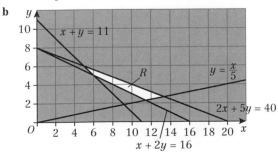

335

c

Vertex	(6, 5)	(5, 6)	$\left(\dfrac{80}{7}, \dfrac{16}{7}\right)$	$\left(\dfrac{40}{3}, \dfrac{8}{3}\right)$
Profit	£153	£166	£139.43	£162.67

So Angie should make 5 mini-packs and 6 mega-packs and the corresponding profit will be £166.

6 a

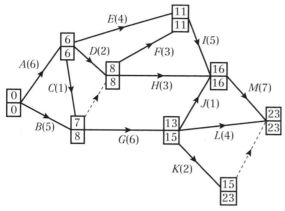

b

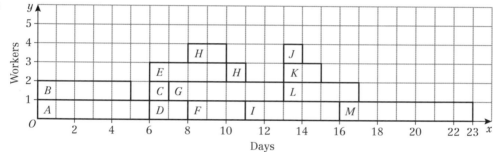

So 4 workers are required.

c Delay the start of activity H by 2 days and the start of activity J by 2 days.

7 a The purpose of the first stage is to provide a basic feasible solution as a starting point for the second stage.

b

b.v.	x	y	z	s_1	s_2	s_3	a_1	Value
s_1	2	1	1	1	0	0	0	50
s_2	1	3	1	0	1	0	0	60
a_1	1	0	0	0	0	−1	1	10
P	−1	−2	−1	0	0	0	0	0
I	−1	0	0	0	0	1	0	−10

c $I = 0$ when $a_1 = 0$, $s_1 = 30$, $s_2 = 50$, $s_3 = 0$, $x = 10$, $y = 0$, $z = 0$

d

b.v.	x	y	z	s_1	s_2	s_3	Value
s_1	0	1	1	1	0	2	30
s_2	0	3	1	0	1	1	50
x	1	0	0	0	0	−1	10
P	0	−2	−1	0	0	−1	10

The pivot is the 3 in the y column.

e $P = 50$ when $x = 10$, $y = 10$, $z = 20$

Index